Sundials of the British Isles

The glorious Ruler of the Morning, ſo
But looks on Flowers, and ſtrait they grow ;
And when his Beams their Light unfold,
Ripens the dull Earth, and warms it into Gold.

Abraham Cowley
1618 - 1667

Sundials
of the
British Isles

A Selection of some of the Finest Sundials from our Islands

Edited

by

Mike Cowham

Cambridge 2005

First Published in Great Britain by M J Cowham

PO Box 970, Haslingfield, Cambridge CB3 7FL

Cambridge 2005

ISBN 0-9551155-0-7

978-0-9551155-0-9

This Edition is limited to

Six Hundred copies of which this is

Number 183

PRINTED BY

Altone Limited
The Old Institute, High Street, Coton
Cambridge CB3 7PL

CONTENTS

The areas covered by the regional chapters include the county, city and borough councils as shown on the map on page vii.

REGIONS OF THE BRITISH ISLES
as divided for this book

SHETLAND ISLANDS

ORKNEY ISLANDS

WESTERN ISLES

MORAY

HIGHLAND

ABERDEENSHIRE

SCOTLAND

ANGUS

PERTH & KINROSS

ARGYLE & BUTE

STIRLING

FIFE

E LOTHIAN

NORTH AYRSHIRE

SOUTH LANARKSHIRE

EAST AYRSHIRE

BORDERS

SOUTH AYRSHIRE

DUMFRIES & GALLOWAY

NORTH EAST

NORTHUMBERLAND

DURHAM

CUMBRIA

DONEGAL

DERRY

ANTRIM

TYRONE

DOWN

FERMANAGH

ARMAGH

MONAGHAN

SLIGO

LEITRIM

CAVAN

LOUTH

MAYO

ROSCOMMON

LONGFORD

MEATH

WESTMEATH

IRELAND

GALWAY

OFFALY

KILDARE

DUBLIN

LEIX

WICKLOW

CLARE

KILKENNY

CARLOW

LIMERICK

TIPPERARY

WEXFORD

KERRY

CORK

WATERFORD

ISLE OF MAN

NORTH WEST

LANCASHIRE

NORTH YORKSHIRE

YORKSHIRE

EAST RIDING (YORKS)

N. LINCS

EAST MIDLANDS

LINCOLNSHIRE

ISLE OF ANGLESEY

CONWY

DENBIGHSHIRE

FLINTSHIRE

CHESHIRE

DERBYSHIRE

NOTTINGHAMSHIRE

GWYNEDD

STAFFORDSHIRE

LEICESTERSHIRE

RUTLAND

NORFOLK

WALES

SHROPSHIRE

WEST MIDLANDS

WORCESTERSHIRE

WARWICKSHIRE

NORTHAMPTONSHIRE

CAMBRIDGESHIRE

EAST OF ENGLAND

SUFFOLK

POWYS

CEREDIGION

HEREFORDSHIRE

BEDS

PEMBROKESHIRE

CARMARTHEN

GLOUCESTERSHIRE

OXFORDSHIRE

BUCKS

HERTS

ESSEX

W. BERKS

GREATER LONDON

LONDON

SURREY

KENT

WILTSHIRE

HAMPSHIRE

WEST SUSSEX

EAST SUSSEX

SOMERSET

SOUTH & SOUTH EAST

DEVON

DORSET

ISLE OF WIGHT

CORNWALL

SOUTH WEST

CHANNEL ISLANDS

GUERNSEY

ALDERNEY

SARK

JERSEY

ISLES OF SCILLY

0   50   100 km

The British Isles have been divided into 12 separate regions as detailed on the map on page *vii*.

These 'regions' were decided as convenient packages for each chapter and follow a fairly self explanatory order, working generally from west to east and from south to north. Several administrative changes have been made over recent years so it was decided to use the new county boundaries as used on maps in 2004, although some of the recent county boroughs may be less familiar to some readers. These county boroughs make listings complex so they are not mentioned in the chapter headings. However, the towns and villages included in this book are indexed with their current counties and county boroughs.

Each region has been allocated a colour, and a similarly coloured box has been added near to the edge of each page to assist with finding the appropriate region.

Dials have been listed in each chapter in alphabetical sequence of nearest postal town or village with the county name added alongside. This mixture within a region simplifies the layout and does not over emphasise counties that only have few dials. It also helps the reader to find these places quickly in a road map index.

Throughout these chapters references are frequently made to either 'Mrs Gatty' to 'Eden & Lloyd' or 'The Book of Sun-Dials'. Further details of this remarkable and still useful book will be found in the Bibliography in Appendix 2.

Other books referred to in the text include Arthur Mee's series 'The King's England' and Nikolaus Pevsner's series 'The Buildings of England'.

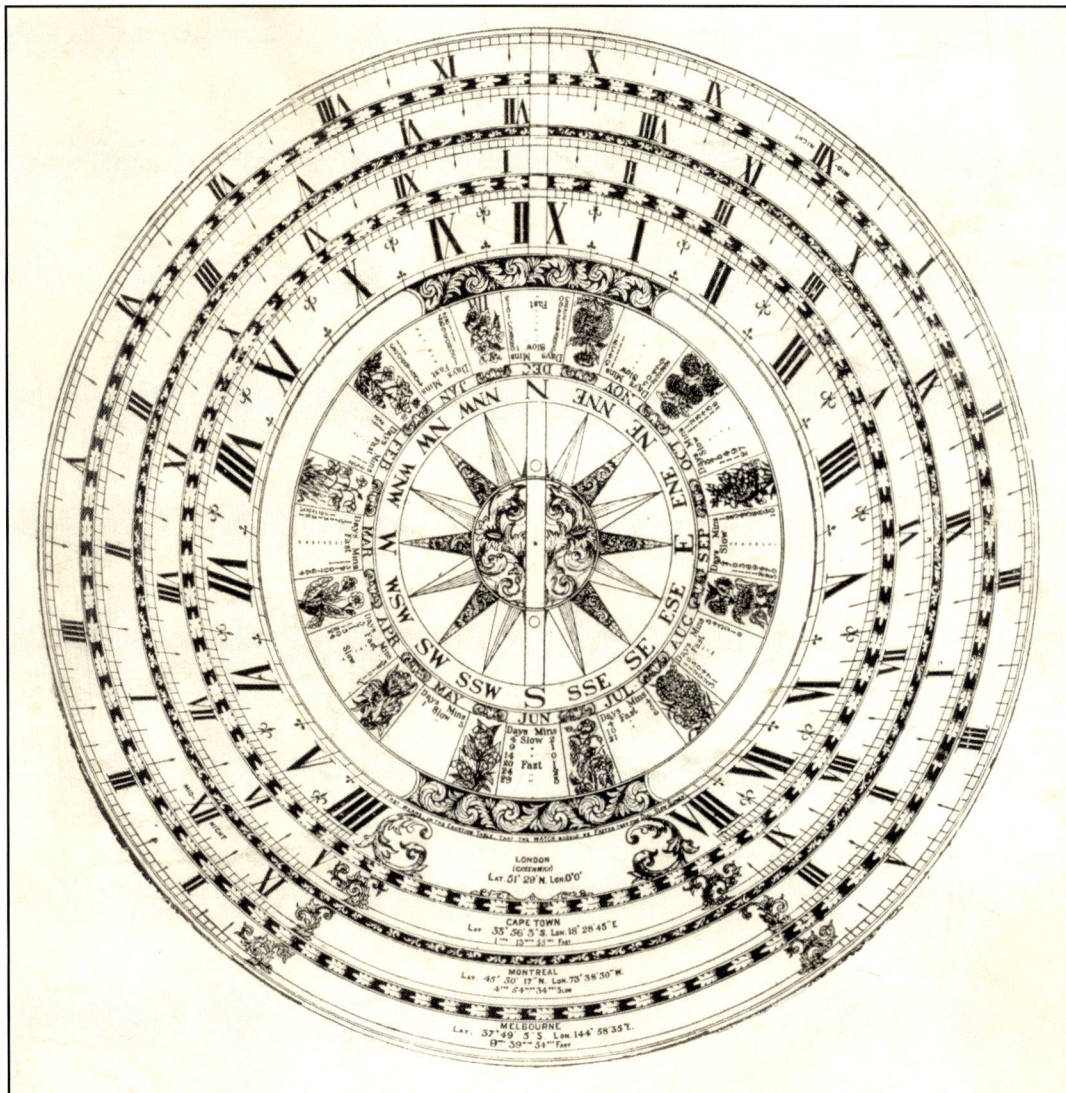

A print of a sundial plate, c1910, thought to be a dial by Francis Barker and Son, Clerkenwell

ACKNOWLEDGEMENTS

In a book of this nature with contributions from so many Authors there are also a number of others who have helped, either directly or indirectly in research into the actual dials or in taking photographs.

Each Author has done a tremendous job, often spending many hours getting together the various bits of information and even more hours travelling around to get the excellent photographs that are essential for a full colour publication of this nature. Up to four main authors have been named for each chapter beneath the titles. However, many of these authors have also contributed to other chapters, where they may not have been named, for which we must thank them.

We would also like to acknowledge the kind help and permission given by owners of many of these dials, who have allowed us to include their dials in this publication. Several have also added further details to the text. Without their co-operation this would be a very slim volume.
Currently active dialmakers have also been invited to add at least one of their works and we thank them for their detailed notes and photographs.

The proof reading too has been quite a task and we need to thank all those who have been involved, particularly Val Cowham who has been very patient with the Editor and has corrected many slips of spelling, punctuation and grammar.

The following list, in alphabetical order, shows the names of many people who have contributed to the book in one way or other in addition to the named authors of each chapter. There will be omissions for which we apologise and hope that these kind persons will forgive this lapse of courtesy.

We would like to thank without reservation the following people and institutions who have helped with producing this book of the 'Sundials of the British Isles':
Wendy Baines (Sulgrave Manor), Jim Bennett (Museum of the History of Science, Oxford), Dr W T Berrill, Kate Bewick (formerly Norris), Sarah

'Dial upon a Glass Globe' supported by Atlas from A Tutor to Astronomy and Geography by Joseph Moxon, 1686

Breach, R D L Brownson (Hawkshead Grammar School), Carmel Byrne (Powerscourt Estate), Lord H Cavendish (Holker Hall), The Marquess of Cholmondley (Houghton Hall), Matthew Constantine (Carlise), James Cox (Gonville & Caius College, Cambridge), Derek Foreshaw (The Argory, NT), Simon Francis (drawing of Monasterboice Dial), Carolin Göhler (Cambridge Preservation Society, Wandlebury), Liz Garrett (Holker Hall), Frank Gate, David Gavine, Elizabeth Green (Lakeland Horticultural Society), Margaret Greg (Powis Castle, NT), Doug Hamon, Ray Hartland, Mrs E M Helme, Sally Hersh, Peter Hingley, David Hodges (Hitchin Museum), Liz Hunter (Hill Top, NT), Ben Jones, Mr D Kaye (Waterton Park Hotel, Wakefield), Alan Langstaff (Packwood House, NT), Sir Mark Lennox-Boyd, David Levitt, William Lloyd, Paul Maher (National Botanic Gardens, Dublin), Clifford

Marcus-Cromwell, Lady Molesworth-St Aubyn (Pencarrow House), Miss Joan Munroe, Mr & Mrs Peake (Sezincote), Peter Ransom, Matthew Richardson (Manx Museum, Douglas), Deirdre Rowsome (Russborough House), Miles C Sandys (Graythwaite Hall), José Spinks (Bollington Arts Centre), James St Aubyn (St Michael's Mount, NT), Thom States (Seascale Millennium Project), Roger W Suddards, Wendy Taylor, Carolyn Taylor, (Carlisle City Council), Dr R D H Walker (Queens' College, Cambridge), Sandra Ward, Philip Warner (Anglesey Abbey, NT), David Warren (Hawkshead Grammar School), Frederick Wayles, John West (Hawkshead Grammar School).

We would also like to thank the owners or administrators of the following properties, whose names we do not have, who have agreed to our publishing pictures of their dials:
Clitheroe Council, University of Manchester, Stonyhurst College, Dial House Farm at Sulgrave, and Tabley Hall.

PHOTOGRAPHIC CREDITS

© The Copyright of all photographs strictly belongs to the photographers

Photographic credits are many and the method used simplifies any listing. Under each photograph will be found small initials; those of the photographer. In most cases these will be the Authors of that chapter. However, a complete listing follows:-

AA	Tony Ashmore
AB	Tony Belk
AJ	Andrew James
AM	Tony Moss
AS	Alan Smith
AW	Tony Wood
AW-	Anthony Watson
BJ	Ben Jones
CD	Christopher Daniel
CM	Carolyn Martin
CMC	Clifford Marcus-Cromwell
DB	Douglas Bateman
DC	David Le Conte
DF	Damian Fennell
DI	Denis Igoe
DL	David Levitt
DY	David Young
FB	Frankie Badcock
FC	Frank Cowham
FE	Frank Evans
JD	John Davis
JF	John Foad
JL	John Lester
JM	John Moir
JT	Janet Thorne
JW	John Wall
KM	Ken Mackay
KT	Ken Thompson
LM	Larry Magnier
MC	Mike Cowham
MC-	Marion Coady
MH	Michael J Harley
ML	Michael Lowne
MN	Mike North
OD	Owen Deignan
PH	Peter Hingley
PR	Patricia Ryan
PR-	Peter Ransom
PS	Peter Scott
RO	Robert Ovens
RS	Robert Sylvester
RT	Richard Thorne
SH	Sally Hersh
SS	Sarah Slazanger
TL	Tom Lawlor
WL	Willam Lloyd
WM	Bill Meminster
WP	Bill Power
WT	Wendy Taylor
WW	Walter Wells
ZK	Zvonko Kracun

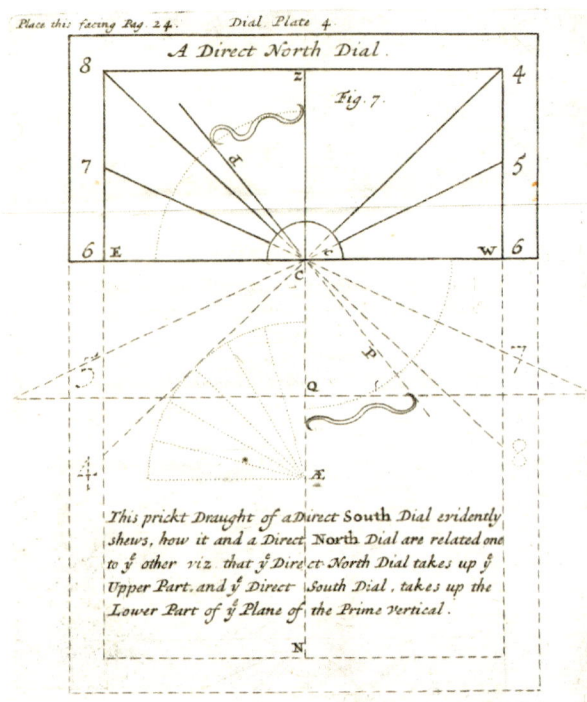

'A Direct North Dial' from The Young Gentleman's Astronomy, Chronology and Dialling by Edward Wells, 1725

FOREWORD

In 1989, some enthusiasts for sundials got together and soon discovered, to their delight and astonishment, that they were not the only people in the world who had an interest in the subject. The British Sundial Society was formed and now has a substantial and growing membership. Well developed societies also exist in the USA, in the rest of Europe and around the world. Some of the contributors to this publication were at that early meeting and all of them are experts and enthusiasts in the subject. The approach in the book is to provide a survey of the British Isles and divides the country into twelve areas. In this sense it is comprehensive, but each section contains examples which are the personal choice of the particular author. Some of the Country's best outdoor sundials are depicted and explained. Modern examples are also included and I am flattered that two of my own designs should have found their way into inclusion. In addition to the area by area examples, there are specialist articles by leading experts on the Early Dials of Britain, Stained Glass Sundials and Portable Dials. The story of measuring the sun goes back a very long way, initially to record a change in season. Newgrange, which is described, is from the Neolithic age and the finest example of such a construction to be found in Europe. Sundials for dividing the day into different intervals were invented in the Middle East in the Classical Age. The word hour derives initially from the Ancient Egyptian. They used rather crude dials, but the first scientific instruments were made by Greek scientists in about 300BC.

Sundials were originally the only timekeepers that Man had and they continued in use into the 19th century. Initially clocks were hopelessly inaccurate, much less than dials, and dials were used to regulate the clocks. Most people today think that dials are inaccurate. In fact it can be argued that the clock is still the less accurate of the two. Clocks tell 'mean time', that is time which is correct *on average*. The clock is trying to reflect the passage of the sun across the sky, but the sun moves at a variable speed and the clock at a constant speed, which explains the discrepancy between the two. Sundials therefore continued in use until the invention of the telegraph and shortly later the wireless, as it was originally called, and in 1924 the Greenwich Time Signal. These famous six little pips were a death knell. For all practical purposes sundials were then forced into redundancy. Gnomons, that is to say the pointers, became bent or disappeared, and columns broken. Although there were a few very fine exceptions, the sundial went down market and more often than not only the crudest examples were to be found on sale in garden centres, as companions to the garden gnome.

Fortunately, they are now back in fashion and regularly employed not so much as timekeepers, but as objects of beauty and interest. They are often commissioned to commemorate some anniversary or event; sometimes they are created as examples of contemporary sculpture which, like Newgrange, remind one of the passing of the seasons. Some modern artists have created beautiful abstractions, which transform the sundial into a sculptural metaphor on the passage of time. Modern sundials are often made to symbolize some important event – a jubilee or the passing of a centenary and all the more so a millennium. So sundials are often about recollection, sometimes sad as well as happy, as are their mottoes. Mottoes are always found on Coats of Arms and sometimes on gravestones. They have often been used on sundials to recall some universal truth or saying, as will be apparent from the examples in this book. The great revival of interest in the subject is due in large part to the enthusiasm and efforts of the British Sundial Society and to the expert contributors to this book. I am flattered to be the Society's patron and delighted to be associated with this publication.

Sir Mark Lennox-Boyd

HOROLOGY, or DYALLING

HOROLOGY, or DYALLING

To the Worshipfull Thomas Stringer of Ivy Church near Salisbury in Wiltshire Esq.

This plate is humbly Dedicated by Ric: Blome

Horology, or Dyalling: may be considered in these three parts

Regular: are such yt are described in a Plaine disposed towards some one determined cheifest part of world. viz.

direct East

of these Regular yt Principall are the

direct West

Horizontall, which is æquidistant towards the Horizon.

South Verticall

North Verticall

are perpendicularly erected aboue the Horizon, & tends directly towards yt Verticall point.

East Meridian

West Meridian

each of which have their severall wayes which are æquidistant from yt Meridian Circle.

The Æquinoctiall aboue.

The Æquinoctiall beneath.

are those which each of them their severall wayes are æquidistant from yt Equator, yt one aboue, yt other beneath yt Horizon.

The Polar above

The Polar beneath

are those which each of them their severall wayes, yt one aboue, yt other beneath, are Parallel to yt Axis of yt world.

The Center of yt hovrs.

yt Elevation of yt Pole, which determines

The Altitude of yt Style.

The ordering yt Dyall it selfe.

Declining from the south either to the East, or West

Declinant, being æquidistant from any verticall circle, & are of two sorts

Declining from yt north either to yt East or west.

Irregular, are such that doe not directly point to those principall parts of yt world, but rather decline from them & are either

to find out yt Meridian line

yt description of yt hours in yt given plain

what things in generall are necessary to be known for the framing sundyalls

Instruments necessary for yt making Dyalls

Rule

Compasses

yt sea Compass

a Plum line

yt Astronomicall Quadrant.

To find yt Center of yt Dyall or hour.

To determine yt Place, & Altitude of yt Style.

To describe yt Arches or yt sines in a

the suns declination

Horizontall Dyall

Verticall Dyall

Meridian Dyall

Æquinoctiall Dyall

Polar Dyall

Inclinant are such yt fall off from yt Verticall point, inclining towards yt Horizon, as not being æquidistant from it

yt contrivance of the Style as to yt forme & Altitude of it.

yt Applycation & disposition of yt dyall it selfe.

The Meridian line

To place yt Gnomon at its true Center.

The Longitude of yt Style.

The Æquinoctiall Line

The Horozontiall line

To find yt hour of yt night by yt rayes of yt Moon.

'Horology or Dyalling'
A print by Ric: Blome from an unidentified book of around 1700
showing in chart form the different types and classes of Sundials that are possible

PREFACE

A book featuring the Sundials of Britain has been lacking for many years. 'Sundials of the British Isles' hopes to redress this point. It respectfully follows that great work 'The Book of Sun-Dials' originated by Margaret Gatty in 1872. However, it does not aim to replace or even update this work but to show some of the more interesting dials that can be found across our islands.

Authors have been chosen for each chapter because they either live in that region or have a detailed knowledge of the dials there. They have now brought before us their own personal choice of some fascinating dials in their areas. Examples have therefore been included from all corners of our Islands, each dutifully photographed and researched by one or more of our team of Authors.

Each sundial is illustrated in colour with its main features described and the entry often includes interesting local stories or historical facts about the region and other buildings worth visiting that are close to the dial.

This is not a highly technical book. There are several of these to be found describing dialling techniques in detail. Therefore specific dialling terms have been limited to essentials, but a Glossary has been included to explain the terms that may be less familiar to the casual reader.

Following the description of many of these dials there is often a brief note of other dials to be seen in the vicinity. Therefore a trip to see one will often be rewarded with the opportunity to see others.

Britain has a great tradition of sundial construction, many of them dating back hundreds of years. Our earliest dials go back to Saxon times and the tradition has continued to the present day. When Railway Time was introduced into Britain towards the end of the 19th century, most sundials became obsolete; time being distributed around the Country by the electric telegraph, and in more recent years by radio. Luckily, particularly in the late 20th century, dialmakers have again taken to the 'Art of Dialling' and have produced some spectacular dials, many of which you will find featured in these pages.

The making of new sundials is now a thriving trade and many of these new designs and ideas have been featured in each region.

Generally speaking there were few regional styles of sundial, much of our Islands having the same types of dials. The exceptions will be seen as the reader goes through each chapter. There will be found the magnificent

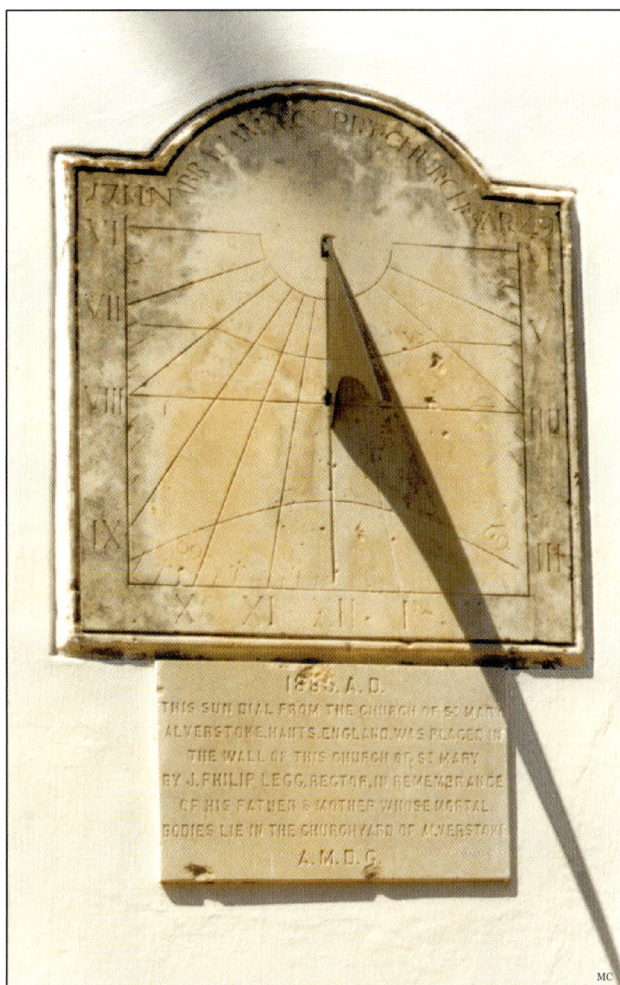

Dial from Alverstoke in Hampshire that is now fitted to the north wall of a church in Stellenbosch, South Africa.
The shadow, of course, travels across the dial in the wrong direction and its delineation will not be correct for the dial's new latitude

Scottish Polyhedral and Lectern Dials, the Saxon Dials found through much of the Northeast and the slate Dials of Wales and the West Country.

The common Horizontal Dial is as English as is the garden. Who could imagine a rose garden without a sundial at its centre? Many of these 'Garden' Dials were made by London instrument makers in brass or bronze. They were made to order, delineated for the customer's latitude and perhaps supplied locally with a suitable supporting pedestal. Vertical Dials, on the other hand, were often made *in-situ*, and would have been done mainly by local makers, sometimes perhaps not quite understanding the finer principles of dialling. Some, as will be seen, have made various mistakes, obviously due to their lack of knowledge of the Art. There were however, several good books detailing how to make a dial on any surface and, if the workman was able to read, he would probably have followed the instructions given very carefully.

Sometimes a clockmaker would provide a sundial in order for his clock to be set correctly. The famous clockmaker Thomas Tompion is known to have provided several. One of his dials that most

Magnetic Compass Dial made in Britain ^{MC}
around 1820 for export to Rio de Janeiro.
Note its low angle gnomon
and anti-clockwise hour scale

people know is at the Pump Rooms in Bath where he supplied the large and accurate longcase clock. The dial still sits just outside the window overlooking the Roman baths, within a few steps of his clock.

The dials chosen for the regional chapters of this book may all be seen by The Public. Although many are privately owned, such as by National Trust or English Heritage, we may freely see them when we visit these properties. There are many excellent and interesting dials in private hands but these have not been included because they are not readily accessible.

The book covers the whole length and breadth of the British Isles with few corners being omitted. It does not cover foreign made dials or British dials that have found their ways overseas, except for the two illustrated in this Preface.

British dialmakers were famous throughout the world, particularly in the 18th century, exporting their superior products, particularly to the British Colonies.

In 1989 a small group of enthusiasts formed the British Sundial Society. Very quickly it was realised that there were many people interested in the various aspects of dialling and the Society grew to a Membership of over 500. Today it is a thriving Society having Members with many different interests, both from Britain and from Overseas. One of its earliest aims was to catalogue all the sundials *in-situ* in the Country. This is an ongoing project but several thousand have been found and a full catalogue has been produced that is available to Members.

'Sundials of the British Isles' is a credit to the many BSS Members whose efforts as Authors and Dial Recorders have each helped to put this work together in celebration of our great dialling heritage.

Mike Cowham
Cambridge 2005

INTRODUCTION
Christopher Daniel

It is, perhaps, remarkable that there are so few modern books on sundials published in Britain today, compared to other countries, particularly France, when Britain can boast some 5,000 'fixed location' sundials scattered around the British Isles. The most famous British sundial book is still the monumental work, originally compiled by Mrs Alfred Gatty in the second half of the 19th century, first published in 1872, which ran to four editions. The last of these, published in 1900, being spoken of in almost hallowed tones and regarded by the cognoscenti as the sundial 'bible', is a large tome principally devoted to sundial mottoes. There were other smaller books, of a somewhat similar nature, by Warrington Hogg, Launcelot Cross and Geoffrey Henslow that were also published around and following the turn of the century. More recent works, primarily concerned with the construction of sundials, include A P Herbert's book 'Sundials Old and New' published in 1967, (better described as 'Fun with the Sun'), Frank Cousin's book 'Sundials, A Simplified Approach', published in 1972, and Peter Drinkwater's book 'The Art of Sundial Construction', which ran to several editions in the 1980s. Nevertheless, the most popular work in this field is that of Albert E. Waugh, a professor for 40 years at the University of Connecticut in the United States of America, whose classic book 'Sundials, Their Theory and Construction' was first published in paperback in 1973 by Dover Publications of New York!

It may seem surprising that America, a country not noted for its sundial heritage, should also have produced several other books dealing with this subject in the 20th century. These include the works of Alice Morse Earle (1902), Roy Marshall (1963), Winthrop Dolan (1975) and, not least, R. Newton Mayall and Margaret W. Mayall, in their well known book 'Sundials, How to know, Use and Make them', first published in 1938 and running to later editions in the 1970s. Furthermore, it is clear, from the fact that the North American Sundial Society was formed in 1994 and that it is an active organisation, that there is an unusual, not to say a remarkable interest in the subject in the United States.

It must be said that there was something of a resurgence of interest in sundials in the second half of the 20th century, both in Britain and abroad, not just as garden ornaments or as having historical significance; but as scientific works of art. Thus, as a sundial designer, I have witnessed and experienced this renaissance, which has brought about the restoration of historic dials and the creation of many fine modern sundials. Indeed, in Northern Ireland in the Carnfunnock Country Park at Larne, one may even find a 'time garden' of sundials in the grounds of a country estate, whilst in south-east London, at the Horniman Museum, there is a recently established 'sundial trail' for visitors to follow! It was in this climate of interest that numerous sundial societies were formed on the Continent and which also saw the foundation of the British Sundial Society in 1989.

Nevertheless, in Britain, one would have expected a somewhat greater interest in sundials and many more publications concerning them. Since the formation of the British Sundial Society it would seem that this is so, since the Society has a membership in the order of 500, has well-attended annual conferences, publishes a fine quarterly 'Bulletin' and a number of specialist works, including a 'Sundial Glossary: A sourcebook of dialling data' (Ed: John Davis), a 'Biographical Index of British Sundial Makers' (Jill Wilson), and a 'Register of Fixed Location Sundials' in the United Kingdom. There are also a number of monographs or descriptive works, including my own 'Sundials on Walls' (1978), Philip Pattenden's 'Sundials at an Oxford College' (1979), Andrew Somerville's 'The Ancient Sundials of Scotland' (1990), Alexis Brookes' & Margaret Stanier's 'Cambridge Sundials' (1991), Margaret Stanier's 'Oxford Sundials' (1994), Carolyn Martin's 'A Celebration of Cornish Sundials' (1994) and my own 'Sundials' Shire Album, an introduction to the subject, first published in 1986. Admittedly, the latter was reprinted five times, with minor amendments, until 2004, when, having sold 24,000 copies, it was upgraded and published in colour as a second edition. In the first six months, the new edition sold 1,300 copies!

Thus, it is evident that there is now a considerable interest in sundials in the United Kingdom; but, by comparison with France, although a larger country with perhaps 11,000 fixed location sundials, books describing our sundial heritage are relatively few. French sundial literature, on the other hand, may perhaps be described as 'prolific', with colourful illustrations and descriptions of their many sundials, produced to the highest quality. One of these, although principally concerning the construction of dials, by one of their best known authorities, the late Réne R J Rohr, is his excellent book 'Sundials, History, Theory, and Practice', which was translated into English as well as German. In this country, we have nothing, as yet, that compares with the scale of this class of sundial book. The various English publications mentioned either deal with a specific aspect of the subject at length or in general with not as much information about individual sundials as one might like. In some measure, this present work 'Sundials of the British Isles' seeks to remedy this situation by describing some of the best or most interesting sundials, both historic and modern, that are to be seen in these islands.

Sundials come in all shapes and sizes, both great and small, from the giant masonry structures in the 18th century astronomical observatories of India to the small 'pocket sundials' that may be found in museum collections. When the 'Mary Rose', King Henry VIII's prized battleship, was brought up from the seabed in the Solent, amongst the treasured artefacts that were found in her, there were a number of little wooden pocket dials. These were circular in shape, per-

Portable Dial from the Mary Rose ^{MC}

haps 40mm in diameter, in an extraordinarily good state of preservation, complete with engraved hour-scales, hinged brass gnomons, and magnetic compasses, used for setting the instrument. The Tudor mariner may well have used such dials for the purposes of navigation; but they may also have been the equivalent of a pocket watch, being the personal possessions of the gentlemen aboard the vessel at the time of her sinking.

The origin of the sundial is lost in the mists of antiquity; but the ancient Egyptians are known to have used shadow casting devices, probably for religious purposes, the earliest reference to such a device dating back to about the year 1300 BC. The Babylonians are credited with having invented the *gnomon*, which took the form of a vertical rod set up on level ground, the tip of its shadow indicating the time. The ancient Greeks were undoubtedly influenced by Babylonian science and perhaps it was from this civilisation that they inherited their knowledge of *gnomonics*, the science of constructing and using sundials, which science they developed to an advanced and sophisticated state. Likewise, Greek civilisation and culture influenced the Romans: many Graeco-Roman sundials are to be found scattered around the shores of the Mediterranean in former centres of the Roman occupation, indeed there is an acquired example to be seen at Hever Castle in Kent.

Following their invasion of Britain, it would appear that the Romans were also responsible for bringing the sundial into Britain. There is a mosaic panel in the floor of the Roman villa at Brading in the Isle of Wight portraying a scene in which there is the earliest representation of a sundial in the country, which may be dated to c 250AD. However, the earliest sundials extant in Britain, dating from the 7th century, include those of the Bewcastle Cross to the north of Hadrian's Wall in Cumbria, the church of St John at Escombe in Co. Durham, and that of the church St Bueno at Clynnog-fawr in Gwynedd, North Wales. Such early dials were not of the sort of which most people think, when they visualise a sundial, namely the common or garden 'horizontal' dial, on a pedestal in a rose garden! The earliest examples mentioned are of the 'vertical' kind, cut into the stonework of these ancient monuments by Anglo-Saxon, Celtic or Cymric sculptors, to provide a means by which the times of church services could be regulated. There are

Roman mosaic at Brading, Isle of Wight

degenerated into the often crudely carved medieval so-called 'scratch' dial or mass dial. This, no doubt, was due to a common requirement for churches to have a sundial that would determine the time; but sculptors and stonemasons, with the ability to carve them, were not so common, at least not in remote country areas. Thus, the sundial was probably scored into the stonework by an individual priest, who, at the time, may not have taken into account the change in the seasons and consequent change in the position of the Sun in the heavens. Churches in those days were usually lime-washed, inside and outside, for hygienic reasons and so as to be visible from afar. These various conditions often resulted in several mass dials being carved close together on the building. Many such mass dials are to be found today on the walls of churches, often quite close to a south porch. These medieval devices were the precursors of what is called the 'scientific' sundial, the instrument that came to be set up on many a church in later centuries to regulate the clock, "which often goes wrong".

many other somewhat later examples of Anglo-Saxon sundials to be found around the country, most notably the one at St Gregory's Minster at Kirkdale in North Yorkshire, dating from about the year 1064, just before the Norman Conquest. It has been shielded from the weather over the centuries by a porch and is in an excellent state of preservation.

The corpus of Anglo-Saxon sundials shows that they were often well executed and beautifully decorated, their function being to indicate the times of religious services, as laid down by canon law. Carved in the stonework of the south facing walls of churches, these semi-circular devices, with radiating lines denoting the times of mass, had a gnomon, in the form of a short metal rod or wooden peg, that projected horizontally out from a hole in the wall, at the centre of the dial's delineation, being also perpendicular to the wall. The tip of the gnomon's shadow indicated the hour; but, whilst they were sufficient for their purpose, such sundials only gave accurate indications of the time at noon, i.e. at mid-day, when the sun was at its maximum altitude. As England sank into the Middle Ages, this form of sundial

The 'scientific' sundial, excluding Anglo-Saxon sundials and medieval mass dials, embraces all sundials as we know them today. The principle may best be described by looking at the common or garden horizontal. It takes the form of a flat usually square or circular plate, the *dial-plate*, engraved with hour-lines to indicate the hour of the day. Aligned with the 12 o'clock hour-line and fixed to the dial-plate, at or just off centre, there is a second triangular-shaped plate, the *gnomon*, the shadow of the inclined straight edge of which indicates the time, as it comes into contact with each hour-line. The delineation of the instrument must be calculated for the latitude of the place at which it is to be used, and the angle of the inclined edge of the gnomon to the dial-plate must be equal to this same latitude. To set the sundial correctly on its pedestal, the dial-plate must be *horizontal*, i.e. it must lie in the plane of the horizon, whilst the inclined edge of the gnomon must lie in the plane of the *meridian*, such that the tip of the gnomon points due north towards the Pole Star. What this means is that the inclined edge of the gnomon will be parallel to, i.e. aligned with the polar axis of the Earth. As a result, as the Earth moves in its orbit around the Sun and regardless of the tilt of the Earth's polar axis, which is the principal cause of the Sun's changing altitude during the year, i.e. the seasons, so, in its apparent diurnal motion, the Sun will always remain in a plane that is at right-angles to the polar axis. Furthermore, as the Earth rotates on its axis, the Sun will appear to move at a uniform rate around the globe, completing the 360° circuit in 24 hours. Since the axis of the Earth and the inclined shadow-casting edge of the gnomon may be regarded as one and the same, in respect of the sundial, the Sun will apparently move about the axis of the dial at a uniform rate of 15° per hour. The knowledge of this fundamental principle allows the sundial to be constructed to indicate *equal* hours, whereby the *local apparent time* by the sun may be determined.

It was this invention, in which the gnomon was tilted from the vertical or the horizontal, to be aligned with the polar axis of the Earth, which enabled sundials to be delineated so as to indicate the time with some degree of accuracy. Thus was the scientific sundial conceived; but when and by whom are still unanswered questions. It is quite possible that the Greeks were the first to consider the idea. In the city of Alexandria, founded by Alexander the Great in 332BC, following his invasion of Egypt, there flourished the greatest trading city and the greatest centre of learning in the ancient world. Under the patronage of one of his generals, Ptolemy, who instituted his own family line of kings in Egypt of the same name, there was established the first great public library in the world, with some 700,000 books written on papyrus. Here also a university was founded, including 'schools' of mathematics and astronomy, where, during a period of some three hundred years, Greek scientists studied and mapped the heavens, observed the apparent motions of the Sun, Moon, planets and stars, and constructed their civil calendars. It was not until the year 30BC, when Cleopatra was Queen of Egypt, the last monarch of the line of the Ptolemies, when the country was annexed by the Romans, that the greatest days of the intellectual Alexandrian 'school' of Greek science began to wane.

The Romans governed the region for nearly 700 years and, during this time, Alexandria remained the principal city of Graeco-Roman Egypt. It was still a centre of culture and learning, and Greek astronomy still had its brilliant exponents, notably Claudius Ptolemaeus, better known as Ptolemy, who lived from about 100AD to about 175AD and who produced the great work on mathematical astronomy for which he is famous, the *Almagest*. After the fall of the Roman Empire, Alexandria became a major intellectual centre of early Christianity; but it was during this era that rivalries with other centres of Christian power brought about the final collapse of its scientific activity and the burning of most of its fabulous libraries. Following the rise of Islam, Alexandria was captured by the Arabs in the year 642, when the city was laid waste and the remains of the great library were either plundered or destroyed. Nevertheless, following their conquests of the many countries, stretching from the River Indus and the region that is now Pakistan in the east, across North Africa to Spain and southern France in the west, the Muslims brought a new and highly civilised way of life. They studied art, astronomy, mathematics, medicine, geography, cartography, botany, chemistry and physics, and brought their culture to the whole of their empire and beyond. Despite the destruction of Alexandria, it was the Arabs who inherited much of what remained of Greek mathematics and science. They, in turn, continued the studies of their Greek predecessors and, indeed, it was they who gave Ptolemy's great work the name *Almagest*, from the Arabic "The greatest".

There is evidence to suggest that a sundial, that indicated equal hours by the use of a pole-oriented gnomon, was in use in the first century AD and, considering the hundreds of years of outstanding astronomical achievement of the Alexandrian 'school', it is difficult to conceive that the inclined gnomon could have been unknown in Egypt at this time. Whether this gnomonical knowledge was lost with the destruction of this centre of scientific learning, one may never know. Nevertheless, Arab astronomers enthusiastically took up the study of gnomonics, produced treatises on shadows and sundials, and it is to Islamic science that the practical development of the theory of equal hours is attributed. The works of the Moroccan scientist, Aboul Hhassan, suggest

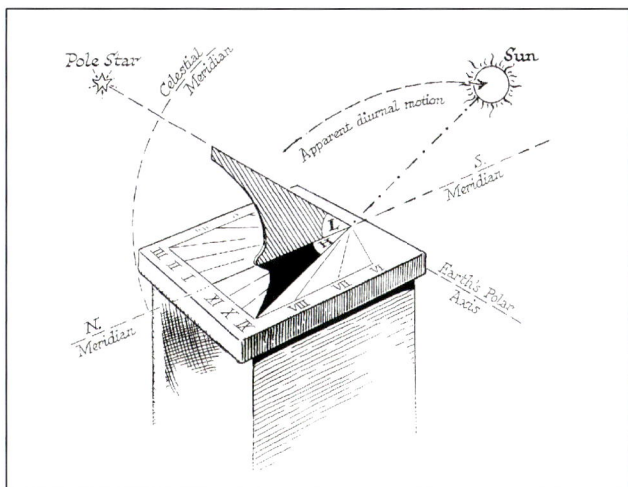

The 'Moorish' Sundial*

dials is that of the *equinoctial* class, sometimes called 'equatorial', where the dial is delineated in the equinoctial plane, parallel to the equinoctial circle of the celestial sphere, which is sometimes called the 'celestial equator'. The gnomon usually takes the form of a metal rod, aligned parallel to the Earth's polar axis, whilst the dial-plate may take the form of a disk, where the hour-angles are all delineated at 15° intervals. There are more sophisticated variations of this class of sundial and the one that most people may be familiar with is of the *armillary* kind, comprising several metal rings. In fact, these rings represent the great imaginary circles of the heavens or the celestial sphere, such as the meridian, the horizon and the equinoctial hour-ring. The gnomon is often represented as an arrow, passing through the centre of the assembly of rings, with the shadow of its shaft indicating the time on the inner surface of the equinoctial hour-ring.

Another kind of sundial is that of the *polar* class, in which not only does the gnomon lie in the

that it was he who invented the 'scientific' sundial, sometime in the late 12th or early 13th century. In time, the principle of this new scientific instrument found its way into Europe, perhaps via the Muslim occupied territory of Spain. Sometimes called the 'Moorish' sundial, due to its considered origin, this invention seems to have made its first physical appearance in Germany in the 14th century or early in the 15th century. Certainly, by the end of the 15th century, the scientific sundial had become an established instrument for use in the determination of time by the equal or equinoctial hour system. Probably in the latter part of the 15th century, the scientific sundial was introduced into England from Europe.

Quite what form this dial took is uncertain, since, as has already been mentioned, sundials come in all shapes and sizes, and they can be placed in various classes, such as the common or garden *horizontal* sundial, discussed earlier, and the *vertical* kind that are carved into or set up on the walls of buildings. Surprisingly, perhaps, the most basic of all

* Adapted from 'Science for the Citizen' by Lancelot Hogben

Armillary Sundial at the Savoy Hotel, London

INTRODUCTION

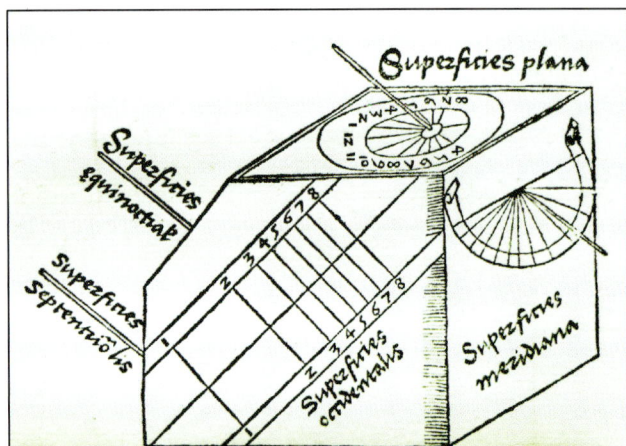

'Kratzer' Multiple Dial*

plane of the polar axis, but the dial-plate also lies in the same plane. However, in order for such a dial to function, the dial-plate and the gnomon must be set apart, at a convenient distance, but still lying parallel to each other, so that the gnomon may cast a shadow on to the surface of the dial-plate. The polar sundial is simple to construct, since the plane of the polar dial-plate is perpendicular to the plane of the equinoctial and tangential to a circular equinoctial dial-plate. Thus, the hour-angles of the polar dial are a function of the tangent of the 15° hour-angles of the equinoctial dial and the distance that the gnomon is set apart from the plane of the polar dial-plate. Such polar dials are comparatively rare; but there are three modern 'Millennium' examples in London and there is a modern 'double-polar' sundial to be found at Otley in Yorkshire. The latter has two dial-plates, both of which lie in the plane of the polar-axis; but which are set at different angles to each other, somewhat resembling a giant open book!

Perhaps the most remarkable form of sundial is that of the *multiple* class, which comprises numerous subsidiary individual component dials of the various other classes. In the late 15th or early 16th century, this form of sundial had become popular on the continent and it was probably introduced into England by Nicolas Kratzer (1487-1550), a Bavarian astronomer, born in Munich. Kratzer had a reputation as being well-versed in the design, calculation and delineation of sundials, and experienced in their construction. In late 1517 or early 1518, evidently through the offices of Sir Thomas More, he was invited to England and welcomed to the court of King Henry VIII, where he became the 'deviser of the King's horologes'. His employment not only

included horological work and the construction of sundials; but matters relating to astronomy and other technical duties. However, sometime in or about 1521, 'at the King's command', probably through the influence of Cardinal Wolsey, Kratzer was given an appointment to lecture in astronomy at Oxford University. He was in Oxford for about three years, during which time he is known to have devised a number of sundials, most notably the free standing pillar dial for St Mary's Church. England has few multiple sundials as such, most of them being of the kind that consist of a 'cuboid' block of stone, with incised vertical dials on the south, east and west faces. However, there are a number of historic examples of multiple sundials, which appear to date from the 16th century, which may owe their origins to Nicolaus Kratzer. The most striking example still extant is that which is in the churchyard of Old Eastbourne Church. It takes the form of a polyhedral 'cuboid' block of limestone, set upon an ancient stone pillar, and, although badly weathered and worn, still exhibits the characteristics that suggest that it dates from the early 16th century. The dial-block has been cut to provide six or seven planes for vertical, horizontal and polar dials, as well as having some five equinoctial *scaphe* dials, in different shapes, hollowed out of the stonework. These characteristics bear a remarkable resemblance to the multiple sundials that are illustrated in early 16th century German gnomonical works, with which Nicolas Kratzer would have been familiar. Multiple sundials of a later period, many dating from the 17th century and of an almost entirely different nature, are to be found in profusion in Scotland; but these Scottish dials were developed at a time when the scientific sundial was already well established in Britain.

* From Sebastian Munster's 'Horologiographia', 1533

Polyhedral Dial at Old Eastbourne Church[CD]

6

Whilst the introduction of the multiple sundial into England may be credited to Nicolas Kratzer, it is probable that the *vertical* class of the scientific sundial arrived independently, perhaps late in the 15th century. Likewise, the *horizontal* class of dial, the so-called 'Moorish' sundial, may well have been brought to this country at much the same time. Certainly, in the grand Tudor gardens of Henry VIII, brass sundials of the horizontal kind were not uncommon. Whilst they served a useful purpose in determining the hour of the day, set upon carved stone pedestals, they were evidently also considered to be attractive as ornamental garden features. The horizontal sundial continued to serve this dual role in the centuries that followed, although in the 17th century it developed into a more sophisticated instrument. This was especially so when this instrument took the form of the *double-horizontal* dial, in which the shadow of the vertical edge of the triangular gnomon, in conjunction with the shadow of the inclined edge, indicated not only the hour of the day, but provided an extraordinary wealth of ancillary information on the dial-plate. Such dials are relatively rare; but even the ordinary horizontal 'garden' dial of the 18th and 19th centuries was quite normally furnished with tables by which clocks and watches could be regulated. However, in the Victorian and Edwardian era, the horizontal sundial probably increased its popularity as a garden feature, although, with the advent of the railways and the electric telegraph, the use of the sundial, as a practical instrument for determining the time, was on the wane. Today, in most garden centres, one may find plenty of so-called 'sundials' of the common or garden kind, some of which may well sport a nice 17th century date; but they serve only as garden ornaments and seldom, if ever, will show the proper time!

The *vertical* sundial, of course, carved into the fabric of or set up on the wall of a building, would normally be more visible than any other form of dial. It replaced the earlier inaccurate mass dial, scored or scratched into the walls of churches. A very early example of the scientific vertical sundial, probably one of the earliest extant, is to be found on the west corner of the south wall of St Mary's Church at Selling in Kent. This dial has been incised as a direct south-facing instrument, but, to achieve this, the stonework has been cut back, to allow for the fact that the wall of the church *declines* from due south slightly towards

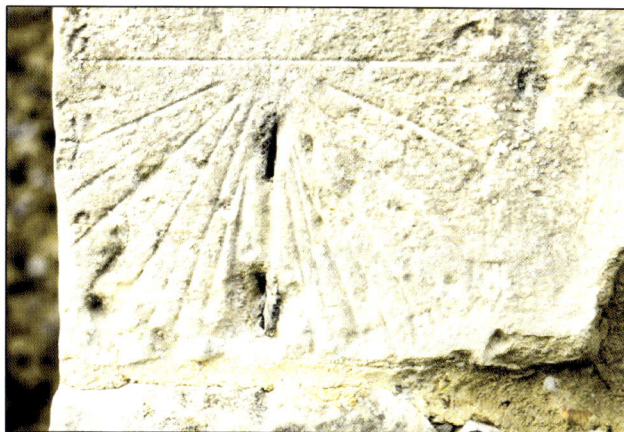

Early 'Scientific' Dial at Selling in Kent CD

the east. The sundial probably dates from the early 16th century, but appears to have been re-incised at a later date, perhaps in the 17th century. Both Roman and Arabic numerals may still be discerned, whilst there the remains of an iron gnomon and old lead-work are still just visible. However, carving a vertical sundial into the stonework of a building, especially if it were to be situated quite high up, where it could be seen from a distance, would have required scaffolding and would have meant that the stonemason would have to have worked for long periods, often in inclement weather. Furthermore, the walls of the buildings would not always face directly towards the cardinal points of the compass. As with St Mary's Church at Selling, a south-facing wall may not always face due south: on the contrary, seldom does one come across a wall that does not *decline* from due south! These factors encouraged sundial-makers to first determine the *declination* of the wall of the building, where a sundial was wanted, and then to construct the sundial in stone or other material in their workshops, unaffected by the weather. When finished, all that was then required was to fix the dial to the wall. Thus, one does not come across a scientific vertical sundial that has been carved directly into the stonework of a building very often, although a fine example may be found in the Tower of London and another may be seen on Ripon Cathedral.

Whilst, in principle, equinoctial and polar sundials are simple to construct, they must necessarily be set for the latitude of the place where they are to be used. Horizontal and vertical sundials, on the other hand, must actually be constructed for the latitude of the place concerned, although, in theory, if not in practice, they may be moved to other locations. Such a move is not recommended, as, in order to indicate the correct time, the dials must be set so that their gnomons remain aligned in the polar axis. In theory, for example, a horizontal sundial made for the latitude of Edinburgh, could be set up as a vertical direct north dial on a north facing wall in Sydney, Australia, and, in a sense, it would work! Thus, horizontal and vertical sundials require some calculation, whilst vertical dials that are to be placed on walls that decline from north or south call for rather more calculation. The advantage of the vertical dial is that it can often be made quite large and that, consequently, it can be adorned with attractive 'furniture', providing ancillary information, such as the height and direction of the sun, the 'sign of the zodiac' in which the sun is situated, the number of hours when the sun will be above the horizon during the course of the day, and much more.

In due course, when clocks were put up on churches and public buildings, a vertical sundial would frequently be set up nearby to regulate the clock, since early clocks often went wrong! As clocks became more accurate, so sundials were required to provide the means to regulate such clocks with greater accuracy. The time indicated by the sundial varies according to the apparent motion of the sun, called 'apparent solar time' or *local apparent time*, whilst the time shown by a clock is an artificial *mean* of this quantity, called 'mean solar time' or *local mean time*. The

difference between the two is called the *equation of time*, caused partly by the varying speed of the Earth in its orbit around the sun and partly by the tilt of the Earth's axis to the plane of its orbit. This difference is a known quantity, which can be tabulated and which was normally a feature of the furniture of well-made horizontal sundials in the 18th and 19th centuries. Furthermore, the equation of time may be incorporated into the construction of a sundial, most effectively with a vertical sundial, in the shape of a figure-of-eight, or *analemma*, on to which the Sun's rays are projected as a spot of light, by means of a small aperture or *nodus* at the extremity of the gnomon, such that the sundial will indicate mean-time or 'clock' time directly. The invention of this form of dial probably dates from the early 18th century and was popularly employed on vertical noon-mark or meridian sundials in France. Whilst there is evidence, in the form of an existing gnomon, with a central aperture of the kind used in France, to suggest that such dials may have been introduced earlier, they only seem to have made their appearance in England in the late 19th century. The only working examples to be found in this country today are modern noon-mark mean-time sundials. One of these is a fine vertical declining meridian dial, carved in black slate, on a south wall at Green College in Oxford. The other is in the headquarters building of the research company QinetiQ in Farnborough. This is a unique noon-mark sundial in the window of the large 'curtain wall' of glass. The dial features

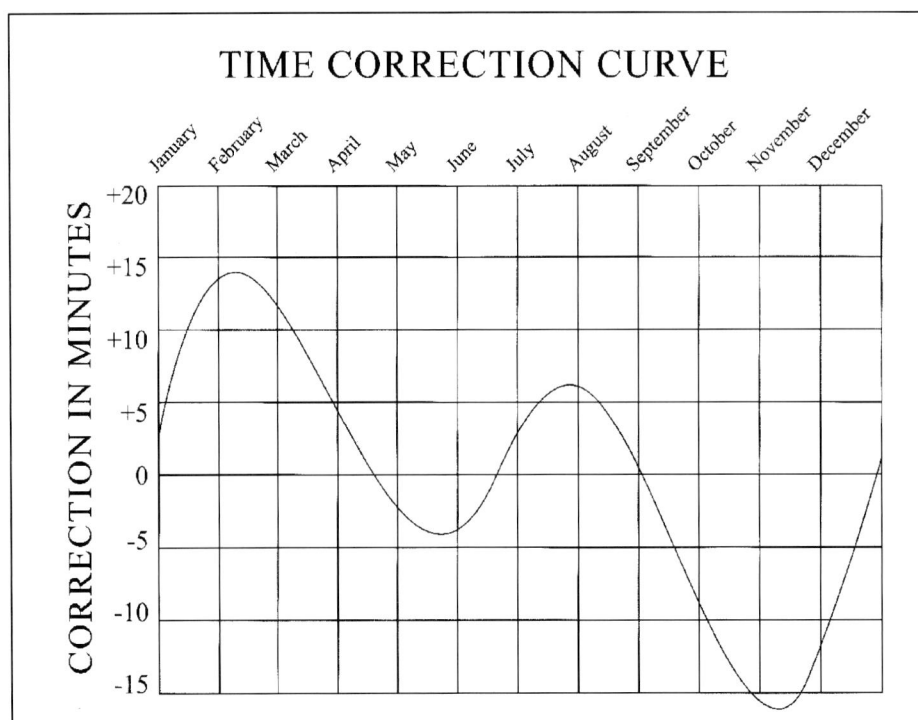

Graph showing the 'Equation of Time'

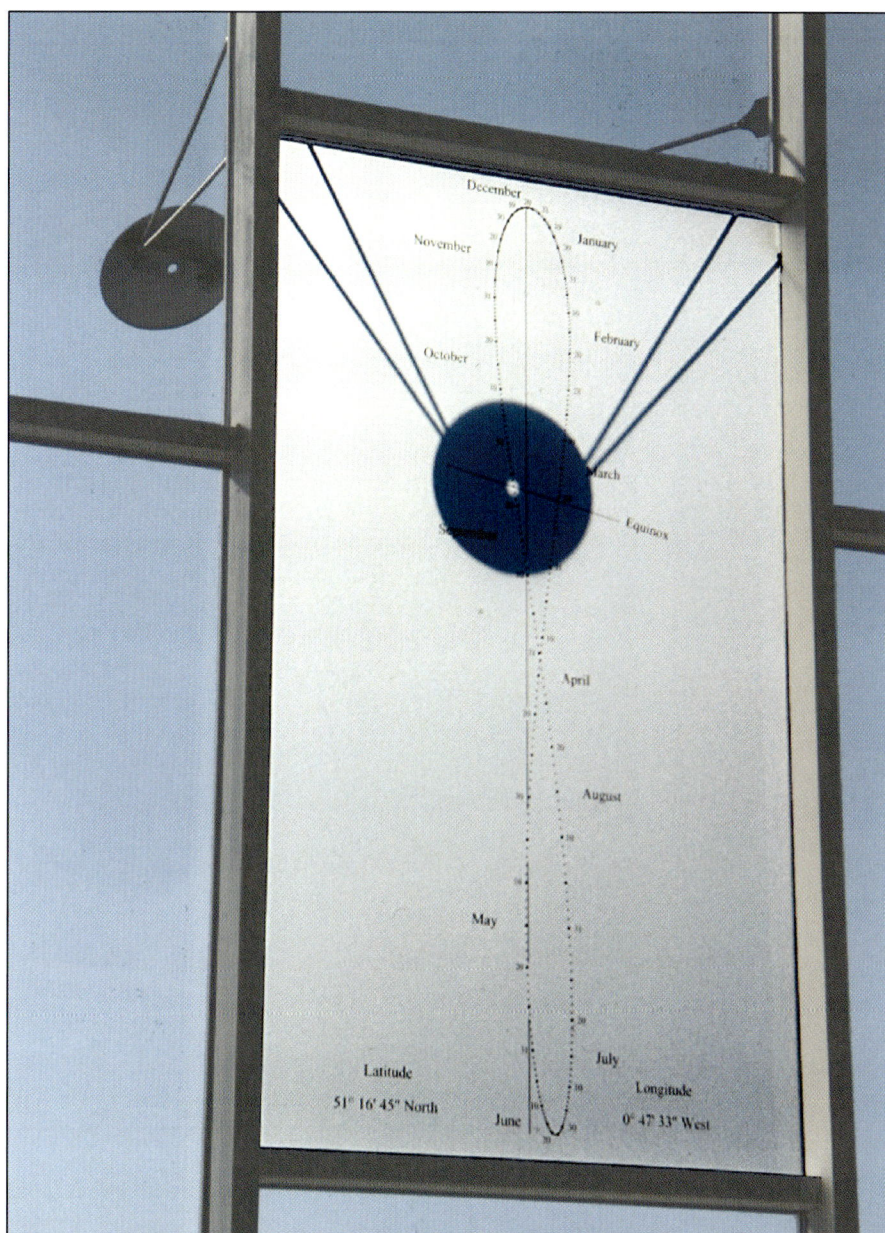

Calendrical Mean-Time Dial at QinetiQ, Farnborough

one. Made in stained-glass, the earliest recorded example of a glass window sundial seems to have been one dated 1518, in the castle of Kurfürst von Sachsen in Altenberg, south of Dresden, in Germany and close to the border with what is to-day the Czech Republic; but which would then have been Bohemia. Altenberg was noted for its lead-work, whilst this region is famous for its glass, so perhaps it is not surprising that stained-glass sundials first made their appearance in this part of Germany. Even so, it was not long before this beautiful form of sundial reached England, possibly from either the German or Flemish schools of glass-painting. However, the earliest extant example, dated 1585, was made by Bernard Dininckhoff, thought to have been a refugee glass-painter from Bohemia, which had been annexed in 1526 into the great Austro-Hungarian Empire. Stained-glass sundials in England mostly date from the 17th century, when glass-painters were obliged to cease producing religious scenes for church glass and so turned to the production of heraldic and civic glass. The Art of Dialling, as the construction of sundials was called in this country, which was a flourishing mathematical science and a popular pastime, offered new opportunities for the glass-painter. There are charming examples of this art in such historic houses as Groombridge Place in Kent, Berkeley Castle in Gloucestershire, Arbury Hall in Warwickshire, and Tredegar House in Gwent, South Wales. There are also dials of this nature to be found in one or two country churches, such as Bucklebury Church in Berkshire and Widdington Church in Essex. However, whilst there may have been many such dials, scattered about the British Isles, at one time, they are of a fragile nature. Vandalism, neglect and natural causes, including such mundane items as tennis balls, must have reduced

an *analemma*, which resembles an elongated 'figure-of-eight" formed by an engraved dotted line, each dot denoting a day in the year. The Sun's rays are projected onto the dial-plate as a spot of light from an elliptical aperture in the gnomon (which is attached to the outside of the building, and elliptic in form to cast a circular shadow at the equinox). As the spot of light crosses the line of the analemma, the dial will indicate both the date and the time of 12 o'clock (GMT) or 1 pm (BST) to within 14 seconds! Regarded as the most accurate sundial in the British Isles, it was designed by Douglas Bateman, Honorary Secretary of the British Sundial Society and made 'in house' in 1996.

The concept of glass window sundials is not a new

**Stained Glass Dial of 1655 and its associated window,
originally from St Clement Danes Church, London**

these delightful works of scientific art to the few that remain in England and Wales today. In modern times, with the 'renaissance' of the sundial, some new stained-glass has been created, notably at Buckland Abbey in Devon, the Merchant Adventurers' Hall in York, and in the church at Toller Porcorum in Dorset.

The British Isles has a wealth of historic sundials, many of considerable interest, such as the famous sundial, said to have been designed by Sir Isaac Newton, on the south wall of the Old Court in Queens' College, Cambridge, or the equally famous 'pelican' or Turnbull Dial in Corpus Christi College, Oxford. In present times the country has also gained many new and splendid modern sundials in all the various classes mentioned. The now famous equinoctial 'dolphin' sundial at Greenwich is just one example, a mean-time dial that makes use of the analemma to indicate the time to within an accuracy of some 30 seconds.

Then there is the superb modified interpretation of the hemisphaerium scaphe sundial of Berosus, the Babylonian astronomer, who is said to have flourished on the Greek island of Cos in about 270BC, made from a flawless block of slate, as a shallow 'dish', over 1.5 metres in diameter. It is to be found in the grounds of Holker Hall in Cumbria and is a magnificent example of a modern work of scientific art, relating the past to the present. The readers of 'Sundials of the British Isles' may prefer to tour the country, looking at these many splendid sundials, in the armchair comfort of their own homes. However, should they choose to venture out, in search of some of these historic or modern sundials, they are likely to find many more dials than are mentioned in this book; but they will surely be equally well rewarded by their discoveries.

THE EARLY DIALS OF BRITAIN

Tony Wood

Saxon Dials & Mass Dials

British sundial history goes back to the 8th century. However, before that we have the remains of some much earlier Solar Observatories such as Stonehenge in Wiltshire and Newgrange in Ireland. Stonehenge was not so much a sundial as we know them but was for observing the sun and was able to record the various seasons, essential

These simple dials are known as Mass Dials or Scratch Dials, and in the past, Saxon Dials have been lumped into the same category as Mass Dials by some authors. It seems however that the distinction is sufficiently clear, with little overlap; a convenient dividing line being provided by the date 1066.

Stonehenge, one of Britain's earliest Solar Observatories

for the late Bronze Age people, c1600BC, to know when to sow their crops etc. It is still well-known for its midsummer celebrations where it shows the rising of the midsummer sun. Some observers also believe that it was used for the phases of the moon and could even predict eclipses.

In Britain, sundial history as we know it started in Saxon times. These date from before the Norman Conquest of 1066. Also, from the Conquest to the 16th century a simple form of dial appeared before the invention the 'scientific' dial with a sloping gnomon as we know them today.

Both Saxon and Mass Dials had their gnomons horizontal, perpendicular to the wall. The shadow cast would therefore not indicate regular hours but would vary considerably from winter to summer.

SAXON DIALS

A small corpus of about 30 Saxon Dials exist, stretching from the oldest British sundial at Bewcastle in Cumbria south to Gloucestershire, Hampshire and Kent. Their design is remarkably varied but is usually distinguished by good quality carving, sometimes ornate and often in relief.

Britain's earliest Sundial at Bewcastle ^{MC}

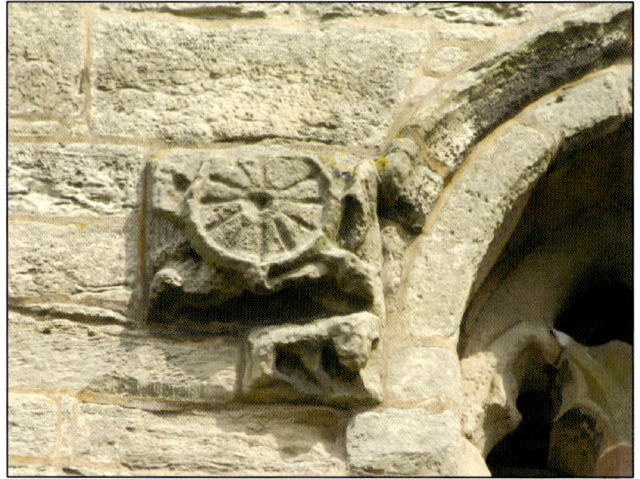

Saxon Dial high on the tower at Dunchurch ^{MC}

Saxon Dial at Daglingworth, Gloucestershire ^{AW}

The time marker divisions may be by three or four lines but some early ones, notably Bewcastle and Dalton-le-Dale, Co. Durham have twelve.

As their origin dates from an era where the churches were built mainly of wood it is likely that many surviving Saxon Dials are not in their original positions; indeed some are now placed very high up as at Dunchurch, Warwickshire and formerly at Stoke d'Abernon in Surrey. Some dials have associated inscriptions, notably Kirkdale in North Yorkshire. The text is Anglo-Saxon but there are other texts in Latin and one dial, in Kent, has Runic characters.

The lines marking morning and afternoon masses are frequently marked by an end cross as is the noon line. Needless to say their condition is usually well worn but those now within covered porches are well preserved as at Daglingworth, Gloucestershire. Also quite well preserved in Gloucestershire but damaged are the dials at Eastleach Martin and Stowell, the latter being regarded as a bit of a mystery as it is unique in being of a square design.

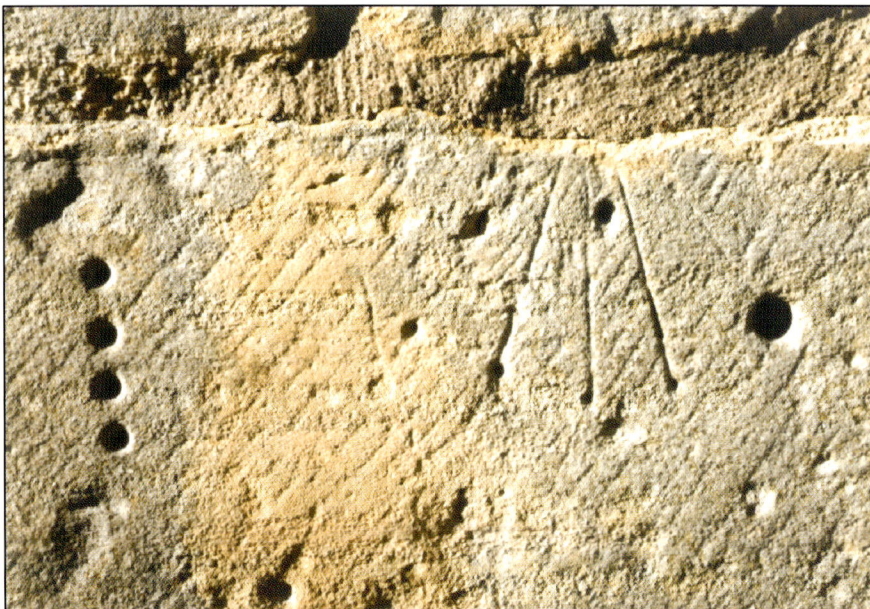

Mass Dial at St Barbara, Ashton-under-Hill, Worcestershire ^{AW}

MASS DIALS

These are normally about 8 or 9 inches across and rather roughly cut. They come in a variety of designs, from semicircles of dots to complete

St Martin, Wootton, Kent AW

St Bartholomew, Waltham, Kent AW

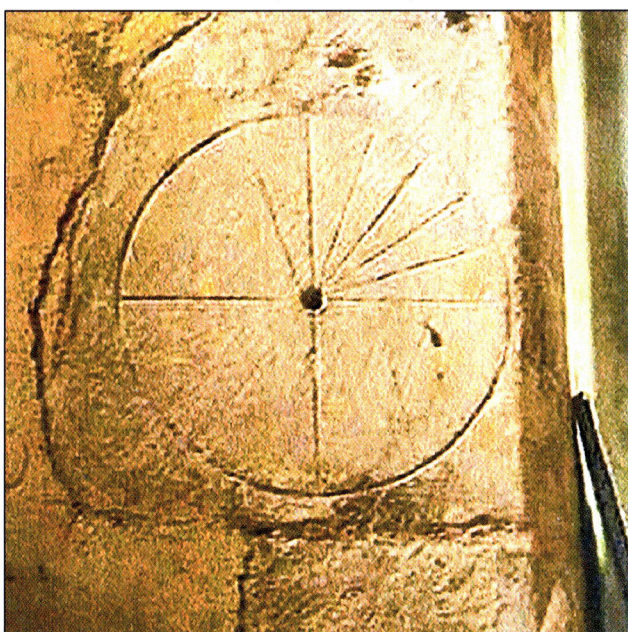

St John the Baptist, Little Missenden,
Buckinghamshire AW

inverted example is at Little Missenden Bucks.

The usual form of later dials has radii principally in the lower half and may be bounded by a circle or an arc of 'pocks'. Waltham, Kent is typical. Lines above the horizontal are a bit of a mystery in practical terms but were probably 'space filling' or alignment lines used if the church was lime-washed over.

The quality of carving is very variable and some very crudely cut dials are found. Great Witcombe, Gloucestershire and Rhossili, West Glamorgan are good examples. Although usually 8 or 9 inches across (200mm), there is actually an enormous range of sizes recorded, from 50mm diameter at

circles with associated radii. Their chronology is difficult to determine but it is thought that the earliest mass dials may be a simple carving of four or five lines from a gnomon hole, possibly with a circle or arc later. Ashton-under-Hill, Worcester-shire and Wootton, Kent are typical examples. The Ashton-under-Hill stone also includes an example of 'four holes in a row'. The row of dots would originally have been horizontal with the gnomon stuck into a mortar line above. As with Saxon dials, the 'time marks' would indicate the church services rather than the time of day.

Re-positioning of Mass Dials is quite common, an

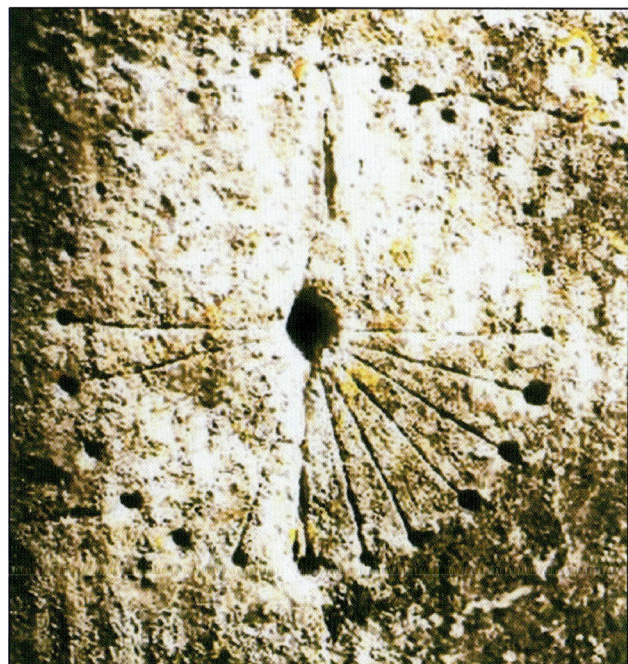

St Mary, Great Witcombe, Gloucestershire AW

Rhossili, West Glamorgan

**Sandhurst, Gloucestershire,
with Arabic Numerals**

**Avening, Gloucestershire,
with dots in place of numerals**

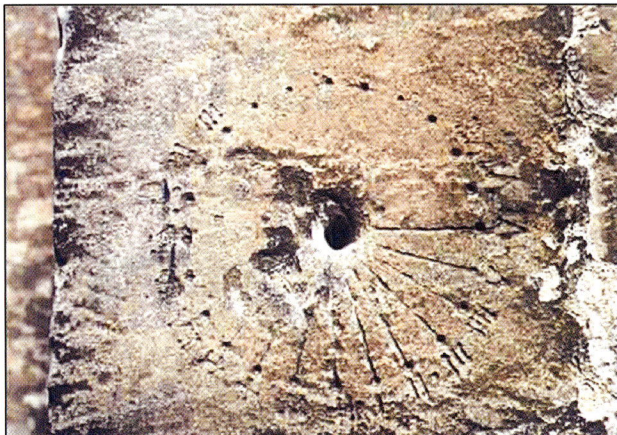

**Bibury, Gloucestershire,
with Roman Numerals**

Lynsted, Kent to around 1000mm diameter at Badsey in Worcestershire. The very large and very small ones being quite rare. Frequently, several dials appear on one church. Why this occurs is not known; possibly there are several reasons. So far no regional variation in design has been established.

The final phase is probably 'dials with numbers round the edge'. The numbers can be in the form of some dot arrangement – Avening, Gloucestershire, Roman at Bibury or even Arabic as at Sandhurst, Gloucestershire. This latter dial probably represents the end of the Mass Dial line as the 'hour line' spacing seems to be variable rather than a uniform 15° as on many dials. Also there is no 'mass marker' at around 9am which is a feature of many Mass Dials.

Within the British Isles they are largely an English device; some English counties have few dials ('dial deserts'), whilst others are well documented to have many. Scotland and Wales seem to have few; Rhossili, West Glamorgan is one of the westernmost. Their boundaries of occurrence are slowly appearing. Cornwall, Lancashire and West Yorkshire have practically no dials whereas other counties, like Gloucestershire, Lincolnshire, Somerset and Kent have many.

It now seems likely that the Mass Dial may have been introduced into Britain from France and research is continuing into their origins on both sides of the Channel.

SOUTH WEST

Tony Belk, David Le Conte, Carolyn Martin & Richard Thorne
Cornwall, Devon, Dorset, Somerset, Wiltshire & Channel Islands

Until the coming of the railways and the Brunel railway bridge across the Tamar in 1859, the south of Cornwall was virtually cut off from the rest of the country, a wild and remote peninsula full of wonders to be explored. Transport by road was slow and hazardous and travellers from other parts of Britain usually arrived by sea. A strong both employment and visitor appeal. Because Cornwall has a mild and damp climate, Victorian plant hunters introduced tender and exotic plants to Cornish gardens and in the spring, camellias and rhododendrons show off cascades of blooms.

The strong, clear light of Cornwall has always

The Cathedral at Salisbury

Celtic influence can still be seen in the villages, with names such as St Winnow or St Wenn and it is said that *there are more saints in Cornwall than in heaven*. The county has many small settlements but no large towns and in the far west, Iron Age field patterns can still be seen alongside granite standing stones and Celtic crosses. The former staple industries of fishing and tin mining have been replaced, in part, by heritage sites or industrial museums, creating attracted artists, notably to Newlyn and St Ives and generally, there is strong support for the arts. The jewels of Cornwall are the ancient churches, almost untouched by history. Many boast fine but unpretentious sundials and most are made of slate, an inexpensive material for a poor county. Often the dials are carved with the names of the churchwardens in the parish; these prominent individuals would have financed the project and controlled the design.

Nearby Devon is a county with similar diversity. To the north, where the rugged headlands push out into the Bristol Channel there are wonderful sandy beaches and waterfalls. Here the coast path provides spectacular walking for both the long distance enthusiast and the summer afternoon stroller. By contrast the southern coast is more rolling and because of the warm waters of the Gulf Stream has been developed as the main holiday area for the county. Exeter the county town which was a settlement long before the Romans found their way so far west, binds together the West Country, its independence, its loyalty to the throne, its pride, its trade, its history, all of these traits abound here.

Plymouth Hoe, where Sir Francis Drake played bowls before attacking the Spanish Armada 1588 overlooks the deep-water naval dockyard of Devonport at the estuary of the River Tamar which for all of its length forms a natural boundary with Cornwall. One other river flows into the Sound, the Plym, and it was from a small harbour at the mouth of this river that the Pilgrim Fathers set sail to the New World in 1620.

Between these coasts Devon can boast not one, but two of England's finest National Parks. Exmoor in the northern part has many steep sided valleys and is vividly described in the story of Lorna Doone.

Dartmoor in the south has been described as England's last wilderness, set in 365 square miles it is a land of myth and legend. The vast majority of Devon's sundials are on religious buildings and, like Cornwall, many are constructed of slate. Many of the makers were unknown and in most cases, whilst being superb craftsmen were often illiterate and would copy the design laid out by the vicar of the parish. Whilst many of these dials are rather plain, some with only the hour lines and numerals, others display a wealth of dial furniture.

In the Wiltshire part of the region the geology provides an appealing downland landscape. The Marlborough Downs and Salisbury Plain exemplify the contrast. It is often referred to as chalk and cheese – chalk downs for sheep and plains for arable and dairy produce. Wiltshire is the richest county in terms of prehistoric remains, and this emphasises also the stone available for monuments and buildings. The Ridgeway, Stonehenge, Avebury, all in the land hitherto known as Wessex. Later came the canals and railways, both

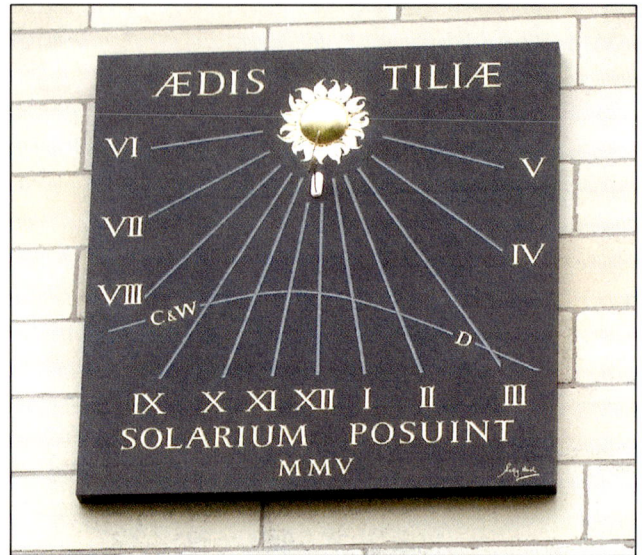

Linden House, Bath

of which are still present and add character to its towns, villages and countryside. The advent of the motorways has cut through the county but also given access to its hidden treasures for those from some distance away. The Army has for long had a major presence in the county and more recently the racing industry has brought added prosperity. Apart from the scenery the villages have their own attraction with many thatched buildings adding to their charm.

The County of Somerset is of considerable beauty with the splendour of the Quantocks, the Mendips, Cheddar Gorge and the wilderness of Exmoor. The cities of Wells and Bath have their cathedrals and the town of Glastonbury is reputedly in the Isle of Avalon, famed for the legends of King Arthur.

The old City of Bath, once a Roman spa town, has many fine buildings. In the Pump Room will be found a longcase clock by Thomas Tompion and just outside is one of his sundials for setting it.

The English Channel coastline of Dorset is famed for its natural features; Poole Harbour, Chesil Beach, Portland Bill and the glorious Lulworth Cove. It is the county of Thomas Hardy who used Dorchester as his model for Casterbridge in his novels. The seaside towns of Weymouth and Lyme Regis are worth a visit. Lyme Regis was the setting for the film, 'The French Lieutenant's Woman' by John Fowles.

The South West is truly a varied and interesting part of our islands, with many special places in which to get away from it all, and a great hunting ground for sundials. Those on the following pages will hopefully inspire the visitor to take a closer look at the region.

BATH

A modern Vertical Dial by Sally Hersh may be found at Linden House in College Road, Lansdown. It was commissioned by the owners to mark their wedding anniversary on 2nd August, hence the line of declination in that position, and their initials. It also commemorates 10 years of their living there. It was felt that the house could be enhanced by the presence of a large decorative sundial placed centrally in the front, and with the inclusion of a raised dome and sunburst chosen to give additional enjoyment. The owners composed the inscription, in Latin:

ÆDIS TILIÆ SOLARIUM POSUINT MMV
(Linden House Sundial Placed 2005)

The dial is 3 feet square and made from Cumbrian blue slate and the numerals are gilded with gold leaf. The lettering and the decorative sunburst are deeply 'V' cut.

BLEADON, North Somerset

The church at Bleadon, just to the south of Weston-Super-Mare, has a set of most unusual Scaphe Dials carved into one of its south buttresses. These dials are now difficult to see having weathered over the ages and are covered in lichens. Their markings are now invisible. However, it is still possible to make out their functions. Conveniently the top of the buttress is tilted such that it lies approximately in line with the Earth's axis making such dials much easier to construct. The upper dial is of simple spherical section with a rusting iron gnomon. This will have shown the hours from 6am to 6pm and probably solstice and equinox lines.

The next dial down is of a triangular 'V' section, the two halves separated by a remaining ridge which acts as the gnomon. The left dial would show the morning hours and the right the afternoon hours. This dial may have had a nodus, the shadow of which would travel along the groove in the 'V' at the equinoxes.

Set of Scaphe Dials on a sloping buttress at Bleadon

The lower dial is a hollow with two ridges remaining, each acting as a gnomon for each of the two outer sections, similar to the triangular dial above. However the centre section is of cylindrical form and would have shown the hours from either gnomon throughout the day.

What is not obvious from the illustration is that there are two further Scaphe Dials in the east and west faces of the buttress in the position of the central triangular one. These are two spherical hollows using the buttress edge as gnomons, east for morning hours and west for afternoon.

Note that in Eden & Lloyd's 'Book of Sun-Dials' of 1900 these dials are shown somewhat differently, perhaps wrongly drawn from information supplied.

The church also has a fine Mass Dial on another buttress.

BREMHILL, Wiltshire

A 15th century market woman who had to walk from her home at Bremhill to Chippenham market to sell her eggs and poultry is an unlikely candidate for commemoration on a sundial. However Maud Heath was so tired of the muddy track that she had to follow and the accompanying falls, breakages and washing that this involved that she established a raised causeway. It is over 4 miles long and consists of 64 segmental arches carrying the footpath beside the road. In 1474 she bequeathed £8 per year forever in money and houses to pay for the causeway and as the charity was maintained it also paid for a

Pillar at Bremhill in Memory of Maud Heath ^{AB}

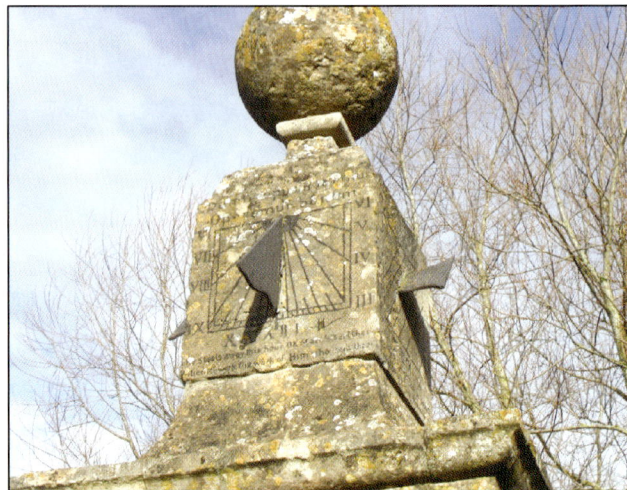

Cube Dial at Bremhill ^{AB}

bridge over the River Avon. The story so impressed the local community that a pillar commemorating the causeway was erected in 1698. The pillar carries the inscription:

To the memory of the worthy MAUD HEATH, of Langley Burrell widow, who in the year of grace, 1474, for the good of Travellers did in Charity bestow in Land and haufes about eight pounds a year forever to be laid out on the Highway and Causey leading from Wick Hill to Chippenham Clift. This pillar was set up by the Feoffees, 1698. Injure me not.

Above the pillar is a cube with east, south and west facing sundials, each with an inscription, topped by a stone ball.

The east facing dial has inscriptions:

Volat tempus
(Time Flies)
**Oh Early passenger look up be wise,
And think how night and daytime onward flies**

The west facing dial:

**Haste traveller the sun is sinking now,
He shall return again but never thou.**

The south facing dial (the legible motto):

**Ere steals away this hour oh Man is sent thee
Hasten the work of Him who sent thee**

The memorial has recently been repaired after having been struck by a car in 1996. To further remember Maud Heath on a site overlooking the causeway at Wick Hill there is another monument. It carries the inscription:

Erected at the joint expense of Henry Marquis of Lansdowne, Lord of the Manor, and Wm. L. Bowles, vicar of the parish of Bremhill, trustees 1838
**Thou who dost pause on this aerial height
Where Maud Heath's Pathway
winds in shade or light
Christian wayfarer in a world of strife
Be still and ponder on the Path of Life
W. L. B.**

BRENT TOR, Devon

Brent Tor church Vertical Dial is included here, not because it is one of the finest examples in the county but more because of its location. It is believed to have the oldest slate Vertical Dial set upon any church in the county. The church is situated upon a large outcrop of rock 1130 feet above sea level and this imposing natural feature can be seen for miles around.

There are two legends associated with its location. One, that it was erected by a wealthy merchant, who vowed, in the midst of a tremendous storm at sea (possibly whilst addressing himself to his patron, St Michael), that if he escaped, he would build a church on the first land he saw. The other, that its foundation was originally laid at the foot of the hill; but that the enemy of all angels, the Prince of Darkness, removed the stones by night from the base to the summit, probably to be nearer his own dominion, but that, immediately on the church being dedicated to St Michael, the patron of the edifice hurled upon the devil such

Slate Dial at Brent Tor

BRIDESTOWE, Devon

In mid-Devon there is a Vertical Dial on the wall of The Royal Oak at Bridestowe. In the Gatty 'Book of Sun-Dials' of 1890, a dial is referred to as being on St Bridget's Church with the motto:

Obrepit Non Intellecta Senectus
(Old Age creeps on unawares)

and dated 1714. This dial is actually on the west facing wall of the pub. As this wall faces the porch of the church, from which it can be seen about 50 metres away, it is interesting to speculate that perhaps the parishioners were instrumental in its positioning, thus ensuring they were not late for either church services or opening times.

an enormous mass of rock that he never afterwards ventured to approach it. However, more prosaically the monks of Tavistock built it in 1319 and dedicated it to St Michael. The building is only 37 feet long and 15 feet wide with a 40 feet high bell tower on which the dial, dated 1624, is located.

Consequently the views from all sides of the summit over Dartmoor and the Tamar Valley are quite spectacular. The church is reached from the car park at the bottom and involves a half mile walk up a grass-terraced path to the summit. This may involve some effort to climb to the top but on a fine day this is well rewarded by the surrounding views and of course the dial itself.

**Dial with date 1586 on
Ilminster Grammar School**

ILMINSTER, Somerset

There is a fine Perpendicular Minster Church here (c1460) with a rather worn Vertical Dial above its south porch. However, behind the church in Court Barton is situated the Old Grammar School which moved here in 1586 but was closed in 1971. It has a fine semi-circular Vertical Dial on its front. The bracket below the dial carries the date 1586, but the dial appears to be much later. It carries the motto across the top:

SIC TRANSEVNT DIES TVI
(So Pass Thy Days)

The dial was recorded in 'The Book of Sun-Dials' in 1900, so it has been on the school for many years.

INGLESHAM, Swindon

Inglesham is in a small peninsular of north Wiltshire bordering on Gloucestershire and Oxfordshire and close to the River Thames. The village has now moved further south, leaving the ancient church of St John the Baptist isolated.

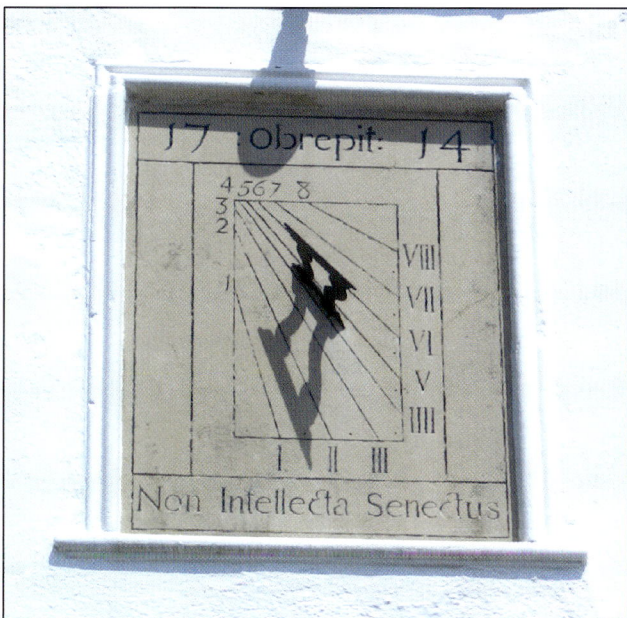

**West Facing Dial at the Royal Oak,
Bridestowe**

Mass Dial at Inglesham AB

Horizontal Dial at Lacock Abbey AB

latitude 51° 24'. The maker's inscription reads:
**Made by Tho Wright,
Instrument maker to His Royal Highness
ye Prince of Wales**

This will have been Frederick, son of George II. Frederick died in 1751 and his son became George III in 1760. Thomas Wright, who having been an apprentice of John Rowley, was active from 1718 and was the leading dialmaker of his day. The dial contains many features including transversals which were probably introduced first by John Rowley. There is a compass rose central to the dial plate with 16 directions marked. It covers hours from 4am to 8pm marked in Roman numerals and one minute intervals. There is a ring with 1/4 hours and another with 1/8 hours. The transversals are diagonal lines around the time scale which allow more accurate time readings than are possible with closely spaced minute marks. Noon is shown at 17 other locations:

The original south wall has a late Anglo-Saxon carving on which a Mass Dial has been inscribed. This is now in the south aisle of the church, a later addition. William Morris carefully restored the church in 1888 and it now looks just as it would have done in the 16th century. Pevsner considers the carving of the Virgin and Child with the Hand of God above deserving of being better known. He points out that it is an unconventional composition with the bodies in profile and the heads full face. It is also a very unconventional place to find a Mass Dial.

LACOCK, Wiltshire

Lacock Abbey was founded by Ela Countess of Salisbury and widow of William Longespée, who retired to Lacock in 1238 and became its first abbess. The buildings date from 1247 but have been much added to over the centuries. After the dissolution the premises were sold and finally belonged to the Talbot family until 1958. W H Fox Talbot did his ground breaking photographic work here. John Ivory Talbot, who inherited the estate in 1714 modified the house in the Gothic style and it is probably as part of this work that the pedestal mounted sundial on the south-east terrace was installed. It is a superb Horizontal Dial with coat of arms of its sponsor, and the

**Horizontal Dial at Lacock Abbey with its
superb engraving by Thomas Wright** AB

Inside, the visitor is met with brightness and light and a truly historic church to explore. Colour streams in through the modern stained glass east window, dedicated in 1973, with a Cornish cross as the central feature, surrounded by several local saints. The original church was founded by St Uny, who is said to have arrived as a missionary from Ireland in the 6th century. Six hundred years later, the Normans built a church on the site and the font in the church dates from these times. Norman pillars can be seen in the nave, together with an arch from the original church. Over the years there has been a continual battle against the sand blown in from the shore which occasionally filled the inside of the church, but the situation improved when marram grass was planted on the towans in 1824. In the hey-day of smuggling, it is said that the church was used to store French brandy!

For art lovers, nearby St Ives has much to recommend it, with the Tate St Ives, numerous art shops and galleries. There is also a Vertical Dial on the Parish church and another on a private house, which can be seen from the road close to the Barbara Hepworth Museum. To the east of Lelant, the saltings at Hayle are of international importance and attract a wide range of migrant and overwintering birds.

The Lelant gnomon supported by a skeleton

Peking, Bengal, Fort St George (Malasia), Isphan (Iran), Surrat (India), Jerusalem, Moscow, Hannover, Iamaico, Constantinople, Dublin, Lisbon, Madrea, Fernando Isle (off Brazil), St Sebastian, Barbadoes and Mexico. There is an Equation of Time scale showing when the watch is fast or slow and by how much throughout the year. A close inspection of this scale shows that the maker was still using the Julian calendar. This in turn means that the dial was inscribed before September 1752.

LELANT, Cornwall

The approach to the church of St Uny, through a gloomy arch of yew trees, is hardly encouraging, especially when faced with a menacing skeletal figure on the Vertical Dial above the south porch. With an hourglass in one hand and a bone in the other, the crowned skeleton supporting the gnomon points to the fragility of life and approaching death. Made of bronze, with an elaborate border and Roman numerals, the dial sits over a niche previously occupied by a religious statue. It can be dated as being between 1714 and 1760 according to the Royal Georgian Arms on the face. A unique sundial, it was repaired and restored in 2002.

Vertical Dial at Lelant,
high above the church porch.

MARAZION, Cornwall

If there is one image that represents the essence of Cornwall, it must be the views of St Michael's Mount, across Mount's Bay from the causeway at Marazion. But unknown to many, St Michael's Mount has another attraction for visitors, a fine 19th century bronze Horizontal Dial, with unrivalled views out to sea, telling the time across the waves. It was made by the firm of Troughton and Simms of London, with much dial furniture and a sturdy gnomon. The sundial is set into the battlements overlooking the English Channel on the sunny south side of the island.

The Mount was originally the site of a Benedictine Priory and the castle itself dates from the 12th century. A place of myth and history, with the giant's stone heart set into the entrance path and

Human Analemmatic Dial at Killacourt

time is read by a person standing on the correct month in the central dial plate and the shadow cast tells the time.

Commissioned by Newquay Town Council to commemorate the Millennium and carved by Sarah Stewart-Smith, the dial has a strong Cornish identity. The illustrations interpret Newquay's past and present, the decline of the pilchard fishing industry and the rise of tourism. As well as coats of arms and a crest, there are a number of Cornish symbols and words. *Tewyn Plustri* is the old name for Newquay, for example and *Ro an mor* means 'gift of the sea'. The verse on the sundial encapsulates the theme:-

The Horizontal Dial at St Michael's Mount looking out across the sea

The pilchards are come
Hevva be heard
The town from the top
To the bottom is stirred

'Hevva' is the shout that used to be given from the cliff-top lookout to alert the townspeople that a shoal of pilchards had been sighted.

Queen Victoria's footprint on the quay. The Mount has been a private residence since the 17th century and the present St Aubyn family have their own private quarters within the castle. The property is now managed by the National Trust and access is limited to opening hours.

Nearby church sundials can be found at Paul, Penzance, Gulval, Madron and St Hilary.

NEWQUAY, Cornwall

The Human Analemmatic Dial at Killacourt has a spectacular view across the golden sands of Towan beach. Considering the proximity to Newquay's town centre and the number of holidaymakers frequenting the area, this unusual Delabole slate Analemmatic Dial has worn well. The

There is a plaque alongside the dial, explaining that the inner ring of numerals is to be used in summertime and the outer ring in wintertime. A time capsule has been buried under the main sundial date plate, to be opened in the year 3000.

In Newquay itself there is a large Aquarium and a Zoo to visit and both are open all year. Trerice, which belongs to the National Trust, has a Horizontal Dial and there are church dials at St Newlyn East and St Columb Minor. Robert Oliver's carving on the latter is exceptionally elaborate and decorative.

Dial plate by Ben Jones at Ottery St Mary

Reclining dial at Ottery St Mary

OTTERY-ST-MARY, Devon

The Reclining Dial at the hospital in Ottery-St-Mary was made by Ben Jones in 2001. It is made of green slate and Ham Hill stone and has a brass gnomon. It was made reclining and not horizontal so that someone standing on the nearby path could easily read it. It was set up in memory of John Sedgman, being a gift from Sally Sedgman who was so grateful for all the care given to her husband at the hospital.

The verse that Sally chose to have carved on the dial is:

**SORROW MAY ENDURE
FOR A NIGHT,
BUT JOY COMETH IN
THE MORNING**

By placing the two lines of the text around the top and bottom of the circular dial-plate sorrow is not so obviously the first word of the verse and the line with joy in it is nearest to the reader.

PERRANPORTH, Cornwall

Set high above the beach at Droskyn Point, with views across the water to Ligger Point and Carter's Rock, is a large Cornish Horizontal Sundial. It is a strong contender for the title of the most scenic setting for a sundial in the county. The dial was commissioned by Perranzabuloe Parish Council to mark the turn of the Millennium in 2000 and is built on the site of the former Droskyn mine.

The sheer size of the dial (some 10 metres across) alone makes it impressive and the four large upright granite posts, marked with the points of

The large monumental dial at Perranporth with its wonderful vista over the bay

23

the compass, together with the 12 smaller posts in between are of some antiquity. These were found under an old bridge in Perranporth when flood defence work was being carried out. The hour posts are sunk deeply into the ground and the whole dial rests on a concrete raft for stability. The solid stainless steel gnomon is approximately 3 metres in length. In addition, the surrounding area has been landscaped and resembles an amphitheatre, with granite bench seating so that visitors can enjoy the dial to the full in this spectacular setting.

Large Horizontal Dial at Plymouth

The site belongs to Perranzabuloe Parish Council who are responsible for maintaining the area and the design was the inspiration of Stuart Thorn, a designer from Perranporth. The whole project is said to have cost around £25,000.

There are three explanatory plaques above the dial, giving details of the funders, an Equation of Time and the third plaque, with some words in Cornish, reads as follows:

> *Dydhyel Howlyek Droskyn -*
> *The Droskyn Sundial*
> *Sited on the former Droskyn mine, this*
> *Millennium sundial points north*
> *across Perran Bay to Penhale*
> *and Gull Rock. St Piran, Cornwall's*
> *foremost saint landed on this beach*
> *from Ireland in the 6th century*
> *and established an oratory in the dunes.*
> *The sundial is calculated to Cornish time*
> *and it shows a difference of 20 mins to GMT.*

Perranporth has an extensive range of sand dunes to explore, where the original church of St Piran is buried in the sand and a short distance away there is an Iron Age fort at Piran Round.

PLYMOUTH

The City of Plymouth was severely damaged during the Second World War blitz and at the end of hostilities the whole area was a mass of rubble with only a few buildings left standing. In 1948 a Plan for Plymouth was proposed and subsequently the reconstruction work was carried out. This involved demolishing the few remaining buildings left standing, and the result is the completion in 1953 of the shopping precinct you see today.

The Horizontal Sundial, commissioned by Plymouth City Council, was constructed in 1988 to commemorate the pedestrian scheme and the celebration of the 400th anniversary of the defeat of the Spanish Armada. It is set in the middle of Plymouth City Centre at the junction of Armada Way and New George Street.

Its stainless steel gnomon, 8 metres high, gives the appearance of a sailing ship on course for Plymouth Hoe. It soars above the two dials over which water flows. The upper dial is 2.2 metres diameter and the lower dial 3.2 metres. The water cascades into a 6.4 metre pool below. Steps surround this sundial, where at the top are granite plinths showing the National flags and the Equation of Time set alternately spaced around the edge of the pool.

PRINCETOWN, Devon

Unusually, the county can boast at least three public houses with dials. At Princetown is a Horizontal Dial in the courtyard at the Plume of Feathers Inn.

This brass dial was commissioned by James Langton to enhance the exterior of the hostelry. It was designed and manufactured by Richard Thorne, an engraver who worked for many years

Horizontal Dial at the Plume and Feathers, Princetown

III, whose family lived there from 1660. He died in 1780 and has a monument in the north transept of the cathedral praising him for his great learning. On the south wall of the house overlooking the close is a high quality Vertical Dial dated 1749. Across the top it carries the motto:

LIFE'S BUT A WALKING SHADOW

Below the dial is a text from St John's Gospel 17:3

THIS IS LIFE ETERNAL THAT THEY MAY KNOW THEE THE ONLY TRUE GOD, AND JESUS CHRIST, WHOM THOU HAST SENT

The hours are marked with gilded Roman numerals from VII through XII to VI and there are ½ hour, ¼ hour and 5 minute

in Plymouth. It is believed to be one of the first sundials engraved using a computerised engraving machine. The dial plate incorporates the Prince of Wales feathers and the motto:

ICH DIEN
(I Serve)

The plate has the Equation of Time and the length of time that the Langton family have owned the Inn. The gnomon incorporates a stylised '**T**' to indicate the makers surname.

Another dial on a pub, an octagonal slate Vertical Dial, is on the wall of The Pack of Cards Inn at Combe Martin.

SALISBURY, Wiltshire

Just in the Cathedral Close in Salisbury through St Ann's Gate on the right stands Malmesbury House. This is considered by Pevsner to be of Queen Anne date, built before 1719. It is based on an existing 15th or 16th century house. The first floor library was made for James Harris

Dial in the Cathedral Close at Salisbury

marks. There are also eleven curved lines marking the sun's declination on certain dates. These dates are according to the Julian calendar that was in operation in England when the dial was first made. Conveniently there is a plaque which was placed on the garden wall in June 1989, when the dial was last restored, which gives a succinct account of the Reformation of the Calendar:

An important timely point of interest to the passerby…. In the year of our Lord 1752, the Reformation of the Calendar took place ~ see the Wall Dial above dated 1749. This Julian Calendar made the year too short, thus the accumulated error amounted to eleven days. England adopted the Gregorian or Reformed Calendar, so the next day after Sept 2nd 1752 became Sept 14th 1752.

In all, the dial and its explanation are a fitting memorial to James Harris III's great learning.

ST BURYAN, Cornwall

Like a lighthouse adrift in the landscape, the tall, square church tower at St Buryan beckons to the faithful across the bleak landscape of West Penwith. The stark grey stone exterior of the church contrasts sharply with the rich interior, which amongst other treasures boasts a well restored 16th century rood screen and a stone archway indicating the position of King Athestan's original 10th century church. Over the years, the church

St Buryan Church

has been restored and refurbished many times, reflecting its importance in the heart of the village. Outside, in the churchyard, there are notable memorials and two fine Celtic crosses; the parish of St Buryan is said to possess more Celtic crosses than any other parish in the county.

The Vertical Dial at St Buryan is essential viewing. It is dated 1747 and set prominently above the door of the south porch. It is said to be the most advanced church sundial in Cornwall, with a calendar and a solar azimuth scale. The latter shows how far the sun has travelled across the landscape and is read from the shadow cast by the nodus (a small cut on the gnomon) on the vertical lines incised on the face of the dial. The nodus is also used to read the celestial zodiac calendar. Here the shadow passes over the curved horizontal lines from left to right, moving down and then up the face of the dial over a 12 month period, beginning with Aries and the spring equinox in March.

When telling the time, the number four in the Roman numerals is shown as **IIII** rather than the more usual **IV** and the craftsman seems to have had problems in squeezing the last few letters of the motto into the allotted space. It reads:

The slate dial at St Buryan

Slate Vertical Dial at St Neot

several claims to fame. The parish is the largest in the county and there is a finely decorated 10th century Celtic cross in the churchyard. Inside, the glory of the church can be seen in the famous 16th century stained glass windows, depicting scenes from the Creation, Noah and the Flood as well as the lives of the saints. Although the windows were restored in the 19th century, much of the original glass remains.

St Neot has two sundials, a slate 17th century Vertical Dial on the south wall of the nave and a modern Scaphe Dial in the churchyard. The Vertical Dial dates from the golden age of Cornish dialling and was engraved by **William Olliuer** (Oliver) in 1682. A geometric border frames the dial, with a half double rose around the top of the gnomon and an intriguing verse at the head of the dial:

JAMBICA UITA
VT HORA VITA PRÆTERIT CITO PEDE
PRIOR BREVIS, SECUNDA VITA LONGIOR
MANENS IN ÆVA SEMPITERNA CUM DEO
ET INDE VITA NOSTRA DICTA IAMBICA

A truly poetic verse, written around the word iambic and can be roughly translated as:-

(An hour of life goes by swiftly
Our first life is short, our second longer
Lasting for all eternity with God
And therefore our life is said to be an iamb)

Pereunt et Imputant[ur]
(They perish and are recorded)
emphasising life's fragility and the final reckoning.

There are several other church dials of interest nearby, at St Levan, Sennen, Paul, Madron, Penzance, Zennor (the church with the mermaid) and Morvah (one of the oldest dials in the county) as well as a dial with a difference, on the battlements at St Michael's Mount. In addition, the surrounding area offers gardens to be visited and the freedom of the whole of the Land's End peninsula where the Iron Age field patterns can still be seen.

ST NEOT, Cornwall
On the edge of Bodmin Moor and close to urban centres, St Neot has a foot in both camps and yet manages to retain the quiet charm of a true Cornish village. In 2004, St Neot was voted the best village in the South West.

The solid grey stone 15th century church of St Neot sits above the village and has

Modern multiple Scaphe Dial at St Neot

Cube Dial on Pillar at Steeple Ashton

In poetic terms, an iamb consists of two syllables, the first short and the last long. The verse thus hints at our short life on Earth and the longer life to come.

In complete contrast, a modern multiple Scaphe memorial dial, designed by S H Grylls can be found on the south side of the church, close to the church wall and gate. A scaphe dial is where the edge of a hollowed out area acts as a gnomon, casting its shadow into the hollow, which is inscribed with the hours of the day.

There is also a holy well about 300 yards from the church, and further afield, there are the Carnglaze Slate Caverns to explore and delightful walks through the woods to the Golitha Falls.

Another church sundial can be seen at nearby Menheniot.

STEEPLE ASHTON, Wiltshire

The view along the High Street of Steeple Ashton is very attractive with Ashton House, many cottages of different styles, and the village green with a windowless lockup and the market cross. The cross, which established Steeple Ashton as a market centre, is surmounted by a cube with four Vertical Dials, a ball, a cross and a crown. The dials face north-east, south-east, south-west and north-west, and have quite elaborate scrolled gnomons. In each case there is a sunburst at the origin of the gnomon and a gilded sun at the end

away from the dial face. The half-hours are marked and the dial faces are in good condition. The inscription on the column of the market cross claims that it was founded in 1071. It is known that in medieval times the local estates belonged to Romsey Abbey and that a weekly market and annual fair were held at Steeple Ashton in 1266. The present column dates from the reign of Charles II, 1679, and the dials are thought to have been added at the accession of George I in 1714. The cross claims to have been repaired in 1783, 1820 and 1887. To mark the coronation of Queen Elizabeth II the dials were refurbished by Ivor Smith, a local blacksmith, and they have stayed in good condition since. The south-east dial covers the hours from 4am to 2pm. The south-west dial covers the hours from 10am to 8pm. The north-east dial covers 4am to 8am, and the north-west dial 4pm to 8pm. The dial designer only needed to do one set of hour angle calculations as the north-east and north-west dials are mirror images and the south-east and south-west dials are also mirror images. In both cases the hours are inscribed on the opposite side of noon. As is often found, noon is marked by +. At any time two faces of the dial are illuminated.

Cube Dial at Steeple Ashton

John Berry dial at Tawstock

TAWSTOCK, Devon

John Berry was one of the better-known dial makers of Devon and much of his work can be found in North Devon. He was a stonemason who was born at Muddiford in the parish of Marwood in 1724 and died in 1796 at the age of 73. He is buried in Marwood churchyard and had six children, only three of whom lived beyond the age of 22. His third son Thomas survived to go into partnership with his father, his name appearing on two dials with John.

There still exist eight dials signed and dated by John Berry. They are located at Tawstock 1757, Marwood 1762, Kentisbury 1762, Bittadon 1764, Landkey 1768, Stoke Rivers 1770, Pilton 1780 and Yarnscombe 1788.

Those made in collaboration with his son are Heanton Punchardon 1795 and Braunton 1795.

Tawstock, Marwood, Kentisbury, Bittadon, and Yarnscombe are very elaborate. The dial at Tawstock is perhaps the best of these dials with furniture showing the signs of the Zodiac as well as the lines of the Tropic of Cancer, Capricorn and the equator. There are also lines indicating midday at locations as diverse as Samarcand, Fort George (now known as Madras), Boston and Babylon. This dial like many of the period shows the correct local solar time, which, as Tawstock is 4° west of Greenwich, is 16 minutes later than GMT.

That John Berry, a simple village stonemason

from North Devon, would have had either the necessary education, or the ability to design such a dial involving all the mathematical calculations is hard to imagine. Dial making in the 18th century was probably too advanced for a local craftsmen of the period and would have relied on educated members of the church and gentry to produce designs that they could copy.

The calculation required for the delineation of a Vertical Dial was well established by this time, and much of the information John Berry needed would have been provided by the rector of Marwood parish. The living of Marwood is in the gift of St John's College, Cambridge and has always been presented to a fellow of the college who would have been educated in mathematics. Apart from the elaborate dial plate furniture of the Berry dials, it can be noticed that the ½ hour lines are represented by stylised fleur-de-lys and the 12 o'clock lines by an abstract design. Cherubs' faces or wings are often depicted on the upper corners of the dial and can be found on the Thomas Berry dial at East Buckland dated 1819.

These dials are some of the best vertical slate dials to be seen in Devon.

URCHFONT, Wiltshire

At Urchfont Manor there is a Vertical Dial declining 17° east. It has circular reliefs with '**UM**' and a phoenix and a gilded Sun at the origin. There are Arabic gilded numbers with half hour marks and a **+** for noon. The latitude is quoted as 51.3° N and longitude 1.9° W. A plaque below carries the Equation of Time graph.

Urchfont Manor was built in 1680 by Sir William Pynsent who died in 1719. His son left it in 1770

Modern Dial at Urchfont Manor

29

Dial in Italian Garden at Pencarrow CM

of pictures, porcelain and furniture. The property is open to the public during the season and is set in an attractive formal garden, surrounded by 50 acres of woodland and meadow walks.

Alongside the house are two prestigious sundials. One is set in the Italian garden, on the south side and the other is close to the entrance, on the east side. The dial in the Italian garden is manufactured by Heath and Wing, in The Strand. They flourished in London between 1740-1771. As well as the months, compass points and Equation of Natural Days, many places around the world are engraved around the perimeter. In addition, the word **CLOWANT** can be discerned on the face, indicating that it could well have been moved to Pencarrow from Clowance, another property at one time owned by the Molesworth-St Aubyn family.

The second sundial is set on a limestone pedestal with a circular granite base, possibly a mill wheel. The details on the face can be read more clearly and the dial is richly engraved, with 16 compass points, months, days and numerals. The dial is made by Thomas Heath, London 1714-1765. With Pencarrow on the dial face, the dial could have been made specifically for this location. There is a latitude degree on the dial but the worn numerals are difficult to decipher and give no indication of the history.

to William Pitt the younger. The house had been remodelled in 1690 by William Talman, who may have added the splendid doorway. Pevsner considered it to be one of the best houses of its type in Wiltshire. After William Pitt it had a number of private owners including the Dukes of Queensberry. The last private owner was Hamilton Rives Pollock, a lawyer, who bought it in 1928 and died in 1940. After the Second World War it was bought by Wiltshire County Council and after some initial pioneering work it has become established as a residential adult education college. Telephone application should be made to arrange to see the dial.

WADEBRIDGE, Cornwall

Pencarrow, near Wadebridge, an elegant Georgian stately home belonging to the Molesworth - St Aubyn family, houses an interesting collection

Dial by Heath & Wing, in The Strand CM

Dial to the east of Pencarrow House CM

Dial by Thomas Heath at Pencarrow

Vertical dials at Wimborne

Other dials to visit in the area can be found on the churches at St Breward, Blisland and Michaelstowe. In nearby Bodmin, the former County Town, there is an excellent military museum and the Bodmin & Wenford steam railway.

WARMINSTER, Wiltshire

The combination of railways and sundials is an unusual one, particularly as the advent of the railways led to the demise of local apparent time (read by sundials) and required the use of Greenwich Time to establish reliable timetables. A British Sundial Society member, who is also a railway enthusiast, lives in Warminster with a railway line at the bottom of his garden. He has

installed a west facing Vertical Dial with a very strong railway motif, made for him by Harriet James. It commemorates the centenary of the house and gives the dates. It can easily be seen from the nearby railway bridge. Its motto is:

NON MOROR
(I do not stop)

The hour lines, which are the sleepers on the railway track, are all parallel because it is a west facing dial. A Midland compound steam locomotive is featured across the top of the dial and up the right hand side is a semaphore signal. This is a unique and very pleasing dial.

Railway theme dial at Warminster

WIMBORNE, Dorset

The minster church of St Cuthberga (sister to King Ina of the West Saxons) is mostly Norman built 1120 and 1180 on Saxon foundations. On a south buttress are to be found a set of three Vertical Dials. The central one is facing almost due south showing the hours from 6am to 6 pm. On either side of the buttress are direct east and direct west dials to show the morning and evening hours in the summer months.

31

CHANNEL ISLANDS
David Le Conte

The Channel Islands, though British, are geographically French, and local churches came under French Bishops until the 16th century. Even in later centuries, when the churches were under English jurisdiction, it seems likely that French limestone was used for the construction of many of the ecclesiastical sundials. Granite is the natural local building stone but its hardness makes it difficult to carve accurate hour lines and symbols. Many churches have sundials but only one Mass Dial has been found, from Lihou Island Priory and is in the Guernsey Museum collection. There are some 50 dials (public and private) of interest in Jersey and 30 in Guernsey, with a handful in the smaller islands of Alderney and Sark.

FOREST, Guernsey

At Forest Primary School, near Guernsey Airport, a gold-coloured direct south Vertical Dial is mounted prominently on the front wall. Its 1829 date most likely marks major refurbishments to the school building following its adoption by the Forest Parish. The school was demolished and rebuilt in 1999, and the dial replaced in its present position at that time.

ST ANNE'S, Alderney

A large south declining (13° west) Vertical Dial is on the wall of the Pharmacy in Victoria Street. It was designed by local architect Doug Hamon, with calculations by Roger Pierpont, and constructed in 1998 when the building was largely re-built. The sundial uses the small round

Pharmacy dial at St Anne's

window as a centre point for the steel gnomon and is constructed using tiles obtained from the local Alderney Pottery. It acts as a striking piece of artwork in this main shopping street. It is accompanied by a plaque showing the Equation of Time curve, an explanation of it and an example of how to determine GMT or BST from the shadow reading.

ST BRELADE, Jersey

A direct south Vertical Dial is above the south door of the church of the parish of St Brelade. It is dated 1837, and is inscribed with a quotation, in French, from Psalm 144:4

L'HOMME EST SEMBLABLE
A LA VANITE: SES JOURS
SONT COMME UNE OMBRE
QUI PASSE

(Man is like to vanity:
his days are as a shadow that passeth away)

Its distinguishing feature is the unusual table of Equation of Time corrections with months denoted by signs of the zodiac, and a stepped-pyramid representation of the Equation of Time offsets from the local noon line.

Vertical Dial at Forest Primary School

St Brelade's Church, Jersey

ST HELIER, Jersey

An attractive declining Vertical Dial is on the Picket House on the north side of Royal Square, and fits into what was formerly a window. Dating from the 1820s, it was made by Jersey-man Elias le Gros, a mathematician, teacher of navigation, cartographer and civil engineer. The fact that its inscription is in English, rather than the French in common use in the island at that time, implies that it was probably intended for use by the British garrison. The inscription, headed by the word 'NOTE' in elaborate letter-ing, exhorts users to refer to 'almanacks' for the Equation of Time:

REGULATE YOUR CLOCKS BY THE SUN DIAL
CORRECTION MUST BE MADE
FOR THE EQUATION OF TIME
WHICH IS GIVEN IN ALL THE ALMANACKS

During a dispute between the civil and military authorities about the tenancy of the Picket House the sundial was plastered over, but was restored in the 1880s, the gnomon being fixed slightly below its true position. It has recently been repainted.

ST PIERRE DU BOIS, Guernsey

The Salle Paroissiale (Parish Hall) near the parish church of St Pierre du Bois, served as a parish school in the 19th century and the dial probably dates from this time, being of the type of Polar Dial known as a 'Book' or 'Open Book' Dial, quite appropriate to its school context. This design is rare. An apparently identical one, in Essex is illustrated in the 1900 edition of the

Picket House, St Helier

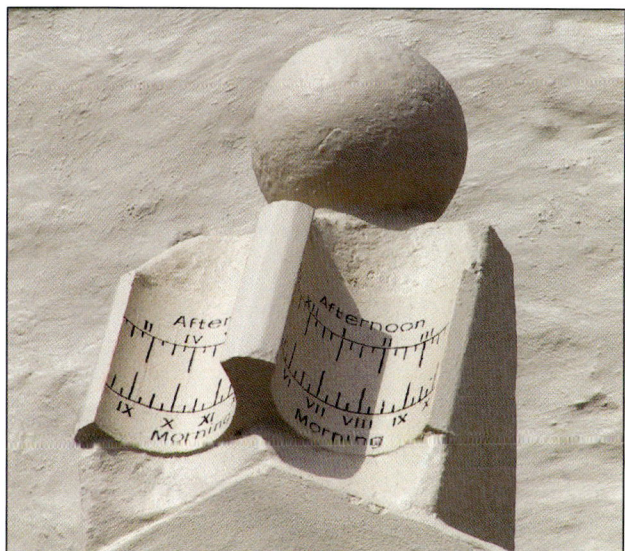
Open Book Dial at St Pierre du Bois

33

'Book of Sun-Dials' by Eden & Lloyd. The illustration clearly shows the original layout of the hour lines, with separate sets of Roman numerals for morning and afternoon on both dial plates.

The sundial has three gnomons: a central one and two side ones. It has two dial plates, one on each side of the central gnomon. It has four 'styles', which are the edges that define the shadow: one each side of the central gnomon, and the inside edges of the two side gnomons. This results in a rather peculiar arrangement for reading the times, with both sides of the dial used for morning and afternoon hours; during some periods of the day, the time may be read on either dial plate.

The hour lines are not incised in the stone of the dial, and have 'migrated' during periodic re-painting during the life of the dial. They have recently been repainted on the basis of the 1900 illustration and a calibration carried out by David Le Conte, which incorporates a 10 minute correction for longitude. The dial is angled slightly off alignment with the building, but does not point directly south, so the hour lines are not symmetrically placed. Also, the angle of the gnomons from the horizontal should be equal to the latitude, 49½° (so that they are parallel to the Earth's axis) but it is about 53°, probably having been designed for a location in England. The effect of these two inaccuracies is that the dial cannot consistently show the correct sundial time throughout the year. It is, nevertheless, a very attractive and unusual dial.

ST PETER PORT, Guernsey

The Guernsey Liberation Monument commemorates the 50th Anniversary, on 9th May 1995, of the Liberation of the Island of Guernsey from Occupation by German armed forces. Designed by local artist Eric Snell, with calculations by

Guernsey Liberation Monument

Guernsey Liberation Monument

David Le Conte, it consists of a 5 metre high obelisk of 50 layers of Guernsey granite, with 40 metres of granite seating, the curve of which accurately follows the path of the tip of the obelisk shadow on the 9th May. The shadow points in turn to inscriptions of events which happened on Liberation Day 1945. The Monument stands at the harbour of St Peter Port, on the spot where the liberating British forces were greeted by the Guernsey people.

It was unveiled by Prince Charles on the 50th anniversary and won a Civic Trust Award in 1997.

SOUTH & SOUTH EAST

Tony Ashmore, Douglas Bateman, Christopher Daniel & Michael Lowne

Buckinghamshire, Hampshire, Kent, Oxfordshire, Surrey, East Sussex, West Berkshire, West Sussex

This region, one of the most heavily populated parts of the country, is made up of expanding towns and gentle countryside with rolling downs and the Weald of Kent. Although the geographical aspects of an area often come to mind, the region has a long coastline with coastal towns and ports such as Chatham, Ramsgate, Dover, Folkestone, Brighton, Portsmouth and Southampton.

remains, a major activity, and the ancient forests provided much of the oak for the shipbuilding towns of Chatham and Portsmouth and the forests of the Weald provided charcoal for the Wealdon iron industry from Roman times to the 18th century. But even the agriculture is changing with increasing soft fruit farming and vineyards. Geologically, the area has relatively undistin-

Marlow Lock on the River Thames overlooked by the Francis Barker Sundial

All of these towns have contributed to the history and development of this part of England and to the British Isles as a whole. Indeed, contributions continue with the extra link to the continent of Europe though the Channel Tunnel.

To the north, the region is more or less bounded by the River Thames and the Chiltern Hills, to the east by the 'Garden of England' in Kent and to the west by the New Forest. Agriculture was, and

guished limestone and alluvial beds, with highest points being the white cliffs of Dover, the South Downs, Butser Hill, Hindhead and the hilly route featuring the Hogs Back, Box Hill and the north Downs. The latter forms part of the route of the ancient Pilgrim's Way to Canterbury.

One might say that the South East has seen its fair share of military history, with the Roman and Norman invasions and the threat from Napoleon;

witness the great naval dockyards at Chatham and Portsmouth, the home of the British Army at Aldershot, and the first man-powered flight in the British Isles at Farnborough, later to become the renowned research establishment. The airfields in Kent and Sussex were central to war in the air during the Battle of Britain. Although defence research continues in different parts of the region, there has been a huge growth in computer related development, software and general information technology. The latter, together with the strong commuter links to London and the expanding leisure industry, are major factors in the overall prosperity of the South East.

Winchester, the old Roman city, became King Alfred's capital in the 9th century, and William the Conqueror was crowned in 1068 in the old Minster. The building of Winchester Cathedral, in the Romanesque style, commenced in 1079. Canterbury was central to the spread of Christianity with the arrival of St Augustine in 597. Chichester with its cathedral, is another city of Roman origins. Whilst Winchester had very early Royal connections, Windsor Castle is one of the principal homes of the Royal Family, and the Castle, river and town form one of primary tourist centres in the region.

Oxford clearly represents the seat of learning in this part of the country (and the colleges are well endowed with sundials) whilst the 20th century saw the emergence of universities in the conurbations in Reading, Southampton, Guildford, Brighton and Canterbury.

The Sundial Society Register, recording nearly 700 sundials in the South and South East, gives plenty to choose from. The following selection emphasises the more unusual, the modern, and some with particularly interesting associations.

ANDOVER, Hampshire

The site of this unusual pair of east and west facing dials is most appropriate for a time sculpture since twice in the last 50 years a floral clock has been located on this pleasant green isthmus. An inclined circular flower bed marks its passing.

For the Millennium year Test Valley Borough Council decided to involve a local artist, Claire Norrington, in the production of a bronze statue. Claire had already produced a variety of animal sculptures in the region. The initial sketches showed 'Walking Man' holding a south facing Vertical Dial above his head. In January 2000, Peter Ransom was invited to consult and calculate the delineation of the dial and it became apparent that since the ideal situation was to have

'Walking Man' Dial at Andover

Walking Man facing Andover town centre, he would have to face east. Matters were further complicated by two trees, one to the south of walking man that would throw him into the shade for an hour either side of noon. The serendipitous solution was to have him holding a pair of dials, one facing east the other facing west. This solved the problem of having the dial perpendicular to the way he was facing as well as countering the tree shadow since the gnomons of east and west dials cast long shadows either side of noon with their edges well beyond the dial plates.

The statue was orientated on site on October 4th 2000 and unveiled the following day. Sharp eyed

East facing dial at Andover

observers may be able to locate the two screws inserted into the tarred path that mark the meridian line used to orientate the dial faces in a true north/south direction. The dials are located at the junction of a minor road and the B3402.

In the town centre, within walking distance, are two Vertical Dials by William Horton Heath dated 1833 and 1846. Both are in London Road, one at number 33, the other at the back of the old Savoy Theatre. They are due east of Walking Man who therefore observes them daily!

ASCOT, Windsor & Maidenhead

Very few of the racegoers at the nearby racecourse will realise that they are near to a rare example of a vertical Reclining Sundial, canted out slightly from its buttress. Instead of the dial plate lying in the vertical plane, it is sloping backwards. Even so, the angle of the gnomon is still parallel to the Earth's axis of rotation and it is merely the geometry of the hour markers that has changed. The dial has a very bold and sturdy gnomon with the hours marked from 7am to 5pm, subdivided into 15 minute intervals. The Equation of Time is given in tabular form at one minute intervals for each month of the year. However, the dial is about 4 metres above ground level, and being inclined backwards, makes the table difficult to read.

The history of the dial is given by Eden and Lloyd

Reclining Dial at All Saints' Church, Ascot [DB]

in 1900 who record the erection of the dial, a lamp, and a memorial inscription in 1896. The inscription says:

**This Dial and this Lamp are dedicated
in memory of Wolfran and Claude Guinness.
Eternal rest give unto them O Lord,
and let perpetual light shine upon them**

In addition, the dial has a very fine biblical motto from Mathew 8:9:

I ALSO AM UNDER AUTHORITY

Both of these men reached the rank of Colonel and Eden & Lloyd state that the motto has also been used in similar military connections. To add to the story of the dial, it was originally fastened in place by some large cap nuts and disaster struck in 1999 when the dial was stolen. The church authorities immediately set about getting a replacement and slides and photographs taken for the records of the British Sundial Society gave sufficient detail for an accurate replica to be made. The only difference is that the modern replica gives the name and location of the church, so that if the dial is taken again, there will be no doubt as to where it came from.

BISHOPSTONE, East Sussex

This church at Bishopstone lies about half a mile north of the A259 coast road, between Seaford and Newhaven. The original village now consists of the church and a few houses, the newer development is just off the main road. The dial is Saxon, possibly late Saxon and shows the sunrise-sunset interval divided into twelfths. The times of services (the canonical hours) are identified by cross-lines with pits at their ends. These are: sunrise (prime), mid-morning (terce), midday (sext), mid-afternoon (nones) and sunset (vespers). The gnomon is modern. It is possible that the dial was originally free-standing. The significance of the inscription **EADRIC** is unknown.

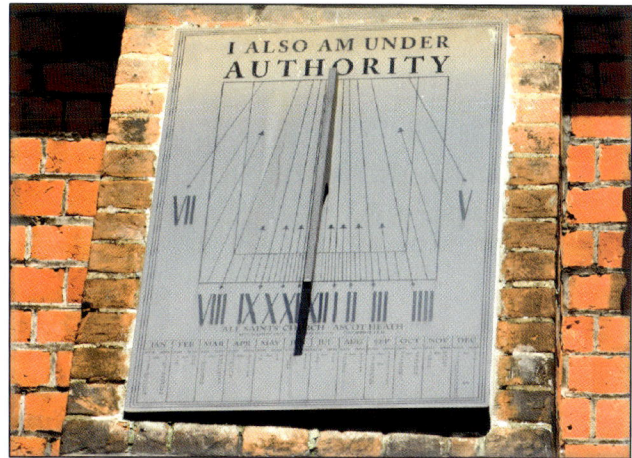

Reclining Dial on buttress at Ascot [DB]

Saxon Dial at Bishopstone

ML

the Prince Regent (later George IV) from 1783, a connection which culminated in the building of the Royal Pavilion in 1822. The high-class visitors and residents demanded facilities such as theatres, libraries, good hotels, a spa and a racecourse, all provided by the entrepreneurs of the day. Development was accelerated by the arrival of the railway from London in 1841.

The Royal Pavilion is restored much as it was in the Prince Regent's day and the Dome Concert Hall (formerly the Pavilion stables) stages concerts and other events. A conference centre built in 1977 completes the status of Brighton as a modern seaside resort.

The church itself retains many Saxon features, with Norman additions and alterations. Little change has taken place since the 13th century.

BRIGHTON, East Sussex

Just off Queen's Road is an impressive slate dial that was mounted in 1896 on a building now known as 'Sundial House'. Although the building is numbered in Queen's Road, the dial is round the corner, in Church Street. Unfortunately it now does not now receive much sun in the winter months, being obscured by a large building opposite.

Above the dial is a quote from the Bible:
Our days on earth are as a shadow
and there is none abiding
from 1 Chronicles 29:15 and below a Latin motto
UT HORA PRAETERITA
SIC FUGIT VITA
(As the hour passes, so life hastens on)

Brighton developed from the impoverished fishing village of Brighthelmstone. The pleasures of sea bathing were discovered in about 1730, and in 1750 a Lewes doctor Richard Russell published a pamphlet extolling the therapeutic qualities of sea-water bathing, and for drinking! Brighton was within a day's coach journey from London and soon lodging houses, terraces and squares of fine private houses were being built near the area called Old Steine where the fishermen used to dry their nets. The expansion gathered pace, and the Royal seal was set on Brighton with the visits of

Vertical Dial at Queen's Road, Brighton

ML

CHATHAM, Kent

The year 2005 marked the 200th anniversary of the Battle of Trafalgar and the death of Lord Horatio Nelson, England's naval hero, on the 21st October 1805. In 1994, mindful of the fact that HMS Victory, Nelson's flagship, had been built in Chatham dockyard, the local authority commissioned a sundial to celebrate the town's association with Nelson and to mark this historic naval victory. Designed by Christopher St J H Daniel and constructed by Ollerton Engineering, the dial is of the vertical kind, declining some 34° from south towards the west, fixed to a frame, situated in front of the southern aspect of the Post Office building. It takes the form of a 3 metre circular metal dial-plate, painted cerulean blue, bordered with a gilded 'rope' surround, in the style of a Royal Naval battleship crest. It is embellished with two 16th century warships, representing the Royal Navy, together with the sword and trident, superimposed over a wreath, representing the Royal Marines. The gnomon is a short horizontal rod, projecting perpendicularly from the dial-plate, the extremity of which comes to a sharp point, which acts as a nodus. The tip of the shadow indicates the time, as it passes over the gilded hour-lines, with numeral scales for both local apparent time and summer time. In addition, there are gilded declination lines for the solstices, the equinoxes and for the 21st October. Thus, the tip of the shadow will track the date of the Battle of Trafalgar, ultimately passing over a small icon, marking the moment of Nelson's death, shortly before sunset.

CROWTHORNE, Wokingham

This vertical declining mean time dial was designed by an astronomer called Alfred Hardcastle in about 1904. He stressed the significance of the method of time telling by putting the letters GMT (Greenwich Mean Time) so boldly on the dial.

The dial cleverly incorporates the two necessary 'corrections' to turn it from an ordinary declining Vertical Dial that tells local apparent time to one that tells mean time. The first is that the hour markers (imaginary in this case) have all been calculated to be positioned to show the time as it would be at Greenwich. The effect is

Vertical Dial at Chatham recording Nelson's death at Trafalgar

to allow for the longitude difference from Greenwich, which is as shown on the dial as Long 0° 48' or 3min 12sec in time after Greenwich. The next step was the calculation of an analemma at each hour, the analemma being the plot of the Sun's declination against the Equation of Time. The gnomon as a whole gives a shadow for the approximate time, but to read the time accurately one has to observe the position of the shadow of the nodus on the analemma. The nodus is the sundial term for a notch or ring on the gnomon that is important for relating the declination of the Sun for the particular season to the position on the analemma. However, the dial does assume that the observer knows which part of the analemma applies in the appropriate season. Looking at the top of the 'figure of eight' and going anticlockwise, the months run from January to

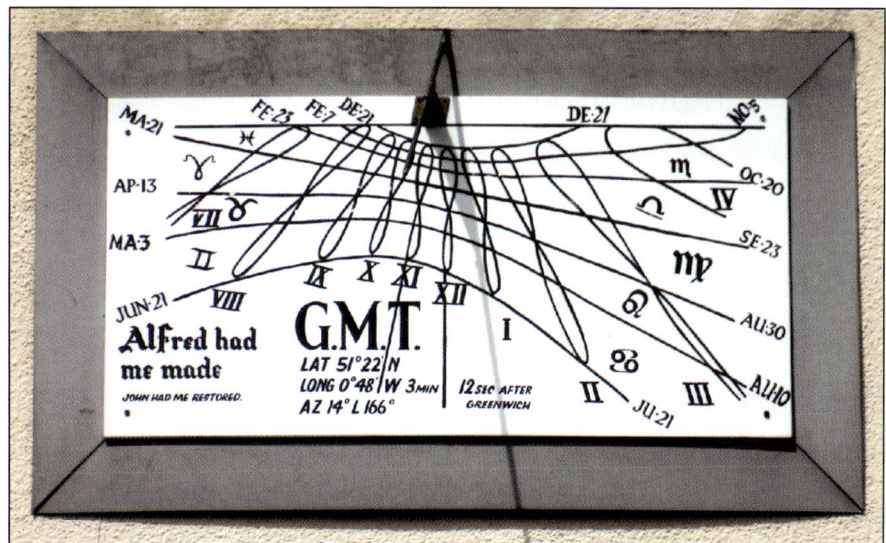

Mean Time Dial at Crowthorne

Shadow of nodus on Mean Time Dial

the crossover in April round to June before turning back up in July and up to December at the top.

The dial also carries lines across it that correspond to the declination of the Sun. The obvious usual line is the straight line at the equinox with dates of MA 21 (March) and SE 23 (September). The other declination lines show the declinations at the summer and winter solstice, and other dates of significance to the designer. The longitude is given on the dial, and unusually, the angle of the dial facing relative to south, in this case 14° east of south. The restorer decided to add the angle of the bearing of the dial relative to north, which is 166°.

The original dial plate was a sheet of iron that will have been repainted several times. There were some errors in the drawing of the signs of the zodiac, and some of the lines lack their original 'mathematical' smoothness, suggesting that the copyist had some difficulty in following what had gone before. Over the years the dial became quite dilapidated and in 1987 it was restored. Unusually the owner chose the acrylic material called Perspex™ for the dial plate. This has turned out to be a good choice - the black paint interacts with the plastic to give a very good bond, and the surface is self cleaning. The appearance after 18 years is as good as new. The only disadvantage is that Perspex™ has a high coefficient of expansion and the dial has warped somewhat by being bonded to inappropriate glass fibre border strips.

EGHAM, Surrey

Nowadays, many great country houses have been converted into fine hotels, which often means the preservation of their character and their history. One such ancestral home is now the Great Fosters Hotel, near Egham in Surrey. The property evidently takes its title from Sir Robert Foster, Chief Justice of the King's Bench, who lived there from about 1643 - 1663. In the garden there stands a tall stone 16th century Multiple Sundial, popularly referred to as 'Sir Francis Drake's dial'. The dial itself takes the form of an octagonal prism, about 610mm in diameter and 260mm in thickness, its shape having distinct similarities to the portable sundial devised by Nicolas Kratzer (c1487-1550) for Cardinal Wolsey. It comprises some 18 individual component dials, including numerous incised scaphes, and is mounted on a pillar, standing on a large rectangular pedestal. Surmounted by a short cylindrical 'capital', which once supported a weathervane, at an overall height of 3 metres the sundial is an impressive monument. As to the 'Drake connection', Sir Francis Drake was the Member of Parliament for Plymouth and thus was often obliged to journey to London. He might have stayed at times with his uncle, Richard Drake, who had bought Esher

Multiple Scaphe Dial at Egham

Dial at Great Fosters Hotel, Egham _{DB}

Oak dial at Charles Kingsley's church, Eversley _{DB}

support for the gnomon. Although made in 1965, the dial echoes the craft traditions of William Morris and Eric Gill. The bare wood will survive because after it gets wet it will dry quickly and naturally there is no lingering moisture to cause rotting. The inscription says that it is replacing a dial of 1852 although a print of the church dated 1834 shows a dial in place. The church has been altered over the years, but retains a warmth and calm serenity, and is well worth a visit.

FARNHAM, Surrey

On the Castle, overlooking the town of Farnham is a pair of declining Vertical Dials. Farnham is midway between Winchester and London and the building of the castle commenced in 1138. It provided a fortification and also hospitable accommodation for the Bishop in his frequent journeys between his cathedral and the capital. From

Farnham Castle _{DB}

Place in 1583, less than ten miles from the property that is now Great Fosters; but the latter was on Sir Francis Drake's direct route from the West Country. So maybe there is some truth in the story.

EVERSLEY, Hampshire

In the north east corner of Hampshire is a small church which has a long history, and is particularly known as the church where Charles Kingsley was rector. He was the curate in another part of Eversley, living in a house called Dial House, for a period of 18 months. Kingsley returned to the parish in 1844 as rector of St Mary's Church where he remained for 33 years until his death in 1874. Kingsley and his wife are buried in the churchyard. Rather discreetly, above and to the left of the porch, is a Vertical Dial that is made from untreated oak. Despite the apparent simplicity, the dial has been well thought out and carved in relief. The noon mark is shown by the small cross, upholding a tradition within the art of dial making. The wavy lines are the only decoration, but even these match the wavy metal

41

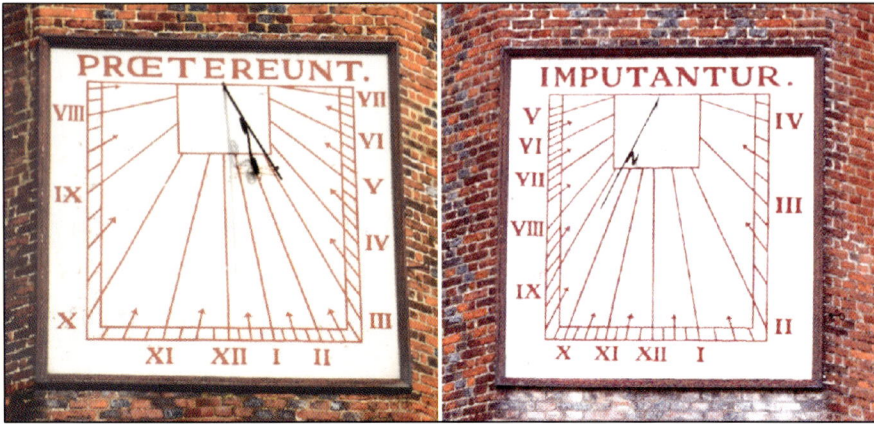

The two Vertical Dials at Farnham Castle

these beginnings, there has developed a close link between Church and town for 800 years. At the foot of the 12th century keep, more spacious and comfortable living quarters were gradually built. Mary Tudor stayed on her way to her marriage to Philip of Spain at Winchester Cathedral and Queen Elizabeth I came to the castle several times.

The sundials were added in the late 1700s, and are large and dramatic in appearance. The ever reliable sundial historian, Mrs Gatty, refers to the mottoes and said *"On the walls of the entrance tower of Farnham Castle, the palace of the Bishops of Winchester, there are two dials which formerly bore the inscription* Eheu, fugaces labuntur anni. *Other mottoes, more appropriate to an episcopal residence, have been substituted..."* Why the earlier motto *'Alas, the fleeting years slip away'* was considered inappropriate is a mystery!

GUILDFORD, Surrey
Guildford has a number of fine sundials, including a traditional vertical dial in stone added to the

Vertical Dial in Tunsgate, Guildford with romantic associations

Elizabethan George Abbot Hospital (actually almshouses) in 1867.

Tunsgate. A charming and beautifully executed vertical dial is situated half way between the High Street and the Castle. The choice of material is unusual; aluminium. For the delineation, Gordon Taylor at the Royal Observatory was consulted; he became a founder member of the British Sundial Society in 1989. The dial has clear romantic associations and a plaque below the dial, quoted literally, gives the following:

Edward I and his much loved princess, Eleanor of Castile, enjoyed their young days with his parents, King Henry III and Queen Eleanor of Provence in Geuildford Castle. This sundial was presented to the people of Guildford by the Prudential Assurance Company Ltd and the Borough Council in 1972. Forged in aluminium at the Quinnell Forge Leatherhead. It was conceived and designed by Ann Garland (1918 - 1979) in whose memory this plaque was erected by the foregoing by the Guildford Society and by her family.

Westnye Gardens. The interlocking bronze rings of the sundial almost conceal the fact that it is a Horizontal Dial. The dial is situated less than 50 metres from the River Wey and an aquatic theme has been used in the engraving with wave-like scrolls and the decorative chain. Some of the links in the chain are loose giving an additional tactile quality to the dial. The motto, in Latin and English states:

Time Reveals All Things

A novel feature is that details of the dial and the method of correction for the Equation of Time are deeply carved into the horizontal slats of a nearby park bench. Westnye was the name originally given to a small island near to the ancient Town Bridge. Apparently the oldest recorded town mill stood there in the medieval period. Twenty-seven local businesses contributed to the cost, and the site is on land that was donated to Guildford Borough Council. The gardens have been neatly landscaped to make the whole a perfect setting for the dial. The designer is Joanna Migdal and this is the second of her sundials to win an award in the 5-yearly scheme run by the British Sundial Society.

Award winning Horizontal Dial near the River Wey, Guildford

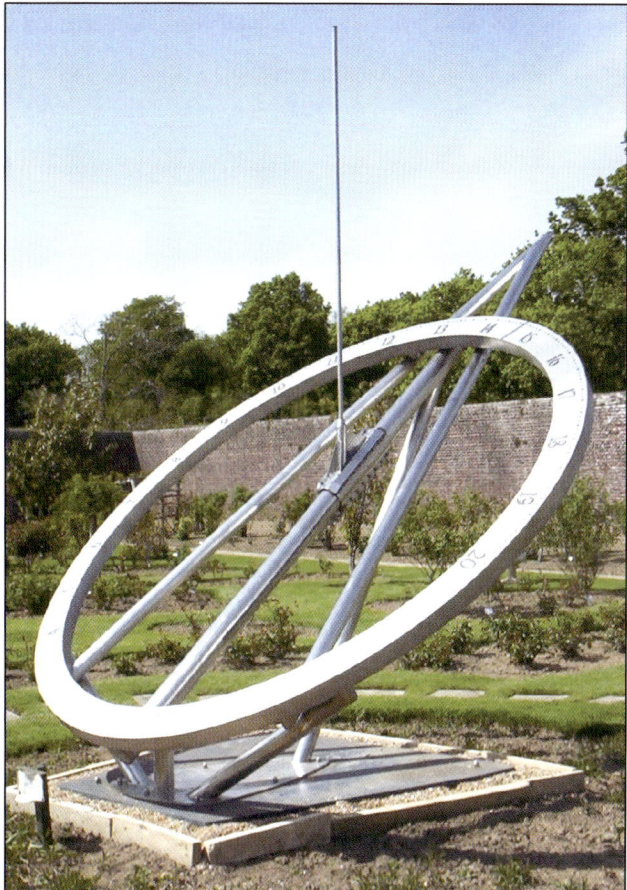

Foster-Lambert Reclining Equiangular Dial at Herstmonceux

HERSTMONCEUX, East Sussex

The unusual stainless steel dial at Herstmonceux Castle was designed by Gordon Taylor. It was installed in 1975 to commemorate the tercentenary of the Royal Greenwich Observatory which was founded at Greenwich in 1675. The Observatory moved to Herstmonceux over the period 1947-1957 and relocated to Cambridge in 1990 taking the dial with it. Upon the closure of the Observatory the dial was returned to Herstmonceux and re-installed in its original position.

The dial is a 'Foster-Lambert' reclining equiangular dial (the time marks are uniformly spaced) in which the gnomon is vertical and the dial-plate inclines towards the north at the latitude angle. Compensation for the changing declination of the Sun throughout the year is

obtained by traversing the gnomon mounting along the central spine and rotation of the annular time-ring in its own plane corrects for the difference of longitude from that of Greenwich, the Equation of Time and Summer Time if in

A sculptural dial in Horsham centre (described on page 44)

43

force, so that clock time is displayed directly.

Nearby is a Cube Dial on a tall pillar, possibly 17th century but restored in modern times.

The Castle was built in the 15th century and is one of the earliest brick-built buildings in England. It was partially dismantled in the 18th century to build a nearby manor house and remained a picturesque ruin until rebuilt in the 20th century. It is now an International Study Centre owned by Queen's University of Ontario: the grounds are open to the public from mid-April to October. Also on the site is the Science Centre, occupying the building and domes still containing the equatorial telescopes.

HORSHAM, West Sussex

It is not every day that a sundial is unveiled by Her Majesty the Queen and HRH the Duke of Edinburgh. This event took place on 24 October, 2003, whilst the Royal party was visiting other facilities in Horsham.

It is a relatively straightforward Equatorial Dial and the gnomon ring is plain with a rectangular cross section. In keeping with this, the gnomon is also rectangular. What distinguishes this dial from others of this type, is the wealth of historical detail that is sculpted into it. The equatorial ring is a 'time line' of a series of cameos starting with the dinosaurs to the present day, covering the topics mentioned above and including such topics as the Crusader Knights, the local iron industry, the horses of Horsham, William Penn and the Quakers, and the author Hilaire Belloc. In addition to all this detail, there is an annular ring of bronze set into the polished limestone plinth with recessed flood lighting. The plinth has hour markers to make the dial work as a Horizontal Dial as well. Credit must be given to the well known sculptor Edwin Russell, his wife Lorne

McKean and Damien Fennell, both professional sculptors. Other contributions were made by Joanna Migdal, Jeremy Knight and Freda Fowler. The dial was sponsored by local businesses.

LASHAM, Hampshire

The airfield at Lasham is the primary centre for gliding in the British Isles, if not the world, and this Armillary Sphere commemorates the life of one of the country's leading glider pilots who was tragically killed in a mid-air collision. Alan Purnell, in a long and passionate career of gliding, had flown over 10,000 hours, covered nearly 400,000km (more than 10 times around the world) and was the first UK pilot to achieve 3

DB

A memorial dial to a remarkable glider pilot at Lasham

Diamond awards. All of these achievements, and more, are engraved on the dial, and on the outer horizontal ring are directions and distances to other notable gliding sites. The gnomon has a nodus ring and to take advantage of this the months for the Sun's declination are engraved on the meridian ring, with an extra date, 26th of April, marking the date of Alan's death.

The dial was made by David Harber, who uses stainless steel to great effect where such dials need to make a strong presence or focal point. This memorial dial is adjacent to the entrance to the club house and reflects, in every sense, the pride and esteem that the members of the club have bestowed on their former colleague. The motto, attributable in various forms, to Leonardo da Vinci, is appropriately uplifting:

Once you have flown, you will walk the earth with your eyes turned skyward - for where you have been, there you long to return.

LEEDS, Kent

Described as the 'loveliest castle in the world', Leeds Castle near Maidstone is set in the middle of a natural lake, surrounded by 500 acres of beautiful wooded parkland and delightful gardens. The Castle is approached by a causeway and the grounds are entered through a fortified gatehouse, leading to what was once a courtyard, where now there is a well-kept circular lawn. Not altogether surprisingly, in a quiet corner of this area, close to the low walls of the once proud towers of what was once a Norman fortress, there stands a fine stone baluster pedestal, surmounted by an 18th century horizontal garden sundial. The dial is made of brass, with an ornate dial-plate about 250mm (10 ins) in diameter and with an interesting gnomon, the underside of which has a scalloped feature. On closer inspection, it

Horizontal Dial at Leeds Castle

Unusual gnomon on dial at Leeds Castle

will be seen that the dial bears the legend 'Made by Tho. Hogben, Land Surveyor & Master of the Free School at Smarden - 1750.' It is also embellished with the owner's coat-of-arms and the words 'Leeds Castle', as well as having much furniture, typical of the period. There is a compass-rose at the centre of the dial-plate, indicating directions by the compass, and a scale of corrections, by which one could set one's watch to the right time. In addition to the finely engraved hour-scale, there is a secondary hour-scale which indicates the time of noon at other places around the world, notably Belvoir in Virginia, USA, with which the owner's family once had close connections, in bygone colonial days!

MARLOW, Buckinghamshire

The town of Marlow, straddling the river Thames, has several sundials and the most noteworthy are a vertical dial in copper, and a large equatorial in granite and bronze.

Marlow Lock. A fine Vertical Dial overlooks the lock on the River Thames and the busy passage of pleasure boats and pedestrians. Many people

45

A Vertical Dial overlooking Marlow Lock

they were listed as mariner's compass and sundial makers. Most of their sundials were horizontal, and the majority of these were exported. The company expanded its range into telescopes, field glasses, magnifiers, aneroid barometers, thermometers, etc. The manufacture of sundials seems to have lapsed in the 1930s, but the company continues to thrive by making the classic prismatic compass for military and civilian use. Given the statement about being a prolific maker, the Register of the British Sundial Society records less than 20 dials by Barker still existing. No doubt there are many more not yet recorded, and some will have been supplied to other makers to add their own name, but it is clear that this dial at Marlow Lock has become a rarity.

will watch the activities in and around the lock, unaware that behind them is this dial made by the company of F Barker & Son, London.

A sundial enthusiast would regard the dial as a bold, well-made and handsome work in embossed copper. The house on which the dial is fixed dates from 1889, and the lettering, numerals and border decoration are consistent with the period when the dial was probably made in early 1900s. In addition there are some nice touches such as the hour lines broadening at the centre, and the small diamonds and triangles either side of the hour numerals. A crescent moon and a couple of stars have been included on the evening part of the dial after 6pm, – quite unusual for a sundial! Equally helpful for the enthusiast and historian is the fact the maker has put his name on the dial.

Francis Barker, trading mainly as F Barker & Son, is recorded as one of the most prolific and important sundial making firms of the 19th and 20th centuries. The firm was established in 1845 in Clerkenwell, London, where

May Balfour Memorial Garden. There is a large Equatorial Dial where the gap between the shadows from the tips of 'horns' falls on the inclined rear surface to show the time. The dial was designed and made by Edwin Russell, FRBS, and has strong similarities with the Dolphin Dial

A massive granite Equatorial Dial at Marlow

46

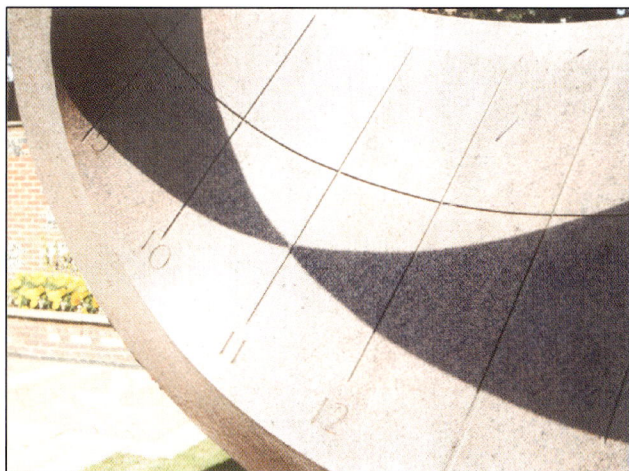

Equatorial Dial at Marlow, the tips of the horns acting as its gnomon [DB]

that he sculpted for the National Maritime Museum, Greenwich, in 1977.

In 1986 Russell was commissioned, by means of a competition organised by the Royal Society of British Sculptors, to create a large dial for the then headquarters of Rank Xerox in Marlow. The dial was hewn from 24 tonnes of Finnish granite with a team of 4 assistants. The project was completed in 1990 and the dial was situated in the front of the building. Rank Xerox moved from the site in 1999 and the building was taken over for refurbishment by the specialist commercial building developer, Akeler. The dial did not fit in with the planned changes to the building, and in 2001 it was generously donated to the town of Marlow. In August 2002, the dial was relocated to the May Balfour Memorial Garden in the centre of the town.

The markings are at hourly intervals from 9am to 5pm, with 1pm for the summer time noon. However, the markings are all offset 3 minutes in time to give the longitude correction for 0° 45' W. (The new location is 1' further west, an 'error' of only 4 seconds in time.) The overall dimensions of the dial are 2.34m high, 2.71m wide, and 1.9m deep. The final weight of the dial is 8.4t.

In the plinth there are four Polar Dials giving the time in New York, Moscow, Tokyo and Delhi. The plinth contains the Equation of Time, rather ingeniously laid out as if a Polar Dial with a ball nodus to reinforce the seasonal nature.

To supplement the dial as a means of telling the time, bronze panels on both sides give representations of the Sun in different cultures. On the east side of the dial the panel shows Ahura Mazda, Persian; Helios, Grecian; Eye of Ra, Egyptian. On the west side there is the symbolism of

Surya, Indian; Flaming Eye, Chinese; Kinich Ahau, Mayan; Xolotl, Aztec; and Nazca, South American.

The final piece of symbolism is a fine relief carving on the rear of the dial that depicts the Twins of Gemini. They stand for the alternate appearance and disappearance of the Sun, therefore night and day, light and shadow. Being twins they are symbolic of the reprographic process and being composed of light and dark are the basis of printed communication. The dial may be seen in the May Balfour Memorial Garden, Institute Road, Marlow.

MIDHURST, West Sussex

A modern Vertical Dial erected in 1995 is located on the Information Centre run by Chichester District Council. It is carved in Welsh slate, with a metal gnomon and gilded numerals. Below the dial is a plaque with the Equation of Time. The dial's declining angle is very high, 78° west of south, and for this reason the dial is constructed as a 'dial without a centre'. In such a dial the point where the hour lines and the gnomon would normally intersect is off the dial face. The gnomon necessarily makes a very shallow angle with the face and, if made in the usual way, the hour lines would be too close together. In the photo the shadow-casting edge of the gnomon can be seen detached from the dial face and making only a small angle with it. The dial was made by Sally Hersh.

Dial by Sally Hersh at Midhurst [SH]

Midhurst is a town at the crossing of two important roads, the long-distance east-west A272 and the north-south A286 between Guildford and Chichester. It has many half-timbered and Georgian houses. To the east of the town lie the ruins of Cowdray, a large manor house destroyed by fire in 1783, which gives its name to Cowdray Park, a venue for polo matches and a popular area for picnics with its many large oak trees.

NORTH STOKE, Oxfordshire

The village stands on the banks of the Thames a little way north of Goring-on-Thames. St Mary the Virgin Church, dating from 1230-1240, replaced an earlier Norman Church (which itself probably replaced a Saxon one). In the churchyard, near the lych gate, is a Horizontal Dial on its original plinth. This dial, erected in 1919, is a memorial of the 1914 - 1918 war. On the dial plate is the motto:

Hours pass Deeds Remain

The dial is delineated from 4am to 8pm with 5 minute markers. In the centre of the dial plate is an 8 point compass rose, the gnomon and its mounting arms forming the four cardinal points. This dial is best visited during the afternoon when it is no longer shaded by large trees to the east of the churchyard.

On the south wall of the church above a blocked up doorway is a 'Saxon' type dial in good condi-

Saxon Dial at North Stoke ^{AA}

tion. The iron gnomon is possibly a replacement. The damaged head and the hands are considered to be representing the figure of Our Lord. There are arms connecting the hands to the head but these were covered up during the church restoration in 1902.

There are the remains of a third dial high up on the south west corner of the church but the stone surface is very badly spalled and the rusty iron gnomon badly bent.

The interior of the church is well worth a visit. It has the remains of wall paintings on three walls (with a board of explanations available) and a carved Jacobean Pulpit.

OXFORD

All Souls' College. This notable dial is mounted high on the building overlooking the North Quad. The dial was designed by Sir Christopher Wren, circa 1653, when he was a Fellow of the College. The College Arms are in the middle of the dial and the hour and half hour lines continue, unusually, outside the Roman hour numerals in the form of narrow gold triangles. Alongside these triangles are a series of dots, arranged as a diagonal scale, which indicate individual minutes, a feature with which Wren was pleased in his design. Its motto is:

PEREVNT ET IMPVTANTVR
(The hours pass away
and are set down to our charge)

As the dial is high up on the wall binoculars are a great help in seeing the finer detail of the dial. It is surmounted by a small but rather nice wind vane.

**First World War Memorial Dial
at North Stoke** ^{AA}

The Dial at All Souls' College

Brasenose College. This dial, painted on a thin stone slab, is seen in a rather colourful summer setting on the right immediately on entering the Quad past the Porters' Lodge. This fine colourful dial, erected in 1719, includes the College Arms and indicates the time to seven and a half minutes. The wall declines slightly to the east thus showing fewer afternoon hours than morning ones.

Although Oxford city maps show the College as being on the High Street, the entrance is from the south west corner of Radcliffe Square, where the Radcliffe Camera is to be found.

Corpus Christi College. In the Front Quad of the College stands one of the best known sundials in Oxford. A round column stands on a rectangular base and on top of the column sits another rectangular block surmounted by a pyramid and ball. On the ball sits a carved pelican, a symbol of piety. This carving gives the whole multi-faced assembly its popular name of 'The Pelican Sundial'. The vertical and inclined surfaces all face the cardinal directions.

Each of the pyramid faces carries a Reclining Dial, three of them with metal gnomons but the west facing one has the dial in the form of a recessed heart, the cusp of which forms the gnomon. On each of the four vertical faces there is a painted heraldic badge surrounded by stone scrolls. In each case a part of the scrollwork projects to form the gnomon for the dial below it.

Vertical Dial at Brazenose College

The 'Pelican Dial' at Corpus Christi College

49

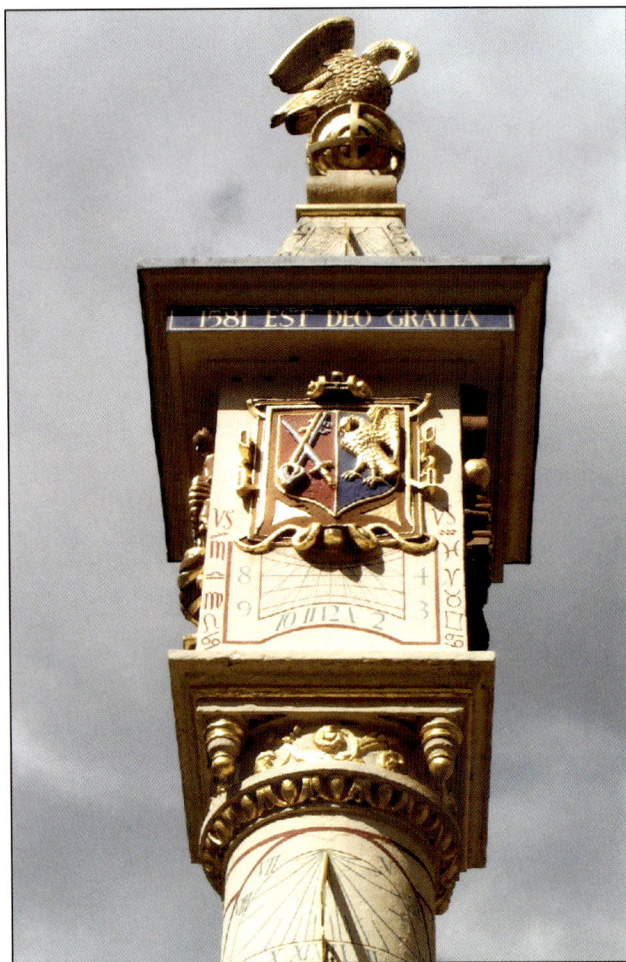

The gilded pelican sitting on the top
of the Corpus Christi dial

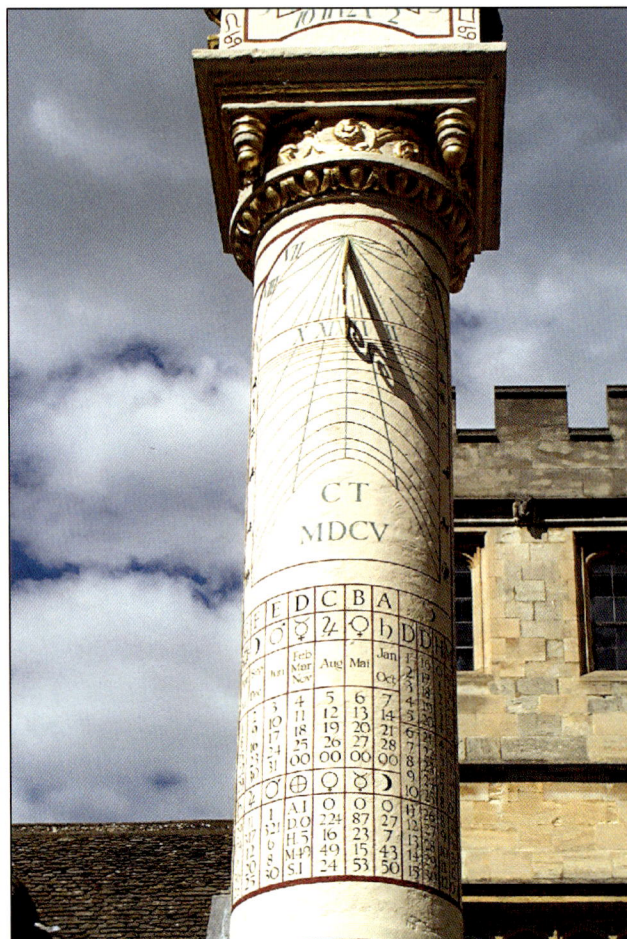

Calendrical Tables on the 'Pelican Dial'

On the south side of the round column is a ninth dial. This is essentially a normal Vertical Dial with a metal gnomon but the curvature of column detracts from the clarity of reading due to the blurring of the shadow except for times close to midday. At the base of the north side of the column is the motto:

HORAS OMNES COMPLECTOR
(I encompass all hours)

Around the column are a number of astronomical and calendrical tables.

This dial was made by the 16th century maker Charles Turnbull in about 1580 and has undergone a number of restorations since then. The most recent and thorough restoration and re-painting was carried out in 1976, supervised by Dr Philip Pattenden, who based the work on old photographs and evidence from the College archives.

Christ Church College. This large but rather plain dial probably dates from the latter half of the 18th century. It is clear and serves its purpose of indicating the local solar time, to a quarter of

an hour very well. It is clearly seen when walking from Peckwater Quad to the college picture gallery.

Green College. A slate Noon-Mark Dial is to be found on the south-facing wall just beyond the Porters' Lodge in Woodstock Road. This college is on the site of the Radcliffe Observatory, the first University Observatory. The dial commemorates the 1994 bicentenary of that observatory and was erected in 1995.

Vertical Dial at Christ Church College

50

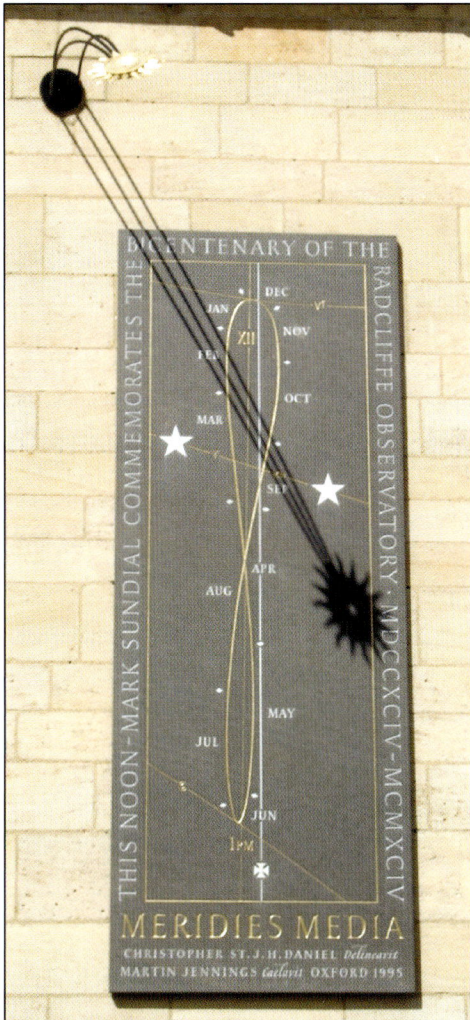
Noon-Mark Dial at Green College [MC]

Vertical Dial at St Catherine's College [AA]

Above the dial is a disc mounted from the wall. An aperture in the disc projects a small image of the Sun onto the dial around noon each day. As the spot moves across the dial it indicates the moment of noon when it crosses the part of the 'figure of eight' curve corresponding to the date of the month.

This curve is known as an analemma and allows for the difference between solar time and clock time throughout the year. The Sun spot crossing the vertical line marked XII and 1pm indicates solar noon at Greenwich, depending on whether GMT or BST is in use, and the spot crossing the line with a cross at the bottom indicates solar noon at Oxford. The three transverse lines marked with the signs of the zodiac show the path of the Sun spot across the dial at the equinoxes (marked with two stars) and the solstices (winter through December and summer through June).

Amongst the observatory buildings in the college grounds still remains a 'replica' of the Athens' Tower of the Winds.

St Catherine's College. The College is slightly away from the main grouping of older colleges but by only a short walk into Manor Road. Its slate dial is on the south wall of the Senior Common Room and commemorates a former Fellow of the college, it being a gift from his widow.

St Giles. Heading north from the city centre is St Giles, a wide but fairly short main thoroughfare. Just outside St Giles Churchyard, is a small, iron, Armillary Sundial. The arrow pointing to the North Pole forms the gnomon and parts of the equatorial ring (carrying the hour numerals) and the meridian ring are omitted to prevent their shadows interfering with the shadow of the gnomon.

This dial was a gift in 1986 to Oxford from its twin city Bonn.

Equatorial Dial near St Giles Churchyard [AA]

Prize-winning Vertical Dial at West Dean ^{MC}

winter (blue diamonds) and summer (red hemispheres). The decorative gnomon is fretted with the Edward James Foundation initials.

Originally gilded with gold leaf, minerals and chemicals exuded from a chimney immediately above the sundial have proved toxic to the copper within the gold leaf which has now been considerably eroded over the years. Hopefully this will be rectified in due course, although the dial is proving most attractive just in its stone state.

A separate, nearby Equation of Time plaque explains to the onlooker the differences between clock and solar time throughout the year.

The gardens at West Dean College, on the A286 north of Chichester, are open throughout the year and the sundial may be seen at any time by visitors and those attending the short courses being run there.

In 1995 Sally Hersh and West Dean College were pronounced the first British Sundial Society Award winners for excellence of design for their sundial.

In contrast to this modern dial, there is a fine large traditional vertical dial on St Andrew's Parish Church, which is just behind West Dean College. The dial was carved in stone in 1729, and is clearly designed to show the time quite accurately by having divisions down to 7½ minutes and there is a 'noon gap' in the line of the gnomon - quite unusual in vertical dials of this period.

WINDSOR, Windsor & Maidenhead

Just a short walk down the steps from the station at Windsor, brings one to the modern frontage of the Windsor Business Park, where, located on the corner of the paved terrace, close to the north and east office blocks, there is a strikingly large and elegant example of a modern Armillary Sundial. Made of stainless-steel, with a diameter of 1.8 metres and standing at an overall height in excess of 2.7 metres, the dial is all the more distinctive since it is entirely gilded, with numerals and

WEST DEAN, West Sussex

When a new wing was built onto West Dean College in 1990 it was decided by the architects to set a sundial into the flint-faced wall to give it a large and an interesting feature. Sally Hersh was chosen to make the designs in collaboration with the architects APP. The dial delineators Frank Smith and Tony Brooks then selected the stone and hand carved the sundial.

This dial, 1310mm high and 940mm wide, has now become a landmark piece in the area and well known for its attractive design. Made of Clipsham stone chosen for its colour and texture to blend with the surrounding flint walling, and with deeply 'V' cut lettering and Roman numerals, the dial shows hourlines, initials EJF - the Edward James Foundation, named after the creator of the Foundation, date 1990, and features a raised dome and deep cut sunburst decoration. Additional unusual features are the blue and pink solstice markers, which, from the nodus on the gnomon, trace the path of the sun's shadow at the

Armillary Dial at Windsor

All Saints' Church dial at Wroxton

All Saints' Church. The dial is located high on the east corner of the south aisle under the parapet. It is painted on the stone wall but this essentially nice dial is suffering from weathering and is in need of restoration. The dial is not quite correctly delineated, possibly due to errors introduced during repainting. The church dates from the 14th century but the south aisle is 15th century.

North Arms. In the rear garden of the pub is an 18th century dial painted on a slab which is canted out slightly from the wall to face due south. It has markers for half and quarter hours. The iron gnomon has a scalloped support and noon is marked by a cross. The spelling of the motto:

Sic Tranfet Gloria mundi
(Thus passes the glory of the world)

has 'transet' instead of the more usual 'transit'. The early morning and later afternoon hours are shaded by the roof overhang.

The Old Post Office. A Vertical Dial is painted directly onto the wall over the front door, showing

markings in royal blue. Unusually, it is constructed of only three rings, comprising a prime vertical ring, a meridian ring and an equinoctial hour-ring. However, the initial impression is that there are a greater number of rings than this, since a central section has been removed from both the meridian ring and the hour-ring. This construction allows the time to be determined in the normal manner by observing the shadow of the gilded gnomon-rod, as it passes across the hour-ring. However, it also permits the disk-shaped nodus, situated at the centre-point of the gnomon, to indicate the declination of the Sun and hence the date, with a correction for the Equation of Time, as its shadow crosses the inner surface of the meridian ring at noon. The dial was designed by Christopher St J H Daniel and constructed by Brookbrae Ltd in 1999.

WROXTON, Oxfordshire
The inhabitants of Wroxton village, near Banbury, claim to have more sundials – 8 in all - than any other village (presumably of a similar size!) in England. All of these dials, except for the Pillar Dial, are to be found in a stretch of less than 500 metres along Church Street and Main Street in the village.

Vertical Dial at the North Arms, Wroxton

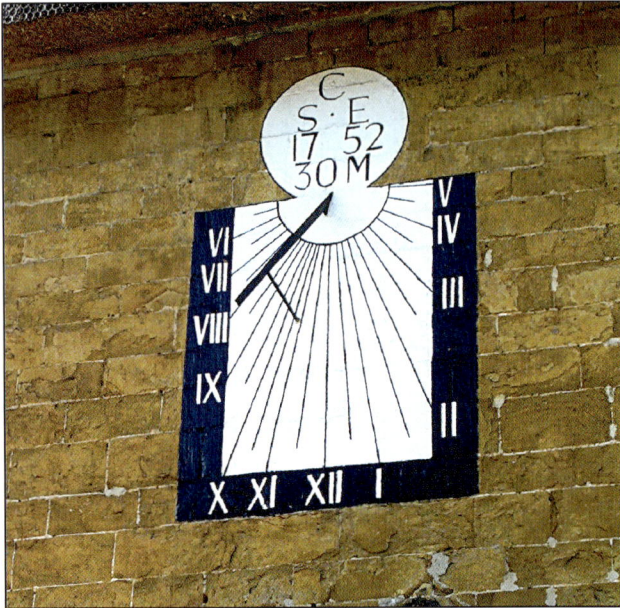

Vertical Dial at the Old Post Office, Wroxton

Pillar Dial at Wroxton

its date of 1752. The dial declines 15° east and is marked every half hour for the period 5.30am to 5pm. Unfortunately the thatched roof puts the dial in shade for much of the summer although the lower angle of the Sun in winter means that the dial is then able to perform its function for part of the time.

Pillar Dial. At the west end of the village at a junction between the A422 and a road to North Newington is a tall ironstone pillar on the grass verge. At the top is an 'elongated' cube having dials on each of the four faces, which are oriented to the cardinal directions. None of the gnomons survive but all the hour lines and numerals are clearly visible as very little weathering of the stone has taken place during the more than 300 years the pillar has stood in its present position. Under the dial cube the rectangular portion of the pillar is a direction indicator being carved with hands pointing to Banbury (south side), London and Stratford (west) and Chipping Norton (north). On the east side is inscribed:

First given by Mr Fran. White in the year 1686

This pillar is a splendid survival and the crispness of the carving and the mason's tool marks shows the very good properties of the ironstone, from which most of the buildings in the village seem to be constructed.

Two other dials at Wroxton may be mentioned. One on a modern bungalow. It is an older sundial slab incorporated above a large window. It is likely to have been transferred from another site but from the same general area as it is the same type of stone as most of the buildings of the village. There is a small slate Vertical Dial at Sundial Farmhouse. The rather thick iron gnomon is slightly bent, perhaps to try to correct an erroneous delineation.

LONDON
John Moir & David Young

For much of its history London has consisted of two cities - the commercial area known as the 'City' to the east and the City of Westminster, site of the Abbey and Royal palace. After the Norman conquest William recognised the financial importance to the Crown of the wealth producing City and allowed its citizens to elect their own governing council, Aldermen and Lord Mayor.

should be used in the rebuilding and that streets should be widened.

Sir Christopher Wren designed and rebuilt 51 of the churches though only 25 now survive. His crowning achievement was St Paul's Cathedral, completed in 1711. Many late 17th century London churches would have had sundials to regulate

The Old Royal Observatory looking out over Greenwich and the River Thames

The City's Guildhall, civic buildings and churches were built of stone, most of which came from Kent as London has no suitable stone of its own. Houses and shops, of timber and thatch were crammed together in narrow streets. Inevitably, there came the Great Fire which destroyed most of the dwellings and 87 City churches. It was decreed that only bricks, stone, tiles and slate

their clocks, though one might question their efficacy after reading the following piece, which appeared in the Athenian Mercury of 1692/3 (iv, No 4). (Note that in this passage 'dyal' refers to a clock.)

I was walking in Covent Garden where the clock struck two, when I came to Somerset-house by that it wanted a quarter of two, when I came to

St Clement it was half past two, when I came to St Dunstans it wanted a quarter of two, by Mr Knib's Dyal in Fleet-street it was just two, when I came to Ludgate it was half an hour past one, when I came to Bow Church it wanted a quarter of two, by the Dyal near Stocks Market it was a quarter past two, and when I came to the Royal Exchange it wanted a quarter of two: This I aver for a truth, and desire to know how long I was walking from the Covent Garden to the Royal Exchange?'

Heavy church clocks would have been made by blacksmiths, but precision clock and watch making was carried out in Clerkenwell workshops.

Instead of individuals making complete clocks and watches, production was organised on the 'outwork' system. Specialist craftsmen worked at home, and in the course of making a watch it passed between workshops all around Clerkenwell. Due to this efficient system London came to dominate the trade. The great innovator of clock and watch design was Thomas Tompion of London (1639 - 1713). He also made scientific instruments including sundials.

During the 18th century wealthy families such as the Grosvenors and the Portlands developed new estates in and around Westminster, whilst the City's overcrowded poor spread into districts to the east. With the coming of the railways in the 19th century thousands of the middle classes sought a better lifestyle by building out of town houses and commuting to the City each day. Their efforts were largely self-defeating. Soon, in the Victorian Music Halls, unkind singers were deriding these would-be country dwellers:
*'Oh, it really is a werry pretty garden
And Chingford to Eastwood could be seen
Wiv a ladder and some glasses
You could see to 'Ackney Marshes
If it wasn't for the 'ouses in between.'*

When the Railways introduced cheap workmen's fares, many more people could afford to commute. This led to the massive expansion of working class suburbs, an expansion which continued well into the 20th century. Greater London now stretches some 40 miles from east to west. However, in most suburban Londoners' minds London is that place you commute to each day, or visit for an entertaining night out.

It is surprising, when studying aerial photos of London and its suburbs, how much green space remains. Apart from municipal parks especially created for public use, many Royal Parks and former private estates are now accessible to the public, and they often contain sundials. In this respect Hampton Court, Kew Gardens, Holland Park, Waterlow Park and Horniman Gardens are just a few worth visiting.

Dials in public parks are usually free standing, whereas wall dials tend to predominate in an urban context. However, there are remarkably few wall dials remaining on churches in central London, in spite of the many old churches to be found there. In the suburbs most churches are of Victorian or later age, and would only have needed a dial for decorative purposes, an opportunity which was rarely taken.

Notwithstanding the above, London has much to offer the visiting diallist. There are sundial trails to follow (www.sundialsoc.org.uk) and the old Royal Observatory at Greenwich is easily reached.

The London Instrument Making Trade
Instrument making in London goes back to at least Elizabethan times. Certainly the trade flourished over the next three centuries and many excellent sundials and other instruments were produced. The majority of Horizontal Dials and Portable Dials to be found throughout the Country will have come from London makers. Many of their dials will be seen throughout the pages of this book.

Some of the finest dials produced came from a small group of workshops. In the 17th century makers such as Henry Wynne and Thomas Tompion produced some exceptional dials. However, the majority of extant dials are from the 18th and 19th centuries. Quantities of fine dials were made in the workshops of makers such as George Adams, John Rowley, Thomas Wright, Troughton & Simms and many others. In Victorian times the use of sundials declined but some excellent ones were made by Barker and Son. Many of these dials were unsigned and therefore their makers are often hard to identify.

BLACKFRIARS, EC4
Just west of the Millennium footbridge, by the City of London school for boys stands a modern Polar Dial, designed by Piers Nicholson for the Worshipful Company of Tylers and Bricklayers. The Company needed their own Millennium monument to reflect their particular trade, so

Modern Polar Dial at Blackfriars made for the Worshipful Company of Tylers & Bricklayers

the Victorian tea merchant. The museum was built on the site of his London house with its beautiful parkland and garden. A trust was set up in the late 19th century with a remit to grant the public free access to the wonderful collections and the surrounding gardens, with their fine views across London.

The Museum has always had a strong connection with the local community and has a flourishing educational facility aimed at helping local schools and the general public to understand the collection and the environment. In the latter part of the 20th century a decision was made to form a sundial trail in the gardens. The British Sundial Society was asked to assist, and there are now about a dozen sundials established around the grounds including a rare ceiling dial in the museum library. Various sundial types are represented and an explanatory leaflet is available at the museum. This Sundial Garden is unique to the British Mainland.

inevitably its plinth is made from bricks, its dimensions being juggled so that exactly 2000 bricks were used.

A Polar Dial is one whose dial plate is parallel to the Earth's axis. Here, the dial plate and gnomon are of stainless steel, with the hours marked in Roman numerals for the winter and Arabic for the summer. The ends of the dial plate are bent up parallel to the gnomon to give the dial a full 12 hour range, and the gnomon has a novel 'ox head' shape to ensure that its shadow falls on the hour scale all year round.

The dial was made by the Royal Engineers' and presented to the Corporation of London by the Company of Tylers and Bricklayers. The inscription on the dial includes the official badges of all three organisations. This dial won an award in the BSS Awards Scheme for 2005.

Two other sundials of the same design have been erected. One stands outside the Royal Engineers' museum in the Brompton Barracks at Chatham in Kent whilst the other is located midway between the Dome and the Thames barrier on the east side of the Greenwich peninsular.

FOREST HILL, SE23

The Horniman Museum in Forest Hill, South London was set up to display the vast collection of artefacts accumulated by Frederick Horniman,

Scaphe or Bowl Dial

This sundial has a perfect inner bowl, contrasting with the disorganised, decaying appearance of the

Scaphe Dial showing the time of 11:10am

Scaphe Dial by Angela Hodgson

Horniman Logo Double Polar Dial

outer base and stem. The designer, Angela Hodgson says that for her it symbolises the perfection of our ideals, which are so unlike the uncontrolled, harsh reality of disease and death which come with the passing of time. The shape of the stem was inspired by the magnificent Magnolia Soulangiana growing nearby. The bowl is part of a sphere inside which sits the pointer, or 'gnomon'. The dial was cast in bronze at the foundry in Central St Martin's College of Art and Design and delineated by Dick Andrewes, who, because of the dial's organic shape had no datum line from which to work. He overcame this problem by rolling a marble in the bowl to find its lowest point, on which the pin gnomon now sits. The shadow of the pin's tip falls on the hour lines and shows solar time.

Horniman Logo Double Polar Dial

A Polar Dial has both the gnomon and the dial plate parallel to the Earth's polar axis. This dial is, in effect, two separate Polar Dials with two gnomons and two overlapping hour scales. In basing the dial on the old Horniman logo the aim was to make no changes other than the addition of the hour marks. Even the words:

The Horniman Museum and Gardens

are preserved - their 27 letters rearranged to provide a motto:

AND HOURS RUN MAD
E'EN AS MEN MIGHT

The sundial, designed by John Moir, has been carved in Hopton Wood limestone and its plinth from Portland stone. The Arts and Crafts design of the plinth was inspired by that of the Horniman clock tower. Richard Klose was the carver who also designed the plinth.

Roman Sundial

Sundials were prevalent in Rome and the Roman Empire as evidenced by the poet Maccius Plautus about the year 200BC. He wrote: *'The gods confound the man who first found out how to distinguish hours! Confound him too, who in this place set up a sun-dial, to cut and hack my days so wretchedly into small portions'.* The hours that Plautus mentions were not the same as the hours we use today. Then, the time between sunrise and sunset was divided into twelve equal parts, and as the length of daylight altered through the year, so the length of the hours altered also. The dial here at the Horniman was

'Roman Sundial'

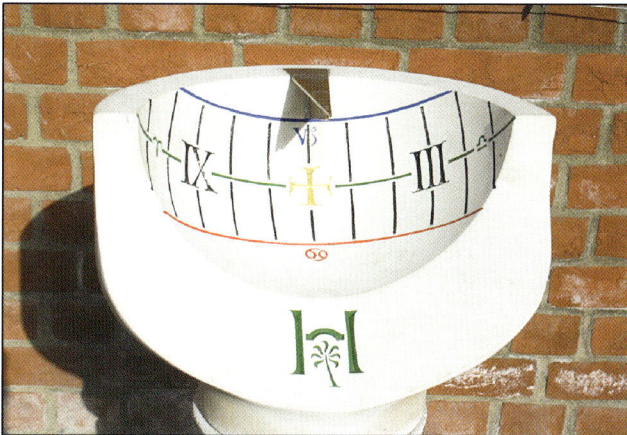
Bowl of Roman Scaphe Dial

Horniman Horizontal Dial

a shadow which can be read all round the centre section, which is coloured green to represent the gardens. Designs on the outer blue panels represent the various departments of the Museum. The top panel shows the museum logo with a fly - a pun on 'time flies'. The dial was designed and

made by David Brown after the style of a Roman Sundial but the hours have been calibrated to read the equal length hours we have all used for about the last six centuries. The dial itself has been carved in Portland stone.

Horniman Horizontal Dial
This dial commemorates the opening of the Gardens to the public in 1897. The gnomon's shape derives from the Horniman 'H' and the name Horniman is written in place of the 9 and 7 hour lines (18**97**). This however may escape notice except for those who know their Morse code! The dial was made by Ray Ashley and designed by John Moir, who seeks to make his dials as quirky as possible!

Stained Glass Dial
On a glass wall of the conservatory, this dial faces slightly west of south. Designed to be read from inside, the orientation of the hour lines are the reverse of the normal wall dial (here morning hours are on the right and afternoon hours are on the left). The gnomon is placed outside to project

The Horniman Stained Glass Dial

The 'Butterfly Dial', a memorial to Noel Ta'Bois

Butterfly Dial

This is a delightful combination of art and science. It is one of a limited edition of 100 originally designed for the Sunday Express garden at the Chelsea flower show many years ago by Edwin Russell and made by Brookbrae. This particular dial commemorates the 20th wedding anniversary of Margaret and Noel Ta'Bois and was given to the museum after his death. Noel was an enthusiastic diallist, Meccano addict, writer and photographer of dials, and a dentist in his spare time.

made by Roselyn Loftin and delineated by David Young.

GREENWICH, SE10

Greenwich began to develop as a cultural visitor destination with Sir James Thornhill's completion of the Painted Hall (1707-26) in what is now the Old Royal Naval College (founded in 1694 as the Royal Hospital for Seamen). Not far from this architectural gem is the National Maritime Museum in Greenwich Park, created by Act of Parliament in 1934. The Museum has the most important holdings in the world on the history of Britain at sea, including maritime art (both British and 17th century Dutch), cartography, manuscripts and scientific and navigational instruments. Its British portraits collection is only exceeded in size by the National Portrait Gallery and its holdings related to Nelson and Cook, among many other individuals, are unrivalled. It has the world's largest maritime historical reference library (100,000 volumes) including books dating back to the 15th century.

The 'Dolphin Dial' at Greenwich

Time shown by the Dolphins' tails, 10:27am

Situated in the narrow strip of garden adjacent to the south face of the building is the 'Dolphin Dial', erected in 1977 to commemorate the Silver Jubilee of Queen Elizabeth II. It is a unique Equinoctial Mean Time Dial indicating clock time, either Greenwich Mean Time or British Summer time. The actual time is indicated by the gap between the shadows of the tails of two diving Dolphins and the curved dial plate is engraved with hour lines that incorporate corrections for the Equation of Time. The dial of bronze and bell metal stands on a Portland stone plinth; it was designed by Christopher Daniel, and executed by the British sculptor, Edwin Russell.

Just a short walk up the hill from the sundial is the world famous Old Royal Observatory with its line of zero longitude and with Christopher Wren's Flamsteed House of 1676. Not to be missed by anyone, most especially those interested in time or astronomy.

LAMBETH, SE1

On the south facing wall of the building that has been the headquarters of The Marine Society (now The Marine Society and Sea Cadets) and the Nautical Institute since 1979, there is a fine sculptured sundial, representing the combined symbols of both these organisations. It comprises a 'sea-dog' holding a 'torch of learning', the symbol of The Marine Society, and a two-dimensional armillary device, being the symbol of the Nautical Institute. Made of copper, it was designed by Christopher Daniel; fashioned by Edwin Russell, the eminent sculptor; and made by Brookbrae Limited of London. The 'torch of learning' acts as the gnomon and its shadow indicates the time. Declining from south by 17° towards the west, it has two hour-scales, one for 'GMT' and the other for 'BST'. It also bears the motto

> **Time, like the tide, waits for no man**

the wording of which, strictly speaking, is not in its original form; but, since it was the expression used in an address to members of the Nautical Institute by the late the Earl Mountbatten, it seemed to be a fitting legend. The dial was unveiled by Her Majesty the Queen on the 5th December 1979.

PETTS WOOD, BR5

Hidden away in Petts Wood near Orpington there is a granite sundial. This is a memorial to William Willett of Chislehurst, a builder who campaigned for daylight saving by moving the

The Marine Society Dial at Lambeth [DY]

clocks forward one hour during summer time. It is said that the idea came to him while he was riding through Petts Wood early one morning and noticed that lots of the blinds were down in the houses and people were still asleep even though the sun had risen hours earlier.

In a long tract he had printed in 1907 he expounded his argument for his case in some detail.

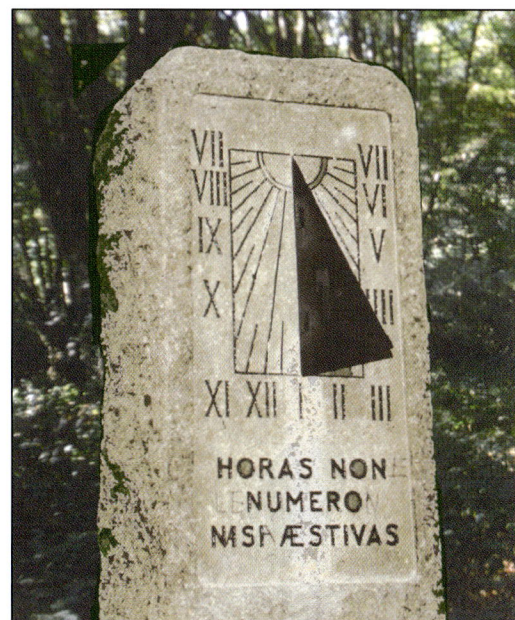

Vertical Dial on Willett Memorial [DY]

Monument in Petts Wood to William Willett originator of 'Summer Time'

He made the point that apart from Sundays, workers' only free time when they could enjoy the open air was for a few hours in the evening during summer time. Advancing the clocks would give everyone seven hours of extra usable daylight each week during the summer months. He also gave in great detail how much money could be saved in lighting alone as this extract shows:

'According to Whitaker the population of Great Britain and Ireland is 43,660,000.

The number of hours during which the cost of artificial light will be saved will be:

During April, 23 hours

During May, June, July and August, 164 hours

During September, 23 hours

Total 210 hours at 1/10 of a penny per hour = 1s 9d, multiply by 43,660,000 = £3,820,250.

Deduct 1/3 to meet all objections and loss of profit by producers of artificial light = £2,546,834

His original suggestion was that the clocks should be moved forward by 20 minutes on each of the four Sundays in April and then moved back in a similar way on four Sundays in September, but this was thought too complicated. Farmers were strongly against Summer Time and many people treated the idea as a joke. However, in the end, William Willett's campaign led to the drafting of a government bill in 1909 but it was rejected on that occasion and on several others. Eventually, a year after his death, Daylight Saving Time (Summer Time) was adopted in England in May

1916. It is interesting that the dial shows BST hours and has an appropriate motto:

HORAS NON NUMERO NISI ÆSTIVAS
(I Number Only the Summer Hours)

SEVEN DIALS, WC2

The original sundial was designed and constructed by Edward Pierce, a leading stonemason, in the 1690s. It had six dial faces - the seventh 'style' being the pillar itself. A story circulated that it was demolished in 1773 by a mob searching for buried gold, but it is now thought to have been removed to rid the district of 'undesirables' who congregated around its base.

The column's remains were purchased by architect James Paine. Unused for many years, they were eventually re-erected in Weybridge, Surrey, as a memorial to the Duchess of York, who died in 1820. The dial itself was replaced by a coronet.

In the 1980s it was decided to erect a new pillar sundial at the London site, using the original drawings held in the British Museum. Since the main sewer and other services now ran beneath the site, the heavy column had to be supported by a concrete raft which straddled the various conduits. The actual dial stone, which weighs a tonne, was designed, carved and gilded by Caroline Webb. The delineator spent three days on site to ensure accurate placement and it is

'Seven Dials' Pillar

'Seven Dials', viewed from the south

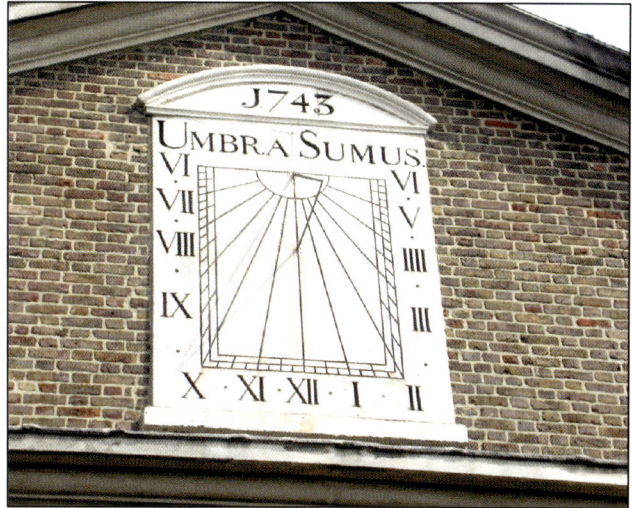
Vertical Dial at Jamme Masjid Mosque

reported that the six dials are accurate to within 10 seconds!

The sundial pillar was unveiled in 1989 by H M Queen Beatrix of the Netherlands as part of the UK's William and Mary tercentenary celebrations. This was appropriate, as the original Seven Dials was built in the reign of William and Mary.

SPITALFIELDS, E1
This handsome dial in Fournier Street, with its motto:

UMBRA SUMUS
(We are shadows)

adorns the Jamme Masjid Mosque in the Spitalfields district often known as 'Banglatown', after

Jamme Masjid Mosque, Fournier Street

its predominately Bangladeshi community. This south facing stone Vertical Dial, dated 1743, decorates a building whose history shows that asylum seeking is far from being a present-day phenomenon.

Following the revoking of the Edict of Nantes in 1685, Huguenots sought sanctuary in the area and by the 1740s their expanding community needed a new church. In 1743 the elegant brick-built 'Neuve Eglise' was built on the corner of Fournier Street and Brick Lane. The Huguenots' main occupation was silk weaving, which in time declined due to mechanisation and cheap foreign imports. Many moved out of the area and by 1819 a Methodist Congregation was using the church.

In the 19th century, Spitalfields experienced great poverty, with refugees from the Irish potato famine joining those Huguenots who had remained. The 1880s brought a further influx, when east European Jews, fleeing persecution, settled in the area and in 1898 they converted the church to a synagogue. As they prospered in the textile trade, many Jews moved to better class areas. In the 1939-45 war many more were killed by the bombing, which also damaged the synagogue.

By the 1970s the synagogue was in disrepair and in 1976 it was bought by the Bangladeshi community for conversion to a mosque. The Bangladeshis were economic rather than political refugees but their presence adds an exotic flavour to Spitalfields.

Not to be missed by the visitor are Christ Church, by Hawksmoor, and the Georgian houses with weavers' windows, in Fournier Street.

Vertical Dial at St Clement Danes Church in The Strand

James Gibbs heightened the tower. This, and much of the walls survived the World War II bombing. After extensive rebuilding in 1958 it became the central church of the Royal Air Force. The floor of the nave is inlaid with 800 Squadron crests, and in the church are Rolls of Honour commemorating wartime casualties.

Although it is partially obscured by trees, the declining Vertical Dial on the south side of the tower is a gem, sitting as it does within an architecturally appealing frame, topped of course with the ubiquitous anchor. A photograph of 1900 shows that the offending trees were not planted then.

Walking around the outside of the church, one finds there is Baroque decoration on the south side but, mysteriously none on the north. The reason is that originally the north side was never seen, due to its proximity to neighbouring buildings. Increasing traffic congestion required the removal of these buildings to provide a relief road, thus leaving the church marooned on an island site.

THE STRAND, WC2

The visitor may wonder why the weathervane, and much of the decoration inside St Clement Danes church depicts an anchor motif. In the 9th century a colony of Danes settled here, naming their church after the patron saint of Danish sailors. A major rebuild, following earlier rebuilds was undertaken by Wren, and in 1719

Whilst the church bells regularly ring out the 'Oranges and Lemons' nursery tune, it is thought more likely that the rhyme referred to St Clement in Eastcheap, which was nearer to the Thames wharves where citrus fruit was unloaded.

Details of sundials in nearby Temple gardens can be found in the City of London Sundial Trail. The Law courts, also close by, are well worth a visit.

TOWER BRIDGE, E1

On a prominent site between the Tower Hotel and Tower Bridge stands an impressive

'The Timepiece' near to Tower Bridge

sculpture by the artist Wendy Taylor. Its elegance and nautical simplicity perfectly compliment its prominent position alongside the Thames. It is named 'The Timepiece' and is in fact an Equinoctial Dial. It consists of a large stainless steel ring, on which the hours are marked with raised dots. The ring is supported on three chains, which spring from a single point on the quayside together with the gnomon that resembles an oversized dockyard nail.

In this type of dial the face points towards the north and it is only in the summer between the spring and autumn equinoxes that the shadow of

'London Underground' dial at Tower Hill

the gnomon falls on the hour marks. In the winter the sun will only illuminate the back of the dial but the gnomon will still cast its shadow on the inside of the dialplate ring. This can just be spotted on the picture showing 8am. The installation of 'The Timepiece' in 1973 marked the beginning of a renaissance in large-scale public sculptures and inspired a revival of interest in sundials for public spaces.

The Sculpture is a well known landmark for the thousands of runners taking part in the annual London Marathon, as it can be clearly seen as they cross Tower Bridge and turn on to the embankment where they follow a devious route round London's dockland to finish up at St James' Park. Visitors not so engaged in such exercise will have plenty of fascinating things to see in the vicinity. Apart from the unique bridge itself there is St Katherine's dock built by the engineer, Thomas Telford on a 27 acre site in 1828.

TOWER HILL, EC3

On an artificial mound above Tower Hill underground station stands a large Horizontal Dial, which was erected in 1992. The mound was built as a viewing platform for the Tower of London, the capital's most popular tourist attraction. This means that the sundial is probably seen by more people than any other dial in Britain.

The dial's gnomon is patterned with wavy lines reminding us of the nearby River Thames. John Chitty, who designed the dial for the London Underground, cleverly incorporated their circle and bar logo into the style of the gnomon. However, the shadow of the projecting part of the circle sometimes falls on the hour scale and at such times the dial can be a little awkward to read. This illustrates how sundial design can very often be a compromise between artistic and scientific considerations.

The dial is calibrated for British Summer Time, so 1 hour has to be deducted in winter. The problem of the Equation of Time is wisely avoided - the dial's plaque merely states that dial time will always be within 16 minutes of clock time.

When you tire of the bustling tourists, tranquillity can be found in the adjacent Trinity Square gardens, which contain memorials to men of the Merchant Navy who died in the two world wars - a colonnade by Sir Edwin Lutyens and a sunken garden by Sir Edward Maufe. A chained off area nearby marks the site of the scaffold where many prisoners in the Tower met their end. Finally , if you walk past the magnificent former HQ building of the Port of London Authority, into Seething Lane, you will discover a bust of Samuel Pepys erected on the site of his garden. It was here that, when the Great Fire threatened, he dug a hole and buried important papers, and his beloved parmesan cheese.

Wimbledon Argos Dial

WIMBLEDON, SW19

Visitors to Wimbledon, who come by train or underground, emerge from the station onto Wimbledon Hill Road. If they are not hard pressed, on their way to watch tennis, and if they turn right and walk as if to proceed up the hill, they will soon catch sight of the Argos store on the corner of the cross-roads. Here they will see a spectacular and unusual mural Vertical Dial on the curved south wall of the building, which, at first, they may not comprehend, since it has no hour-lines, as such, and no numerals to indicate the particular hour of the day. Instead, it comprises thirteen 'Argos-red' stove-enamelled stainless-steel hour-markers, uniformly diamond-shaped to resemble the short swords or daggers of the ancient Greeks. These appear to be randomly spread out on the sand-coloured surface of the reconstituted stone frontage of the building. At the centre of this 'display', there is a polished stainless-steel disk, with a broad circular 'Argos-red' border, resembling a clock face. A polished stainless-steel rod, projecting downwards from the centre of the disk, is the gnomon, the shadow of which indicates the time, as it passes over the dagger-like hour-markers. Since the visitor is likely to have *some* idea of the approximate time of day, a glance at the dial should suffice to confirm this, regardless of whether or not Summer Time is in force! A modern work of scientific art, the sundial was designed by Christopher St J H Daniel and constructed by Brookbrae Limited in 1996.

WOOLWICH, SE18

A modern Vertical Dial in Hare Street was designed for 'Seventh Sun' who deal in new age fashion-wear and paraphernalia. It was decided that its sundial should reflect this by the use of more modern and trendy materials. It is true that the gnomon is made of brass but the dial plate utilises acrylic and vinyl to depict the riotous display of the seven suns.

The shop faces towards the west so that the dial only catches the Sun's rays in the afternoon and evening periods. In some ways this was serendipitous for the designer, since the hour marks 1 to 7, (Greenwich Time), correspond with the 'Seventh Sun' theme.

Visitors should note that shadows from the buildings opposite mean that the dial is best appreciated in the summer months. As the year progresses, the Sun's height increases and the gnomon's shadow can be seen on the face for a longer period each day until the solstice, after which its height begins to decrease again. The dial was designed by John Moir and was constructed by Ray Ashley.

In nearby Beresford Square is the gatehouse of the Old Royal Arsenal and across the road is a handsome Vertical Dial.

Vertical Dial for 'Seventh Sun' in Woolwich

EAST OF ENGLAND

Mike Cowham, John Davis & Margaret Stanier

Bedfordshire, Cambridgeshire, Essex, Hertfordshire, Norfolk, Suffolk

When we think of the East of England we normally envisage large areas devoted to agriculture, a flat land and those big open skies. The area is indeed noted for its fenland with black peaty soils with outcrops of clay but there are also large areas of gently rolling country and a wonderful coastline.

The area contains some interesting architectural clunch (a soft white limestone) from Cambridgeshire, and flint from the coastal regions of Norfolk, Suffolk and Essex. These local materials have all been well used for the building of homes, churches and sundials.

There are numerous dials in the region and we have picked out a few of the more interesting ones to describe. Naturally there are many in

The magnificent cathedral at Ely, truly known as 'The Ship of the Fens'

gems. We have the great cathedrals of Ely, Norwich, Peterborough and Bury St Edmunds, and the incomparable colleges of Cambridge. The area is also famous for its many quiet lanes and small villages. These often appear as if time has overlooked them and they remain little changed by the march of time.

Building materials vary greatly from the hard and durable Barnack limestone from the north,

Cambridge, one of our mightiest seats of learning but there are others, often simple dials, spread throughout the region that are just as interesting in their own right.

The visitor to Cambridge should also see the Whipple Museum which has a great collection of mainly pocket dials and associated astronomical instruments. These too have an important place in our timekeeping history. The Fitzwilliam Mu-

67

seum in Cambridge is a world-class museum and should not be missed by the visitor. It houses fine collections from Egyptian, Greek and Roman times through medieval art to the 20th century. It also contains a fine collection of clocks.

ALDEBURGH, Suffolk

The 16th century Moot Hall was once in the centre of Aldeburgh but erosion by the sea has completely removed one of the three streets which ran parallel to the sea so that it now stands perilously on the edge of the shingle beach. It currently houses a museum of archaeology and has a most attractive Vertical Dial on its southern end. Many tourists view the dial as they watch the remaining fishing boats being hauled up the beach and it even featured briefly in the famous 'tell Sid' British Gas TV commercial! The dial carries the date 1650, although the style and its excellent condition makes it doubtful that this is the original dial for the location. Aldeburgh was at its most prosperous at this time as a leading port although Dunwich, its northerly neighbour, was already losing its battle with the sea. The dial is quite close to a direct south facing one, set just 11° to the west. The motto:

HORAS NON NUMERO NISI SERENAS

is the Latin equivalent of the rather twee, 'I only count the sunny hours' to be found on Victorian dials and poor modern reproductions.

There are several other sundials visible from a stroll around the pleasant streets of Aldeburgh. There is, for example, a painted Vertical Dial on the front of a house in Oakley Square and another

The Aldeburgh Moot Hall dial

in the appropriately named Dial Lane. Even the local rest home is named 'The Sundial'.

BURY ST EDMUNDS, Suffolk

Next to the cathedral, with its new tower, are the delightful Abbey Gardens. As well as the remains of the old abbey, these contain an interesting

The Moot Hall at Aldeburgh

The Pillar Dial at Bury St Edmunds

The unusual equation of time at Bury ^MC

CAMBRIDGE

Queens' College. In the colleges of the City of Cambridge are to be found many impressive sundials. Perhaps the best known is that at Queens', facing approximately south in the Old Court. Its complex lines and unusual table beneath make so many tourists ask their purpose that the College has printed a special leaflet. It is a Sun-and-Moon Dial and is one of the most remarkable wall dials in Britain, well known among gnomonicists for the intricacy and interest of its dial furniture.

The first dial recorded at the college was made in 1642 at a total cost of £3 7s 6d, though it is not known if this was related to the current dial. Then, in 1733 the antiquary Cole reported '....on ye Wall of ye Chapel and over ye door wch leads to it is also lately painted a very elegant Sun Dial with all ye signs. This is no small ornam^t to ye Court to enliven it'. This was the Moondial which has often been attributed to the Cambridge scholar Sir Isaac Newton even though he was dead by then! Since then, the dial has been repainted at least seven times, including a period

Pillar Dial dated to 1870 by a plaque on the base. There is only a direct south face to the dial, with a mix of Roman and Arabic numerals for the hours, but on the west face of the stone block is a unique form of engraving for the equation of time. This is in a very early graphical form and it also corrects for the longitude of Bury, allowing Greenwich Mean Time to be obtained. This would have still been something of a novelty in 1870.

Its motto on the west face reads:-

MONET ANNVS ET ALMUM
QUAE RAPIT HORA DIEM
(The year and the hour which snatches away the nourishing day warn you)
Horace, Odes 4.7

There is a much older declining Vertical Dial, in traditional form, on the cathedral itself. Additionally, there are dials to be seen on St Mary's Church and on the Unitarian Meeting House in Churchgate Street. Whilst in Bury, be certain to visit the Manor House Museum where, amongst the many clocks on display, there is a good variety of portable dials to be seen. Gershom Parkington, the 20th century musician whose collection forms the basis of the museum, is buried locally and has a sundial as a monument, though it is now sadly vandalised.

Queens' College Dial ^MC

The Sun and Moon Dial at Queens' College, Cambridge

HORIZON. The green lines show the Sun's declination and are labelled with the zodiac signs and symbols, and with the associated planetary signs, in two broad strips inside the main numerals; from mid-winter to midsummer on the right and midsummer back to midwinter on the left. Note that the declination lines do not project above the horizon line (which indicates that the sun has set) so the positions of the zodiac signs have to be judged by eye. The months of the year, in Latin and for the old Julian calendar, are set just outside the signs. The column on the left labelled **ORTUS SOLIS** gives the times of sunrise for each of the green declination lines. The corresponding column **LONGITUDO** on the right gives the length of the daylight hours throughout the year. The curved red lines give the elevation (or altitude) of the Sun in degrees above the horizon. The vertical black lines show the azimuth or bearing of the Sun. Finally, the straight black lines which fan out from the centre of the horizon line give the time in temporary, or seasonal hours, i.e. in twelfths of the period between sunrise and sunset.

in the 19th century when it became derelict and gnomon-less. As a result, the accuracy of its delineation has slipped although there is now a computer-generated design available to correct it. The outer Roman numerals of the dial give the local apparent, or solar, time from the shadow of the gnomon in the standard manner. The rest of the dial furniture is read from the shadow of the nodus which is set level with the line marked

The table of numbers underneath the dial is for reading the time at night by the Moon's shadow. The top and bottom rows (1-15 and 16-30) are the age of the Moon in days since the last New Moon, which the observer is expected to know. The times in the middle row, in hours and minutes, then need to be added or subtracted from the time indicated by the shadow to give the time of the night. The complexities of the Moon's orbit mean that this time is likely to be rather inaccurate, even if the shadow can be seen.

Gonville & Caius College. There is a large six-faced Vertical Dial above the gateway between the College and the Senate House Passage. It is

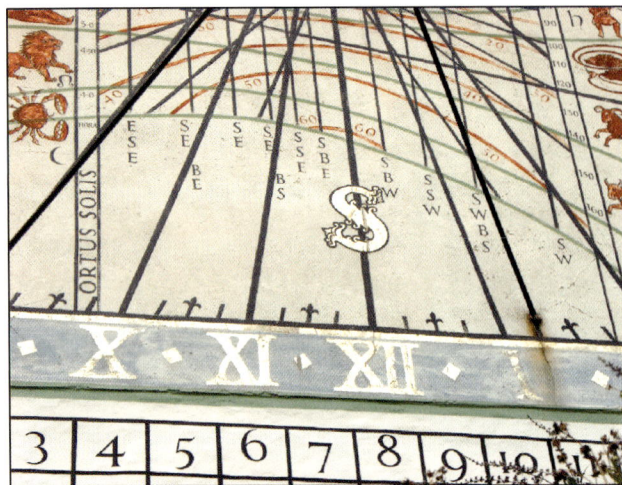

Detail of Queens' College dial

perhaps the most elegantly beautiful of all sundials in Cambridge. The precision of its lines, the restrained colouring, and the clarity and simplicity of the dial faces make it a delight to see and to read. Four of the dial faces (east, south-east, south-west, and west) can be seen from Senate House Passage: the north-west and north-east faces are visible from within Caius Court. It is pleasing and perhaps surprising that despite the nearby presence of several tall buildings all six dial-faces catch some sun, at least

Dial above the Gate of Honour, Gonville & Caius College _{MC}

in the summer months. This dial, though in the position of a much earlier one, dates only from 1963. It was constructed soon after the extensive restoration of the Gate of Honour itself, undertaken by the College as part of a series of restorations marking the 4th centenary of the re-founding of the College by Dr Caius in 1558. Dr Frank Powell, then Junior Bursar, made the first sketches for the design, taking as a guide an engraving dated 1688 from Loggan's Cantabrigia Illustrata, which shows the earlier sundial. Measurements were undertaken by Dr P J Message (Fellow and Astronomer) of the exact corner angles of the hexagonal stone superstructure to which the dial faces are attached, and also of the orientation of the eastward and westward facing sides, which are not precisely parallel with meridians. These measurements allowed great accuracy in the drawing of the hour lines and design of gnomon angles for each face. The dial faces are of bronze, coated with vitreous enamel. The working drawings were made by the Ancient Monuments branch of the Ministry of Works, and the dial faces were made by Birmingham Guild Ltd. An earlier set of sundials was inscribed in the stone of the Gate when (or soon after) it was built in 1575. The designer may have been Theodore Haveus of Cleves, who is known to have designed another sundial, no longer extant, which stood in Caius Court. The hour-lines for this earlier Gate sundial were incised and painted directly on the stone, and the gnomons were

mounted on iron brackets inserted into the stonework. This set of dials needed frequent repair and restoration, and gradually fell into decay. When the thorough restoration of the stonework of the Gate was carried out in 1958-59, it was still possible to see traces of some of the hour-lines, and the fixing holes for the gnomon brackets. The present dial faces will be more durable than their predecessors. They are a handsome adornment to the Gate and a pleasure to passers-by.

Downing Site. The Polyhedral Dial at the University's Downing Site has an unusual 17 faces. It stands on a lawn close to the entrance from Downing Street. This sundial is in the form of a stone cube from which portions of the surface have been cut to form 17 dial planes, each with its own gnomon. This form of dial was quite common in the late 16th and 17th centuries, more as a mathematical conceit than as a practical timepiece. On the upper (horizontal) surface Roman numerals mark the hour lines; on the eight vertical and 8 sloping surfaces the numerals are Arabic. The dial stone stands on a low stone pedestal, round the base of which is the inscription:

SOLI HORISQUE WILELMUS ET UXOR
LUCIA RIDGEWAY POSUERUNT A[nno]
S[alvationis] MCMXIII
(For the sun and the hours, William Ridgeway and his wife Lucy placed this in the year of our salvation 1913)

The support of the uppermost gnomon is in the

71

Polyhedral Dial at Downing Site MC

form of a seated camel; the same inscription with the omission of the first two words is engraved on this gnomon. The sun rarely strikes this sundial in the winter months as there are now high buildings to the south and west of it. The dial has a function in garden conservation however: its presence has saved this portion of lawn from becoming a car park or the site for a temporary laboratory. It is unique in having protection by University ordinance: the first charge on the Ridgeway-Venn Travel Fund of the University is the upkeep of this dial. Professor

Camel on gnomon of Polyhedral Dial MC

Sir William Ridgeway was Disney Professor of Archeology in the University of Cambridge, from 1892 until his death in 1926. He had wide interests in archeology, mainly of the eastern Mediterranean countries. He was instrumental in obtaining recognition by the University of the study of Anthropology as an academic subject.

Magdalene College. The tradition of installing dials in Oxbridge colleges is still current, as can be seen by the modern Vertical Dial high on the wall of Benson Court at Magdelene College (on the west side of Magdelene Street). It is unique amongst Cambridge dials in that it reads GMT directly, having the corrections for the equation of time and the local longitude incorporated in its delineation. The dial is split into two halves, one for the period from the winter to the summer solstices (labelled January-June on the dial) and the other for the period from the summer solstice to the end of the solar year (labelled July-December). The hour lines for the first half are long thin figure 'S's and for the second half they are backward 'S's. If combined onto a single figure they would produce standard figure-eight analemmas but adjacent hour lines would overlap confusingly.

The time markers are the spots of light in the centres of the sunburst shadows, cast by the accurately positioned stainless steel sun disks. At the winter solstice, these spots track along the top edges of the hour curves. The path slowly moves downwards through the year to reach the bottoms of the 'S' curves at the summer solstice.

The design of this unusual sundial arose from the annual design competition run by Magdelene College for its first-year engineering students. The subject for 1986/7 was a sundial and Mr W-F Ng was the prizewinner. The stonework was carried out by an Honorary Fellow of the college, Mr Will Carter, and the metalwork made in the University Engineering Dept. The motto:

FACILIUS INTER PHILOSPHOS QUAM INTER HOROLOGIA CONVENIET
(It is easier to gain agreement amongst philosophers than among timepieces)
is a line from a satirical work by Seneca.

Downing College. To celebrate the bicentenery of Downing College a Horizontal Dial made of slate was commissioned from local stonemason Quin Hollick.

Twin Dials for January-June and July-December at Magdalene College, Cambridge ^{MC}

Downing College, Cambridge ^{MC}

It is inscribed around its edge:

**1800 - 2000 DOWNING COLLEGE
BICENTENARY**

and carries the Latin motto on its face:

QUÆRERE VERUM
(Search for the Truth)

EAST BERGHOLT, Suffolk

East Bergholt is near the Suffolk/Essex border in the heart of the Dedham Vale. The painted wood Vertical Dial with its arched top on the church porch is typical of many church dials and is set just 10° east of true south. It warrants inclusion here because it appears in one of John Constable's earliest paintings, 'The Church Porch' (1809) now in the Tate Gallery. Constable knew the church well having met his wife here. In his painting, he included people of three generations – an old man, a woman and a girl – talking in the churchyard as an allegory of the passing of time. This is also reflected in the motto on the dial:

Time passeth away like a Shadow

which is also found on the Grundisburgh dial.

73

Dial above the porch at East Bergholt ^{MC}

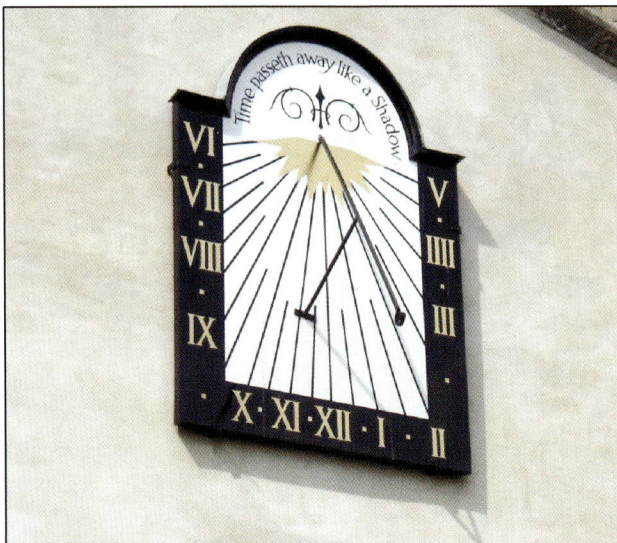
East Bergholt Dial ^{MC}

ELY, Cambridgeshire

The great Norman cathedral at Ely with its charming town stand proudly atop a clay mound, rising out of the fens. The cathedral can be seen from great distances, often looking like a great galleon in the calm sea of fens. It has been described as 'The Ship of the Fens' and truly lives up to its famous title. On its south transept is mounted a Vertical Dial with its Greek motto:

ΚΑΙΡΟΝ ΓΝΩΘΙ

literally 'Know the Season'. It may be more freely translated as 'Recognise the Right Moment'. This relates to the Royalist Bishop of Ely who had the dial made in 1690, after the restoration of the monarchy, when he felt it appropriate to show his allegiances once more. The Dean and Chapter originally paid a Mr Rider £10 for making the dial and even in 1962 it only cost £36 for a partial restoration. The dial is brightly painted in blue with gold lettering and black scales. In theory, it also shows lines for the entry of the sun into the various zodiac signs (the season) but the small nodus on the gnomon is currently missing. The dial was referred to by Mrs Gatty in her 1871 Book of Sun-Dials where she writes that the dial 'is placed on a buttress on the north (sic) side of Ely Cathedral. The face of the sun seems to emit the lines to the surrounding figures, as well as the gnomon; and between the lines are the signs of the zodiac'. The typographical error was corrected in later editions; the dial is, of course, on the south side!

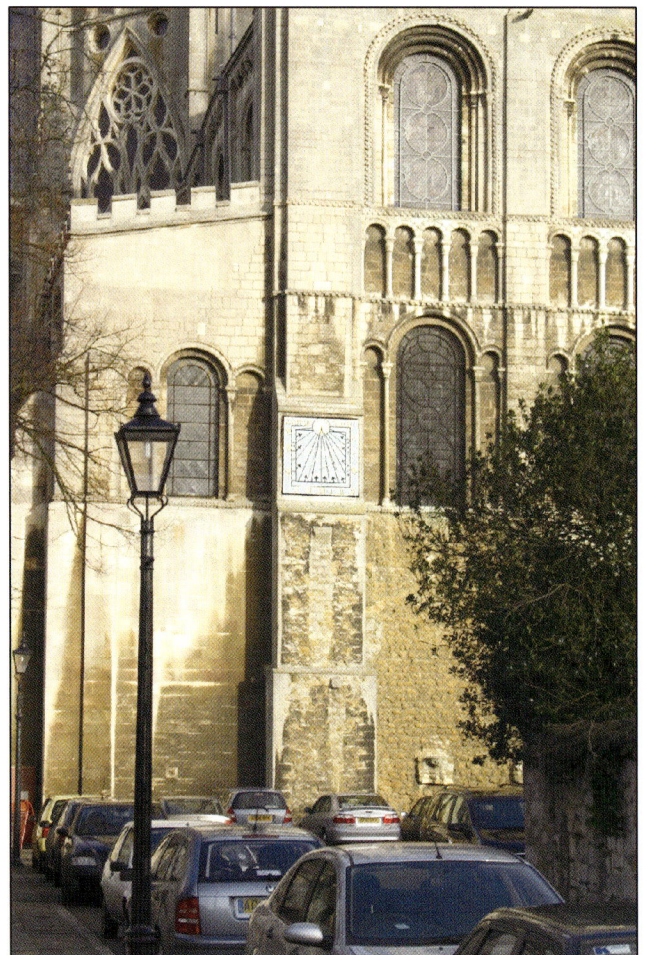
Dial on the south transept of Ely Cathedral ^{MC}

74

No visit to Ely is complete without entering this truly magnificent cathedral. Look out for its maze (more properly a labyrinth) on the floor just inside the entrance door. Its length is supposed to be equal to the height of the Cathedral's tower. See also the many examples of church windows in its Stained Glass Museum.

In the Market Place is an Analemmatic Dial made in 2001 by Quin Hollick with an apt caption for Ely:

'A city built on a hill cannot be hid'.

Another dial with the same motto as the cathedral,
ΚΑΙΡΟΝ ΓΝΩΘΙ
can be found on the Methodist Church at nearby Haddenham.

ΚΑΙΡΟΝ ΓΝΩΘΙ - (KNOW THE SEASON), at Ely Cathedral

GREAT STAUGHTON, Cambridgeshire

On the outskirts of Great Staughton the main A428 road between St Neot's and Higham Ferrers passes a rather unusual Cube Dial set atop an old stone pillar supported by a later brick base. The monument is usually described as the Village Cross. It consists of three separate dials, a direct south dial, a direct east dial and a direct west dial delineated on three faces of a stone cube. Above this is mounted a sphere, possibly once a crepuscular Spherical Dial. The iron gnomons are a little rusty but the original markings are still visible. Little is known or recorded about this dial but on its north side are engraved the date, 1637 and the arcaded letters **E.I.** These are the initials of Edmund Ibbott of Beechhampstead, one of the manorial areas of Staughton. He is described as a carpenter but must have been a master-builder as he possessed a 15 room house and left £137 in his will when he died in 1641; a sizeable sum at the time. It is likely that he paid for the dial though whether he made it with his own hands is unknown.

A short detour to Great Staughton church will be rewarded by the sight of a fine ornate gnomon placed above the porch, but the dial markings

Cube Dial at Great Staughton

75

Cube Dial at Great Staughton

MC

MC

Cube Dial at Great Staughton with date, 1637, and arcaded initials, E I

have long since disappeared due to weathering.

The nearby town of Kimbolton has a large and fine Vertical Dial between the two upper windows of a house in East Street (parallel to the main street). It is usually missed by visitors.

GRUNDISBURGH, Suffolk

The dial on St Mary's church, Grundisburgh, is thought to date from 1731, making it contemporary with the brick tower, although the church itself dates back to the 14th century and also has a well hidden Mass Dial. The dial sits immediately below a good tower clock and the church is in a very pleasant environment beside the village green, complete with a stream flowing through the middle of it. This east declining Vertical Dial with a nodus on the gnomon, has two interesting items of furniture in addition to the rather gloomy motto:

LIFE PAS'S LIKE A SHADOW.

The first is the set of declination lines running approximately diagonally across the dial. These are calculated for the hours of daylight (8 to 16 hours) rather than the more common zodiac signs. These are crossed by vertical straight lines which are for the Sun's azimuth, although they are unlabelled. In the photograph, the nodus shadow indicates there will be about 14¼ hours of daylight and that the sun is currently about 15° east of due south.

The second feature is the pair of extra chapter rings around the foot of the gnomon. The inner one, with Arabic numerals, is 4 hours behind local time and so indicates the time for a longitude of about 60°W. This could be the West Indies (probably Barbados) or Newfoundland. Several

JD

Vertical Dial on Grundisburgh Church tower

merchants from the nearby port of Ipswich traded with the sugar plantations in the West Indies but it is thought that the local squire at Grundisburgh Hall had connections to Newfoundland. The outer ring with Roman numerals is 2½ hours ahead of local time and is thus for a longitude which would fit with Jerusalem.

HITCHIN, Hertfordshire

A most unusual sundial is to be found in the Physic Garden attached to the Victorian chemist shop museum in Hitchin's Paynes Park. It is a Scaphe Dial in the form of a most appropriate pestle and mortar. Although the pestle which forms the gnomon is tapered, because it has a circular cross section it can still accurately indicate the correct solar time as long as the centre of the shadow is estimated. The dial is modern but inscribed in the bowl around the base of the gnomon is:-

DISEASE DOTH OFT RISE
ABOVE MEDICINE
WM. DRAGE   HITCHIN APOTHECARY
1632   1668

The museum includes memorabilia of the medical pioneer Lord Lister who began his education in the town, as well as many of the products, which gained Hitchin the name of 'Lavender Town'.

Unusual Pestle & Mortar Dial at Hitchin

Scaphe Dial at Hitchin where the time is estimated from the centre of the shadow

The Physic Garden was opened in 1990 and the dial itself is in memory to Jimmy James and Douglas Whittet, this dedication being inscribed around the outside of the bowl.

A visit to the nearby St Mary's Church will be rewarded by the sight of two good Vertical Dials on the tower, one south and the other south-east. The south-east dial is engraved:

Anno Salutis
1660.

KING'S LYNN, Norfolk

This old market town lies on the eastern bank of the Great Ouse, approximately two miles from The Wash. 'Lin', as it was originally known, became an important trading port that was at its height in medieval times (when it was known as 'Bishop's Lynn'). It was not until 1536 that it became King's Lynn and was granted a charter by Henry VIII.

One of the most imposing rows of buildings includes the Old Gaol with its sundial and the Town Hall with its stonework pattern facade. The Vertical Dial towers high above the square next to St Margaret's Church. The dial was made by S Bunnett and carries the popular motto:

TEMPUS FUGIT
(Time Flies)

There are still many interesting buildings to see in this busy market town today. They have been carefully restored and looked after by the King's Lynn Preservation Trust. One such building is

Old Gaol at King's Lynn
with its dial high above

Vertical Dial over the Old Gaol at King's Lynn ^{MC}

the Custom House, now a tourist office and small museum containing mostly maritime artefacts, including a small brass Universal Equinoctial Ring Dial, and a fine collection of old Customs & Excise weights and measures.

At the same time that the Custom House was built (1683), Thomas Tue, a clockmaker and church-warden, presented a clockwork driven Moon Dial to St Margaret's Church. This dial would have been of great use to a community dependent on the sea.

Along the South Quay is to be found a modern Equatorial Dial.

LEIGHTON BUZZARD, Bedfordshire

The church of All Saints' in Leighton Buzzard has an incredible number of Vertical Dials; five in all. There are two direct south facing, both on the south transept, an east and a west dial, also on the south transept and a north facing dial on the

north wall. North dials are fairly uncommon, especially on churches and can only function during the summer months and then only before 6am or after 6pm. A north dial is characterised by its strangely inverted gnomon, but this is correct because it has to be aligned with the earth's pole. According to the church's Guide Book, these dials are relatively recent but were certainly in position according to Mrs Gatty in 1888. The two south dials together with the west dial have weathered over the years but the east and north dials are still in excellent condition. It is not clear why so many dials were fixed to the church exterior but it may just have been a dialling exercise for a local mason.

Four of the dials have Latin inscriptions but some are now virtually unreadable. Mrs Gatty's 'A Book of Sun-Dials' records these mottos as follows:

South-west;

Vigila, oraque: Tempus fugit
(Watch and pray: Time flies)

South-east;

Brevis ætas, Vita fugax
(Time is short, Life is fleeting)

Direct west;

Dum spectas fugit
(While thou art looking
[the hour] is flying)

Direct east;

Deus adeft laborantibus
(God stands by those who labour)

The north dial has no inscription.

The church is worth a visit for its beautifully

The five dials on All Saints' Parish Church, Leighton Buzzard. On the left are two Vertical South Dials mounted on the south-west and south-east buttresses, in the centre a Vertical North Dial and on the right a Direct West Dial and a Direct East Dial

carved and gilt angels, high up in the nave, chancel and transept. The church unfortunately suffered a devastating fire in 1985 but sympathetic restoration has been able to restore most of the losses.

Nearby in the High Street there is another fine vertical dial above a shopfront with the motto

WATCH AND PRAY
TIME FLIES AWAY

This dial still retains traces of blue paint around

its chapter ring and has a sunburst pattern at its gnomon root.

LODE, Cambridgeshire

Anglesey Abbey near Lode in fen country to the north-east of Cambridge was left to the National Trust in 1966. It is a country house built out of

Vertical Dial in Leighton Buzzard, High Street

Old Father Time at Anglesey Abbey

Dial plate by Watkins & Smith, London 1769 ^{MC}

one of a series of at least three by sculptor John van Ost (or John Nost) who was born in 1686 in Flanders but moved to a location in the Haymarket in London. His most famous works are the statues in the pediment at Buckingham Palace. The Kronos statue is an attractive piece of the Formal Garden and is surrounded by white and blue hyacinths each spring and dwarf dahlias in summer and autumn. Unfortunately the dial was not correctly set up in its new home so that the statue obscures the sun for part of the day. The circular bronze Horizontal Dial is rather more recent than the statue and is signed:

Watkins & Smith, LONDON 1769

This was a well-respected partnership of mathematical instrument makers working towards the end of the 18th century.

the remains of an Augustinian priory. The site was purchased in 1926 by Huttleston Broughton (1896-1966) the first Lord Fairhaven. In the space of forty years he filled the house with art and treasures, and developed the 98 acre landscape garden adorned with over 100 items of statuary. There are several sundials in the gardens including one showing Kronos holding a dial. Kronos was a Greek god created by the poet Hesiod, writing around the time of Homer. He is now better known as Father Time. The statue was originally at Stowe in Buckinghamshire and is

Another splendid dial is in the Herbaceous Border. It is a large stone Cube Dial surmounted by a statue of the Saxon deity 'Tiw' (for Tuesday, one of seven deities that our weekdays are named after). While the dial is 19th century the statue dates to the mid 18th century and again originally comes from Stowe in Buckinghamshire. The main south facing dial carries the message:

**FEARE GOD,
OBEY YE THE KING**

**Snake supporting the gnomon at
Anglesey Abbey** ^{MC}

Tiw standing on a Cube Dial ^{MC}

Cube Dial supporting the statue of Tiw

Comberton Meridian Dial

Other dials in the grounds include a Horizontal Dial by the south front of the house on a carved stone pedestal. Its dial is inscribed **B. Cox. Kew. 1785** although there is doubt over this and it may be a 20th century copy. There is also a Vertical Dial on the side of the old Chauffeur's house overlooking the Rose Garden which can be seen through a gateway, but it is not directly accessible to the public. It has a similar motto to the 'Tiw' dial

**FEARE GOD
OBEY YE KING**

Inside the main house is an extensive accumula-

tion of art collected by Lord Fairhaven including some interesting and complex clocks.

MELDRETH, Cambridgeshire

Dials have been set up along the line of the Greenwich Meridian in various places. The dial at Meldreth was placed at the roadside by the Parish Council to celebrate the Millennium and was unveiled by Sir Martin Rees FRS, Astronomer Royal, 4 December 1999. The Meldreth dial cleverly has both Vertical and Horizontal Dial planes sharing the same gnomon.

A similar dial a few miles north lies between Comberton and Toft where it was erected in 1984 to commemorate the opening of the Meridian School and the planting of the tree that stands behind the dial in 1968. Unfortunately the close proximity of the tree will eventually shade the sun from the dial. This dial was made by Quin Hollick, a Comberton stonemason.

NEW HOUGHTON, Norfolk

Houghton Hall was built by Robert Walpole, Britain's first Prime Minister, in the 1720s. The house, standing in grounds of 4000 acres, which are home to a herd of white fallow deer, has a central block in the Palladian style with north and south service wings connected by curved colonnades. The north service block was severely damaged by fire sometime in the late 18th century. It was not until 2001 that serious attempts were made to restore the block which includes an octagonal tower with cupola. When it was inspected, four large iron rods were discovered projecting from the original stonework. These turned out to be the gnomons of four lost sundials which would have faced very approximately

Meldreth Meridian Dial

Cupola Dials at Houghton Hall JD

Detail of East, North and West faces of the JD
Houghton Hall Pillar Dial

north, south, east and west. As no signs remained of the original dial faces and no pictures of them could be found, new dials were designed from scratch. The designs drew on the fact that the mirror-image south service block has a tower clock with four faces in the equivalent positions.

Pillar Dial at Houghton Hall JD

Although a 1½ metres in diameter, the dials look small from the ground so a very bold design was made with large numerals. It uses the restored and reset original iron gnomons together with faces of vitreous enamelled steel with gilded numerals.

The recently restored gardens also have a multiple Pillar Dial. The four faces of this dial are arranged to face exactly to the cardinal points of the compass, simplifying the design considerably. Walking round the pillar shows clearly that the four gnomons are all parallel to each other and pointing to the north celestial pole. Whilst the south face of the dial indicates the time from 6am to 6pm (the maximum period any south-facing wall can be illuminated by the sun) the east face operates between the earliest sunrise and mid-morning, and the west face from mid-afternoon to the latest sunset. This is rather academic, though, as the pillar is rather shaded by the surrounding pleached limes. The main attraction of the dial is its excellent decorative stonework featuring lions faces, vine swags and acanthus leaves. Houghton Hall also has a Horizontal Dial, nicely hand engraved but not signed.

NEWPORT, Essex
Above the porch of St Mary's Church in this attractive Essex village is a somewhat unusually carved small wooden Vertical Dial. Its lines and inscriptions have all been carved in relief. This is an uncommon and time consuming technique,

Wooden Vertical Dial at Newport ^{MC}

but several other dials of this type are known, such as at Stoke-by-Clare in Suffolk and at Hertford. Originally the dial would have been painted to make its markings much clearer. This dial at Newport is perhaps the best preserved of its type and it is marked with the motto:

Many will be well paid for abusing Time
a familiar theme. Its maker is only known by his initials, **JC,** but he may have been a local carpenter or clockmaker. The dial was designed for a location declining about 15° east but at Newport it has been canted out by a similar amount towards the west, so it is likely that the dial was originally set up at another church.

NORWICH
On the church of St Peter Mancroft, close to the market place, is a painted Vertical Dial high on the south transept. Note the attractive use of St Peter's keys for supporting its slender gnomon and the fine gilding. The church was rebuilt from

Vertical Dial at St Peter Mancroft, Norwich ^{MC}

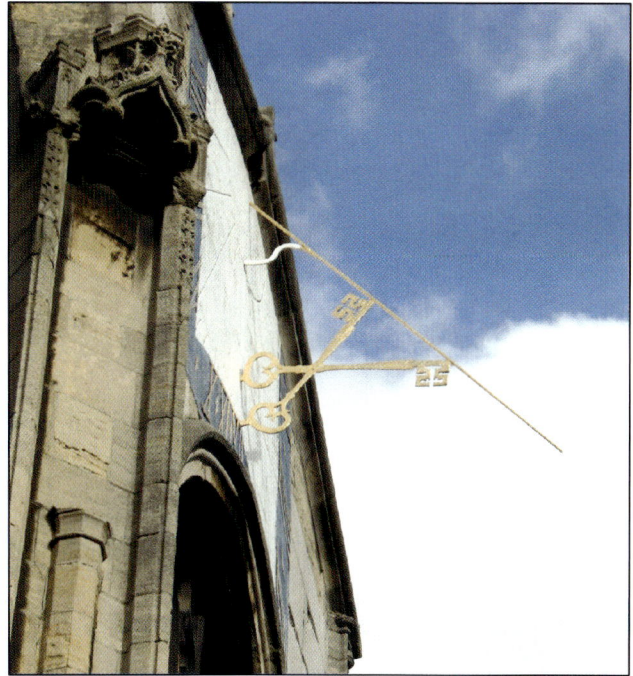
The crossed keys of St Peter holding the ^{MC}
gnomon at St Peter Mancroft

an earlier Norman one in 1455. It has a large peal of 14 bells in its tower that are frequently rung.

Other dials in Norwich include a painted vertical dial on a house on the south side of the Cathedral Close and a small vertical dial on St Andrew's Church.

When visiting Norwich take time to visit the Cathedral with its fine spire.

SILSOE, Bedfordshire
Wrest Park, located some 10 miles from Bedford, was the home of the Grey family from the Middle Ages until the early 20th century. Lying very close to the A6 at Silsoe, it is now a showcase garden featuring large areas of both formal and informal landscape, a canal, and much classical statuary, the latter managed by English Heritage. Halfway down the central axis of the garden is a very large Horizontal Dial surrounded by copies of statues including the Mattei Ceres, the Minerva Guistiniani from the Vatican and the Medici Venus.

This dial is of a rare type known as a Double Horizontal Dial. It was made in 1682 by the mathematical instrument maker Henry Wynne. (Actually, the dial on display is a convincing replica of the original which is now too valuable to risk further exposure to the elements.) Invented in the early 1600s by the mathematician the Revd William Oughtred, who also invented the slide rule, the design uses shadows from both the usual, sloping edge of the gnomon and also the vertical knife-edge in the centre of the plate to provide a wealth of astronomical information. Only about forty such dials, made from approximately 1620 to 1730, are known. The entire face of this dial is covered with fine engraving. The outer circular rings indicate the time at a range of locations around the globe, with some delightful spellings to some well-known places, such as Pequin in China. The centre of the face is dominated by a grid of lines similar to those to be found on an astrolabe. They indicate the date and the time, the time and direction of sunrise and sunset on every day of the year, and the altitude and azimuth of the Sun. With the aid of other information engraved on the dial, it can even be used to indicate the time at night by means of the moon and stars. A form of Perpetual Calendar around the base of the gnomon allows the dates of Easter, and the days of the week, to be calculated between 1683 and 1835.

The gnomon carries the arms of Anthony Grey,

Double Horizontal Dial by Henry Wynne [JD]
at Wrest Park

11th Earl of Kent (1645-1702) together with his motto, **Foy est tout**. The sundial has several other mottoes, including one hidden on the underside of the tail of the gnomon which reads:

Nulla dies sine linea
(not a day without a line)

This is Pliny the Elder's description of a particularly industrious painter and is very appropriate for this sundial with its intricate engraving.

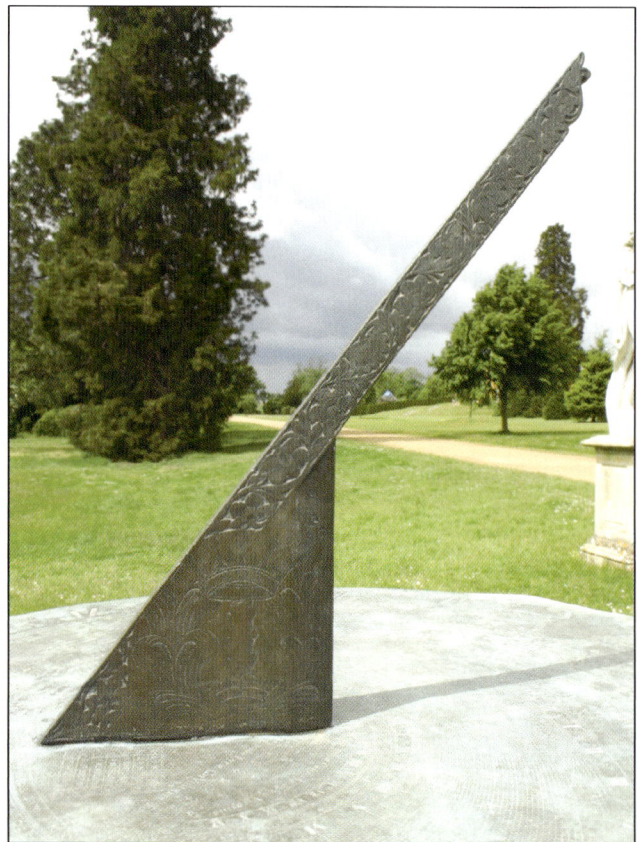

Double Horizontal Dial at Wrest Park [JD]

Decorated gnomon of Wynne Dial [MC]
at Wrest Park

84

The maker of the dial, Henry Wynne (1640-1709) was one of the greatest 18th century mathematical instrument makers. His premises, 'Next the Sugar Loaf' in Chancery Lane, escaped the Great Fire of London by only around 200 yards. He was regularly visited by the diarist Samuel Pepys to whom he sold both books and instruments. He was also an acquaintance of the scientist Robert Hooke. He made a very similar sundial to this for Windsor Castle, and was also responsible for popularising the barometer.

WANDLEBURY, Cambridgeshire

At Wandlebury Country Park on the Gog Magog hills, about four miles south-east of Cambridge, a raised lawn marks the site of the old Gogmagog House, demolished in the 1950s. On this lawn stands an excellent 18th century Horizontal Dial which was made for the original owner of the house, Lord Godolphin. He was a devoted supporter of horse racing and his name is still commemorated by the memorial inscription to the Godolphin Arabian stallion who was one of only three stallions from whom all modern thoroughbreds descend. Wandlebury has been owned since 1954 by the Cambridge Preservation Society and is designated a country park and wildlife reserve with five miles of footpaths giving all year access to the public.

The large sundial is signed by Joseph Jackson, a London mathematical instrument maker who worked from 1735 to 1760. He was a member of the Worshipful Company of Grocers, the London livery company (or Guild) to which many instrument makers belonged. Jackson held the appointment of a mathematical instrument maker to the Office of Ordnance in the 1740s and examples of his work can be found in several museums. The detailed design of the dial was handed down from master to apprentice through many generations. The central compass rose with its fine decorative engraving is surrounded by a series of rings giving the equation of time, or **Æquation of Natural Days,** as it is engraved on the dial, for every day of the year. A study of the values shows that it was drawn up for the old Julian calendar which England abandoned in 1752 and which placed the equinoxes and solstices 11 days earlier than their present dates. The calculations for the values were performed by John Flamsteed, the first Astronomer Royal at Greenwich.

Dial sitting close to the house at Wandlebury

Dial by Joseph Jackson at Wandlebury MC

Although the date is not engraved on the dial, it is believed to be after 1740 because of the ducal coronet on the coat of arms which relates to the marriage of a Godolphin daughter to the Duke of Leeds in that year. The dial does show the latitude that it was designed for, 52° 15′ north, which is very close to the accepted modern value for Wandlebury.

The dial is mounted on an elegant pedestal of Venetian stone. Another Joseph Jackson dial is in a garden just over the Suffolk border and it is on a pedestal of the same design, supporting the view that there was often a strong link between particular instrument makers and sculptors in stone.

Equatorial Dial at Woodbridge JD

Brick carved dial on the Melton Grange JD
Hotel, Woodbridge

WOODBRIDGE, Suffolk

Elmshurst Park in Woodbridge is not far from the Deben estuary, a famous working tide mill and all the yacht building and repairing activity around the harbour. The town, with its crowded old streets and buildings, is one of the most pleasant in Suffolk in which to walk. The park was given to the town in 1935 by Lord Woodbridge as commemorated by the small Horizontal Dial to be found there. It carries the motto:

Docet Umbra
(The shadow informs)

and is quite nicely engraved for a commercial 20th century dial.

A much larger and more interesting dial is also to be found amongst the flowers. This is of the Equatorial or 'Bow-String' design and it was made in 1988 by a local maker, R S Simon, who specialised in this type. It is robustly made in stainless steel as is necessary for a public park. The whole structure of the dial can be rotated about its sloping axis so that it is adjusted for the equation of time, and also for British Summer Time. All the viewer needs to do is to set an indicating mark against the current date so that the shadow indicates clock time directly.

Woodbridge also has a rather unusual Vertical Dial on the front of a house in Cumberland Street and a rather splendid dial in carved brick on the Melton Grange Hotel.

WALES
Val Cowham & David Young

Wales is a country with a rugged coastline, majestic mountains and green lush valleys. It is a country where the inhabitants speak their own language and hold on tenaciously to their traditions. Each year an International Eisteddfoddau is held at Llangollen with poetry and singing the order of the day. All over Wales choirs spring up, usually male voice and some have gained International fame.

The Welsh had turbulent times in the Middle Ages as the Normans fought to take over their

fact the total number of dials in the principality is not large. The reason for this is rather curious, although the likely explanation is the predominance of the chapel over the more English orientated parish church.

The church in England has for many centuries been concerned that the times of its services should be prescribed. Sundials in one form or another have been found on or about such churches, both to indicate the time directly and to correct the tower clock as it became fast or

The Old Bridge at Llanrwst near Conwy with a sundial at its centre

land and built a ring of castles to keep the Welsh in check.

Many parts of Wales are sparsely inhabited as the bulk of the populace has migrated to the large towns and cities. The greater number of people are Nonconformist and many small chapels are to be found with their biblical names.

As far as sundials are concerned it has to be said that there are none particularly 'Welsh' and in

slow. The very presence of these dials would encourage local people to be more aware of 'the time' and so more dials started appearing on public buildings and in the gardens of the rich. In Wales, the parish church barely prospered and the ubiquitous chapel was a much plainer building with little or no adornment and without a churchyard in the accepted sense. Wales has produced many eminent clockmakers, the small market town of Llanrwst being one centre of their industry and they would need a sundial to set their clocks. There is evidence that it was they who made many of the small horizontal dials that can still be found on the remaining parish churches.

Visitors to North Wales inevitably find themselves drawn to Snowdonia National Park. This tract of land has been taken over since 1947 to make sure that its natural beauty remains unspoilt. The park includes most of the Snowdonia Massif and is a climber's paradise. Mount Snowdon itself is impressive as it stands 3560 feet above sea level.

The best-known beauty spot of South Wales is the Gower Penninsula, which runs 18 miles west from Swansea. It is amazing that this area has remained unspoilt, as it is so close to industrial regions. Visitors will come across a local delicacy, laver bread, an edible seaweed high in iodine. On the North Gower coast, cockles are gathered for sale in local markets.

Cardiff, the Capital of Wales, is in the south. It has been a settlement since 75AD when the Romans built a fort on the River Taff to control the Welsh tribesmen. As befits the capital, Cardiff has several fine museums. For a general view of local life the Welsh Folk Museum at St Fagan's, five miles west of Cardiff, is a must. The castle is in the centre of the city and has had many modifications during its life. It was rebuilt in 1871.

For a nation of music lovers' Cardiff has St David's Hall, which hosts classical concerts and the city is home to the Welsh National Opera. Singing of a different, but nevertheless stirring, calibre can be heard at Cardiff Arms Park.

The counties bordering with England have had a turbulent history. Many castles and Offa's Dyke are reminders of these times.

Wales also has mineral deposits. Gold is mined here and our present Queen's wedding ring is made of Welsh gold. For many years there was a thriving coal industry but that has declined recently. Slate quarrying still continues, especially in the north around Blaenau Ffestiniog. Slate is plentiful and is used for many roofs of houses. In some country areas in North Wales slate is used to create a simple fence. From our point of view, most importantly, slate has been used as the basic material for many sundials. It is easily worked and is resistant to erosion by the elements.

Wales, though a small country has much to offer the visitor. A quote from the Welsh National Anthem probably sums up the feelings of the Welsh for their homeland:

The Land of my fathers,
how fair is your fame,
Entwined are proud memories
about thy dear name.
Wales, Wales,
Sweet are thy hills and thy vales,
Thy speech, thy song, to thee belong,
O may they live ever in Wales.

BETTWS CEDEWAIN, Powys

St Beuno's church occupies a prominent position overlooking the village of Bettws Cedewain on the northern edge of the small valley that carries the Bechan Brook to the Severn, about 9 miles to the south west of Welshpool. It is a single-chambered structure with a western tower, set in a near-circular churchyard. The tower was rebuilt in the early 16th century, the nave and chancel in the second half of the 19th century. The building contains the important brass of Rev John ap Meredyth, the only pre-Reformation brass in Powys, a chest, a bier and a small amount of pre-Reformation stained glass. Little else survived the 19th century restoration.

Its Horizontal Sundial is located alongside the south path and stands on a slender sandstone plinth of four clustered shafts. This small dial

Horizontal Dial at Bettws Cedewain

Bettws Cedewain Dial by Thomas Wright ^{MC}

with a solid gnomon is inscribed *Tho Wright London* and *John Harris Esq*, but is undated.

CLYNNOG-FAWR, Gwynedd

This is an attractive village and is growing fast because of its discovery by tourists. One of its impressive features is the church dedicated to St Beuno. Near to the church is a well, which served as a place of baptism and source of water for the Saint. When Bueno arrived in the early 7th century he built a cell not far from the well and the present church is probably built on that site. Inside the church is a pair of iron tongs that were used to expel rowdy and fighting dogs during the service. There is also the Chest of Bueno, an ancient oak strong box, credited with miraculous powers. Sundials of the insular Celtic type as at Clynnog-fawr have also been discovered in areas where ancient Irish colonies settled elsewhere in the British Isles. See those included in the Chapter on Ireland for similar examples. This sundial at the rear of Clynnog church is still standing. The design of the dial is very similar to a dial found at Inishcaltra on Loch Derg in Ireland and the carving is very simple with only the lines for the third, sixth and ninth hours, with the usual forked ends. There is no firm evidence of actual age but if compared with those in Ireland it would appear to be c.10th century.

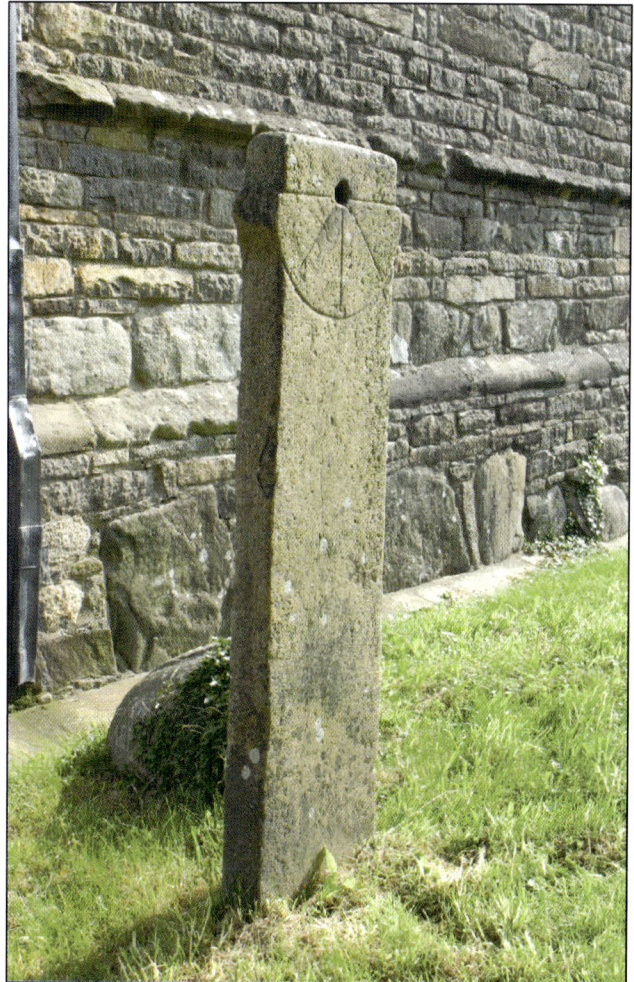

'Irish' style Dial at Clynnog-fawr ^{MC}

The Early Dial at Clynnog-fawr ^{MC}

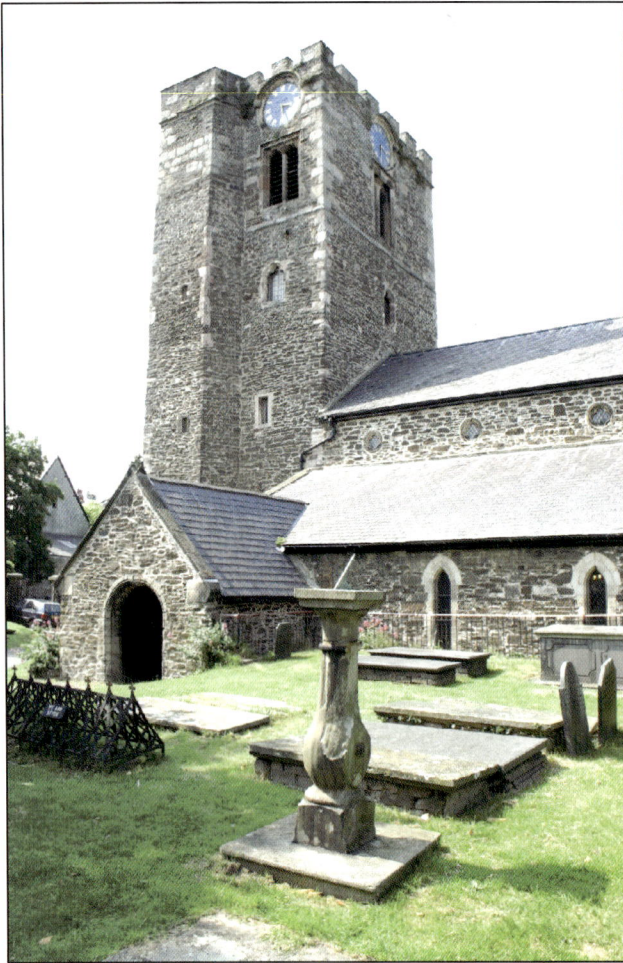

The Horizontal Dial at Conwy MC

It was built between 1576 and 1585 for a Welsh merchant, Robert Wynn and is an architectural treasure. It is probably the finest surviving Elizabethan house to be found in Britain. Wynn's house is noted for the quality and quantity of plasterwork. The house is furnished with many of the original items based on an inventory of the contents in 1665.

The River Conwy is crossed by three bridges. Thomas Telford's suspension bridge was opened in 1826 and was in use until the modern road bridge was finished in 1958. Driving east from Conwy the road travels alongside the old suspension bridge, and beyond that is the railway bridge built by Robert Stevenson in 1846.

There is a Horizontal Dial in the churchyard of St Mary and All Saints'. It was made by Meredith Hughes of Conwy in 1765. It is now rather pitted with exposure to the elements but it is still possible to make out some of the details of its Equation of Time ring.

HARLECH, Gwynedd

The town is dominated by its castle. It was the stronghold of Owain Glyndwr and was the last castle to fall in the Wars of the Roses.

Inside the castle today are preserved six centurial stones which are records of the Roman command nearby. Their faces are 15" x 8" and bear the

CONWY

(*Aberconwy*). The town of Conwy is dominated by its castle built on a high rock with a backdrop of the Snowdonia Range. Conwy Castle was constructed by Edward I between 1283 and 1287 as one of his fortresses in the 'iron ring' to contain the Welsh. From its battlements visitors can look down on Conwy and see the town walls. The walls are over ¾mile long and have 22 towers and 3 original gateways. Conwy is a town of narrow streets and at its heart is Plas Mawr (Great Hall).

Dial by Meredith Hughes, 1765, at Conwy MC

Memorial Dial at Harlech MC

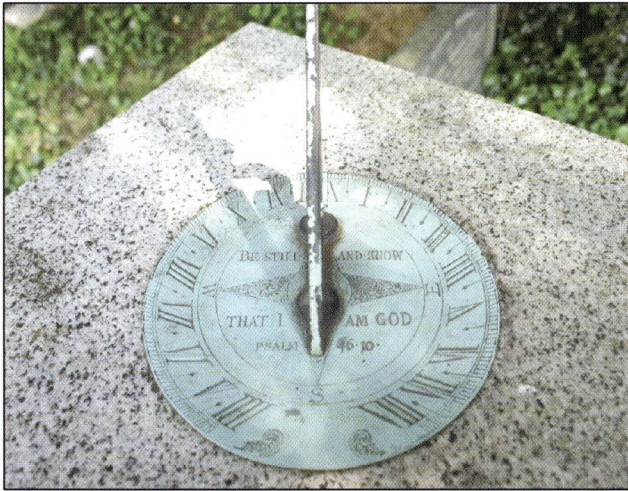
Detail of Dial at Harlech

commander's name. In the town a looped pal-stave (a chisel like tool in a split handle) and a bronze axehead of the type known as Celtic were found. These items show that there was Bronze Age habitation in the area.

The church is built on a slope, and under some trees at the top is a white stone memorial pillar to the Kilmister family with a small brass dial on top. This dial is 20th century and unsigned but carries the inscription:

BE STILL AND KNOW THAT I AM GOD
PSALM 46·10

LLANBRYNMAIR, Powys
The community of Llanbrynmair - St Mary's on the Hill - is so called because of the position and dedication of its Parish Church, which stands on the very summit of a rounded hill. The church (Llan) and its surrounding hamlet lost its domi-nance in the community with the opening of a new turnpike road in 1821 and the coming of the railway in 1861. St Mary's in the centre of town is a single-chambered church that dates back to

Horizontal Dial at Llanbrynmair
by Samuel Roberts

the 15th century if not earlier. There is an early bell turret supported by four oak uprights, a 16th century south porch and a north transept of unknown date. Inside is a 13th century font, a small fragment of medieval glass, and some post-medieval woodwork. In the churchyard is a rather worn octagonal sandstone pillar on an asymmetric base, which supports the sundial. The brass plate is inscribed *Samuel Roberts*, **LLANVAIR, 1754** and *Lat 52:35*. See also Llanfair Caereinion.

Years ago the town was a market centre for the sheep farmers. It also played its part in the Civil War and one of the older houses has mementos of Cromwell's days. In it, beneath later panelling two portraits were found, one of the owner and the other of his wife (both Royalists). The two pictures were riddled with bullets, presumably from Cromwell's men giving vent to their rage.

LLANFAIR, Gwynedd
The village was inhabited in Bronze Age times as was proved by the find of a bronze palstave here. In St Mary's churchyard a simple stone pillar supports a small brass dial made by D Wilson, Nevin. The dial is recessed into the top of its pillar and has a scroll gnomon.

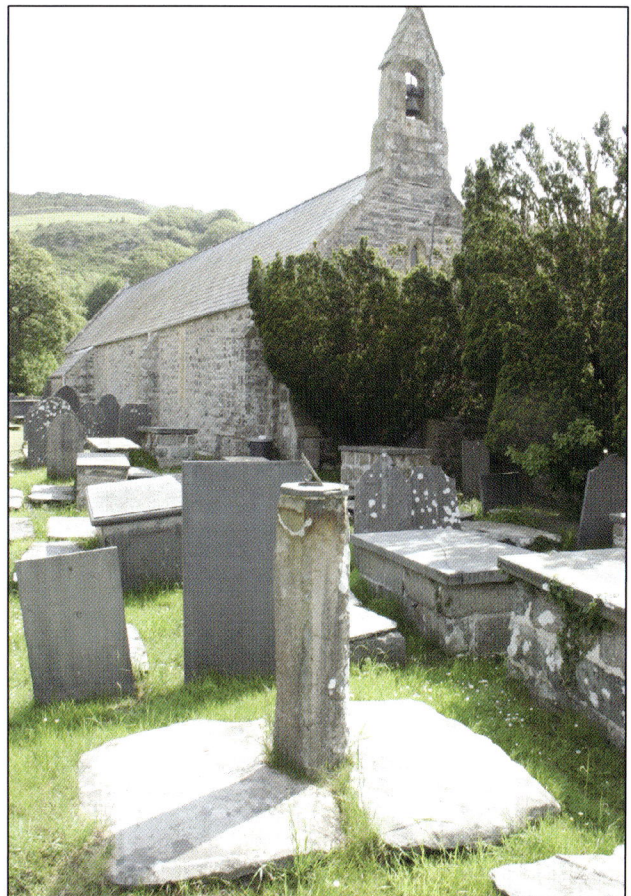
Small Horizontal Dial at Llanfair

**Dial by Samuel Roberts
at Llanfair Caereinion**

LLANFAIR CAEREINION, Powys

St Mary's church in the heart of Llanfair Caereinion was completely rebuilt in 1868 to the plan of the earlier church. Amongst the features that survived the 'Victorianisation' are some of the 15th century roof timbers, a fine south doorway of the early 13th century, a font of similar date and a recumbent medieval effigy. The church occupies a large churchyard on raised ground above the River Banwy and just to the north west of the church is St Mary's Well. The Horizontal Dial, by the church porch was one of three made by Samuel Roberts, a local clockmaker. It has recently been remounted. It sits at the top of a fairly short square-section tapering pillar of white stone. The worn inscription reads *S. Roberts, Llanfair* with a 1755 date, and the plate is set on a tapering square white marble plinth. An inscription in Welsh and English is to the memory of Rev Richard Jones of Llanfair who died in 1868 aged 66 years.

Samuel Roberts, clockmaker, is referred to by Mr Charles H Humphreys in his 'Llanfair Caereinion a hundred years ago' as a member of *'a very old family'* and as the maker of a sundial in Llanfair churchyard. According to a note in the account book in, presumably the present owner's hand, another of Roberts's sundials may be seen in Manafon churchyard. The same note states that Samuel Roberts lived at Pant-y-Tan farm near Llanfair Caereinion and was a farmer who carried on the craft of clockmaking, in which he was self-taught. A third sundial, inscribed *Samuel Roberts, Llanfair, 1754*, is to be seen in Llanbrynmair churchyard.

Slate dial by D Wilson at Llanllechid

Horizontal Dial at Llanfair Caereinion

Flower decoration on Llanllechid slate dial

92

Large Slate Dial at Llanllechid made by D Wilson in 1795 ^{MC}

centre is a 32 point compass rose with each point annotated. The dial stands on a sturdy pedestal made with slate bricks.

LLANRWST, Conwy

Llanrwst is an old market town famed for its 17th century bridge, which is believed to have been designed by Inigo Jones. Pont Fawr (big bridge) is narrow and humped and was a source of congestion until traffic lights were installed. At the centre of the bridge on its southern side is a small horizontal sundial that was placed there for the tercentenary of the bridge, August 27th 1936. Its numerals are in the 20th century Arabic style. It is unusual for dials to be found on bridges but this is quite a sensible place when people walked to and from Llanrwst. Other dials on bridges are known near Ross-on-Wye, Herefordshire; Sinnington, North Yorkshire; Corbridge and Berwick on Tweed both in Northumberland. From the centre of Pont Fawr it is possible to

Roberts died in 1800, and was buried at Llanfair Caereinion. He also made sundials that stand in the churchyards of Llanbrynmair and Manafon.

see a group of standing stones. These are Gorsedd Stones and are connected with bardic

LLANLLECHID, Gwynedd

The churchyard in this small village has a wonderful large slate dial, made by D Wilson. It records that **Robt Pritchard** and **Willm Pierce** were **Wardens** there in **1795**. This large slate dial is 878mm x 858mm with a substantial thickness, lightened by tapering away the slate underneath towards the edges. The dial is also inscribed with its latitude of **53° 20'**. The dial is unusually divided down to 5 minute intervals. In each of its four corners are engraved flowers apparently growing from a bulb. At its

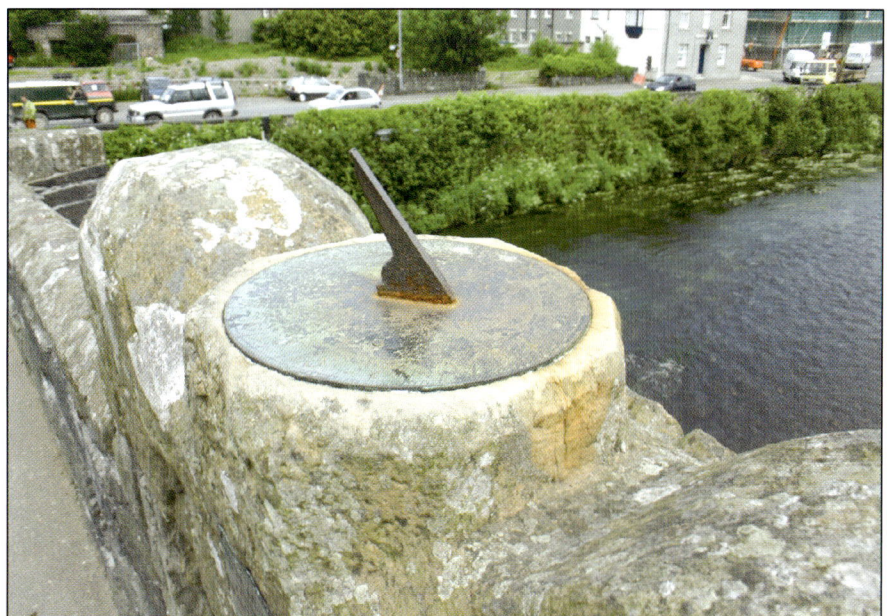

Horizontal Dial at the centre of the bridge in Llanrwst ^{MC}

93

Dial on wooden pillar at Manafon DY

ceremonies and the National Eisteddfodau. At one end of the bridge is a delightful teashop, 'Tu Hwnt I'r Bont', which means 'beyond the bridge'. The building started life as a courthouse for the Wynn family who lived at Gwydir Castle, half a mile away. There was no bridge at that time so drovers and farmers used a ford 100m upstream at Glan-y-Borth. Eventually the building was no longer used as a courthouse and in the 18th & 19th century was probably used as a dwelling. In the early 20th century the house was occupied by Catrin Roberts and when the river flooded she

Horizontal Dial by Samuel Roberts at Manafon DY

refused to evacuate and would have her food passed on a rope pulley from the bridge to an upstairs window. In the 1950s it became a teashop and continues to this day.

MANAFON, Powys

Excerpts from 'The Powys Montgomeryshire Village Book' state that *'Visitors and residents who enjoy driving along the scenic Rhiw valley should be grateful to all those who opposed Liverpool Corporation's plans when they were considering, in 1889 and again in 1966, the flooding of the valley for use as a reservoir. The small villages of New Mills, Manafon and The Green in the parish of Manafon would have been lost and all the interesting facts of other lives and past years would have been gone forever, with no markers to remember them by. There has been a church at Manafon since at least 1254, but the church as it is known today has been restored many times over the years.'* The parish is an upland area of small farms with sheep on the hills. In Victorian times just about everybody in the parish would have been involved in working on the land or like the miller and the blacksmith providing a service to the farmer. Many of the farmers were tenants of the Powis estate. It is recorded that in 1841 there were 741 people in the parish.

St Michael's church at Manafon, about 8 miles to the south west of Welshpool is a simple, structure with a timbered belfry and south porch. The walls almost certainly date back to the 15th century, if not earlier, and the 15th century roof remains. The Horizontal Dial is situated near the south porch and has an octagonal dial plate. It is also made by Samuel Roberts of Llanfair Caereinion. Its really unusual feature is its wooden pedestal fabricated from four shaped wooden legs. This is almost certainly a replacement pedestal but may be a copy of the original.

MERTHYR MAWR, Bridgend

Merthyr Mawr is a picturesque village situated on the west bank of the river Ogmore near to its mouth. It is famous for its miles of sand dunes; in fact the largest sand dune system in Europe. Scenes for the film Lawrence of Arabia were shot here. The elegant church dedicated to St Teilo stands near the village green, built only in the middle of the 19th century but on ancient foundations. The carved stone Vertical Dial on the wall of the church is dated 1720. It faces south east so that it catches the first rays of the summer sunshine at 4am but at the end of the day the sun has long departed from the dial face; in fact the last recorded hour is 4pm. The motto above the dial:

'Tranfit Hora Sine Mora'
(The hour passes without delay)

The gnomon root originates from a Sun's face and around the edges of the dial is an attractive arched pattern.

NORTON, Powys

Norton, near to Presteigne, is one of the few Welsh villages to be mentioned in the Domesday Book of 1086. A church has stood on this site since Norman times. The present Church of St Andrew, although medieval, was restored and refurbished by architect Sir Gilbert Scott in the 1860s. Scott's other works include the Foreign Office and the Albert Memorial in London. At St Andrews, Scott has carefully preserved the old timbers in the tower, altering the fabric as little as possible. On the wall below the tower is a direct south facing bronze Vertical Dial. It was made in 1947 and carries a motto:

DISCE·BENE·VIVERE·ET·MORI·
(Learn well to live and die)

Vertical stone dial at Merthyr Mawr

and extensive inscription, as follows:

GILBERT DRAGE CAUSED THIS DIAL TO BE MADE TO CHRIST AND ST. ANDREW IN THE DAYS OF GEORGE VI KING AND HIS CONSORT QUEEN ELIZABETH
Vicar Canon Clement E. Thomson
IN MEMORY OF MILDRED FRANCES DRAGE.

PRESTEIGNE, Powys

(*Llandras*). This attractive border town between England and Wales lies on the River Lugg. It has a large church which is of Norman origin but has been remodelled in the 15th century. It has two Vertical Dials mounted mid way up its castellated stone tower on its south-east and south-west corners. These are simple Vertical Dials but provide the service of showing the time throughout the day, unlike a single vertical dial that can only operate for a maximum of 12 hours. However, their height above ground makes them quite difficult to read without optics. The original hour lines inscribed on the stone had become virtually unreadable but someone has recently gone over these with a sharp tool to make them visible again. However, the numerals have long since disappeared. It is quite likely that these dials were once painted on the stone, probably in bright colours.

Brass Vertical Dial at Norton

95

Vertical Dials at Presteigne. South-West and South-East ^{MC}

To the south-west of the church stands an interesting memorial stone to Mary Morgan, an unmarried mother who had murdered her child in 1805.

In Tudor times priests and officials visited frequently to enquire into the operation of white witches. In parts of Wales it is thought that white witches were operating into the 20th century and people affirm that a raw potato in the pocket alleviated rheumatism and a handful of earth cures warts.

Memorial stone to Mary Morgan in Presteigne Churchyard ^{MC}

Nearby in the town centre is a brightly painted building shared between a florist and a bookseller. It is decorated with flowers and coloured patterns with large sunbursts on its western side.

RHUDDLAN, Clwyd

There has been a habitation on the site of Bodrhyddan House for at least seven hundred years. Some traces dating from 15th century are still to be found but the main buildings date from 1780. However in 1875 the Victorian architect William Eden Nasfield undertook a major reconstruction which completely changed the look of the building. The original south entrance was relegated to a side door to the garden and the west wing was rebuilt in the style of the Queen Anne revival. The Vertical Dial here, set high in the Dutch gable three storeys above the new main entrance, would have been incorporated at this time. It is a good example of a vertical south dial heavily declining to the west and typical of the period. The motto below the dial is:

DVM : SPECTAS : FVGIO
(Whilst thou lookest I fly)

– another version of 'Time Flies'. The house is the ancestral home of Lord Langford and his family and the house and gardens are open to the public at various times throughout the summer months.

There is also a castle at Rhuddlan which dates back to 1277 and is part of Edward I's ring of castles. The town is set on the banks of the River Clwyd. Nearby on the A5151 to Dyserth there is a spectacular 60ft waterfall.

SWANSEA

(*Abertawe*). This is a large industrial town in South Wales but has wonderful views of the Brecon Beacons and nearby is the most attractive Gower Peninsular. There are many sights of interest in Swansea. Perhaps the most original feature in the Maritime Quarter is the amount of public space decorated by sculpture and carved-stone panels, set into and around the new buildings. This was all part of a Council-sponsored programme that operated between 1985 - 1990 and was designated 'architectural enhancement'. It was conceived to focus attention on the main public walkways and its aim was to arouse interest and give pleasure. The 'enhancement' was not intended to be simply decorative. It is an

West Declining Vertical Dial at Bodrhyddan House ^{MC}

Maritime and Industrial Museum of Wales with a large collection of historic vessels. The Glyn Vivian Art Gallery houses a large collection of Swansea Pottery. Just one mile from the city centre towards the Gower Peninsular is The Brangwyn Hall, a concert hall named after the local worthy, Sir Frank Brangwyn.

TALLREUDDYN, Gwynedd

The church at Tallreuddyn lies close to the railway station of its nearby neighbour Dyffryn Ardudwy. In the churchyard is an unusual reclining Cruciform Dial lying on the top of the rectangular tomb of the Griffiths family. It was erected in memory of Ann Griffiths who died in March 1863 aged 68 years. The top of the dial is pointing to the southern zenith and shows the hours 9am to 12noon and 12noon to 3pm on the top of both horizontal arm surfaces taken from the shadow of the top of the cross. The horizontal arms likewise form shadows on the lower shaft of the cross from 6am to 10.30am and from 1.30pm to 6pm. In the summer months there will also be shadows along the sides of top section from 3am to 6am and 6pm to 9pm, again from the shadows of the horizontal arms. The thickness of the cross is sufficient to allow the shadow to fall along the lower shaft around the two solstices when the sun is at its two extremes of altitude. The dial is inscribed with **Tallreuddyn** on the cross-bar and **1863 ANN GRIFFITH** [sic] on the vertical shaft. Cross Dials of this type are very uncommon and this one is

important element of the whole project and every developer had to agree to its incorporation as part of the planning conditions for the site. Once agreement had been reached, the City Council then organised the design and manufacture of the artwork for the developers to install as part of their building programme.

The first example can be found in St Vincent Crescent and is the Globe Sundial by the artist Wendy Taylor, commissioned by Swansea City Council in 1987. It is a large cast bronze sphere whose axis lies parallel to that of the Earth and reproduces in miniature the way that the Earth is bathed in sunlight. The time is indicated by the fin that casts the least shadow.

While at the Maritime Quarter do not miss the

Globe Dial by Wendy Taylor at Swansea ^{WT}

Side of Cross Dial showing the time ^{MC}

Cross Dial at Tallreuddyn on tomb of the Griffiths family ^{MC}

WALES

TAL-Y-CAFN, Conwy

Bodnant Garden is situated above the river Conwy with extensive views of Snowdonia. The house is owned by Lord and Lady Aberconway but the garden is open to the public through the National Trust. In the rose garden on one of the upper terraces stands a Heliochronometer; a type of very accurate sundial invented at the beginning of the 20th century by G J Gibbs. This dial is a 'late starter' in sundial terms but its great advantage, apart from its accuracy, is that due to the use of a rotating cam it reads true clock time throughout the year. Thus it was used to set clocks in country areas and throughout the Empire. Its demise only came when radio time signals were able to be received almost everywhere.

This Heliochronometer is mounted on an elaborately decorated pillar and is showing some signs of wear from a constant stream of visitors.

TENBY, Pembrokeshire

This favourite seaside resort derives its name from *Din,* a hill fort and *Bach,* meaning small (the same as Denbigh in North Wales). However, Tenby is really an English town, in the old county of Pembrokeshire - known as 'Little England beyond Wales'. Right in the centre of the town

particularly well preserved, being constructed of Welsh slate. However, the cross has developed a split in the lower shaft and is now in urgent need of restoration before winter frosts do more damage.

Heliochronometer at Bodnant Garden ^{MC}

Vertical Stone Dial at Tenby ^{DY}

stands the parish church of St Mary, the largest parish church in South Wales. The original church was built in the 12th century. The one we see today dates from 1470. It is a splendid church with no less than 169 beautifully carved bosses and has a tall spire which is a landmark far out to sea. The town is also the birthplace of the painter Augustus John and of Robert Recorde, the mathematician who invented the equal sign.

Above the south porch is a direct south Vertical Dial that has been slightly canted out to correct for the porch alignment. The dial has been carved into the stone with Roman numerals, the hours divided into quarters. It was restored by John Leech in 1903.

TREGYNON, Powys

St Cynon's Church stands overlooking the small village. St Cynon was a 6th century missionary who came from Brittany. At the end of the 12th century or early 13th century a church was given by a local Welsh lord to the Knights Hospitallers. A moiety of the church was recorded as being in their possession in the Norwich Taxation of 1254

Fine Vertical Dial on the church tower at [MC] **Tregynon**

and by the time of the Lincoln Taxation in 1291 the whole church belonged to the Knights Hospitallers.

After the reformation the church was returned to the local lords of the manor at Gregynog.

Visitors to the church can see its unusual wooden tower which is the mounting place for a large direct south Vertical Dial carved into a light coloured stone. It is a bold dial with hours divided into halves and quarters and its simple but substantial iron gnomon is secured by large support brackets. Apart from a slight crack the dial is in good condition but the wooden frame surrounding it is in need of replacement.

WELSHPOOL, Powys

(*Trallwng*). Powis Castle, a short walk from the centre of Welshpool, was built around 1200, overlooking the picturesque valley of the River Severn. It is now looked after by the National Trust. It was owned by successive generations of Herberts and Clives for four centuries. Its outstanding garden terraces, laid out in the 17th century, are influenced by French and Italian styles. They are overhung with large clipped yew trees and are a shelter for many rare and tender plants.

Powis Castle and the Horizontal Dial by John Bennett [MC]

Powis Castle Dial by John Bennett ^{DY}

The castle itself is perched high on a hillside above the gardens and contains one of the finest collections of furniture and paintings in Wales.

On the lower terrace is a fine sundial by John Bennett of Crown Court, Soho, made around 1765. This large diameter dial (600mm) stands on a fluted pillar. Its motto is:

Audacter et Sincere
(Boldly and Honestly)

West facing Scaphe Dial at Powis Castle ^{MC}

Double Horizontal Dial by Henry Wynne on top of a Cube Dial with east and west scaphes ^{MC}

Also at Powis Castle is a famous Double Horizontal Dial made by Henry Wynne of London and is dated 1664. It is shown here with a bent gnomon but this has recently been restored. This interesting dial is situated on a raised terrace behind the castle but this is not an area currently available to the public due to its being unfenced with a dangerous drop to ground level. However, views of it can be obtained from some windows in the Castle. The dial stands on a large stone cube with a vertical dial at the south and scaphe dials east and west. A further horizontal stone dial with an iron gnomon may be seen on the wall to the east of the Castle entrance but this is in rather poor condition.

Welshpool itself is a small market town with the Montgomery Canal passing through. There is a small Powysland Museum near the canal. There is also a steam railway operation from Welshpool to Llanfair Caereinion.

Outside the door to the Parish Church is another Horizontal Dial of 1745 on a finely shaped stone pillar.

WALES

WEST MIDLANDS

John Lester, Jill Wilson & Tony Wood

Gloucestershire, Herefordshire, Shropshire, Staffordshire, Warwickshire, West Midlands, Worcestershire

The counties of this region show a remarkable variety of both scenery and architecture. Both of these are underpinned by geology and the contrast between the millstone grit of the Staffordshire Moorlands and the Cotswold limestone is

It is hardly surprising that parts of the region are now predominantly urban with the huge conurbation of Birmingham and the smaller one of The Potteries making up the greater part of this increasing urban area.

The picturesque church at Berkswell, Warwickshire

extreme. Between north and south a great variety of rocks dictates the appearance of the countryside and its buildings. The counties of this region have contributed significantly to the history of Britain both politically and, because of the availability of coal, iron and clay, industrially as well.

The distribution of sundials throughout the region is very uneven and as a broad generalisation it can be said that the further north you go the sparser they become. Gloucestershire, with its big wool churches, has a superabundance while Warwickshire, Herefordshire and Worcestershire have far fewer. Shropshire, Staffordshire and the West Midlands trail a long way behind. While old dials tend to disappear due to theft, vandalism or the erosion of the walls on which they are carved, it is heartening to note that new ones are still appearing all over the region, often as a Millennium project and sometimes of an interesting and unusual design.

Staffordshire in the north of the region is regarded by many as an unattractive county. Only those who live there have discovered the delights it offers and are seldom anxious to reveal them to outsiders. Lichfield is its most attractive city though the town of Leek has some pleasant surprises for the visitor too. The Black Country area of Staffordshire is entirely urban but retains a few unexpectedly picturesque scenes in the old centres of its constituent towns. Inhabitants of the Black Country still regard themselves as distinct from and superior to the Brummies.

Shropshire, in spite of the new town of Telford and the continued expansion of Shrewsbury, is predominantly rural. It is paradoxical that Ironbridge Gorge should have been the cradle of the Industrial Revolution and we should look to A E Housman for a truer picture of the county. Many of its villages remain undiscovered by tourists and walkers who confine their attentions to Iron-

Vertical Dial and clock at Berkswell

bridge, Ludlow and the hills around Church Stretton.

Warwickshire is traditionally described as 'leafy' or identified as Shakespeare's county. It is both but it is much more. From its less attractive and formerly industrial north where coal was mined, through its splendid county town to its rural south bordering on the Cotswolds, it never lacks interest. By some it is regarded as the most typical of English shires.

Although both Romans and Saxons were active in Worcestershire they have left little behind them; Norman remains are more abundant, nearly half the churches in the county containing some masonry from that period. There is a better concentration of sundials and Mass Dials here than in the counties already described and the villages surrounding Bredon Hill provide a particularly rewarding starting point for the interested visitor. Herefordshire, like Shropshire, is a rural border county, its place names turning Welsh as you travel west. The Normans have left their mark here too and there is a wealth of half-timbering in both houses and the towers of country churches. Sundials are not abundant but the search for them in such a pleasant county cannot be other than enjoyable.

In the British Sundial Society's Register of sundials, Gloucestershire has more dials than any other county; more than the total for the rest of the West Midlands region put together. It has at least six Saxon Dials and a number of 'Prism' Dials peculiar to the county with faces on the southwest and south-east of a cube cut vertically along a diagonal. Its many fine medieval churches carry numerous mass dials while Chipping Campden probably contains more sundials than any other town in Britain. There is much else of interest in the county apart from sundials and this includes Roman sites and many National Trust properties of various ages.

BERKSWELL, Solihull
The pleasant village of Berkswell is named for the water that still flows near the church, contained in a stone trough or tank perhaps once used for baptisms. Walking towards the mediaeval tower of St John's Church the visitor is surprised to realise that the sundial is larger than the church

Painted Dial at Berkswell _{PH}

clock. The handsome Vertical Dial is made of painted wood with a dramatic golden sunburst on a black background providing the appearance of sunlight even on a cloudy day. It faces slightly west and so cannot show the hours before 8 in the morning. The date of the dial is uncertain but its design suggests it to be perhaps some centuries younger than the tower. On the south wall of the church can be seen two Mass Dials.

BLOCKLEY, Gloucestershire

The Church of St Peter and St Paul has a vertical sundial that is not in too good a condition but full of interest. Considerable effort has gone into installing the dial; in addition to the inset square there is a single stone above with dedication details. The offset gnomon indicates an east declination and the delineation is from 5.30am to 5.30pm although the hour lines and numerals are

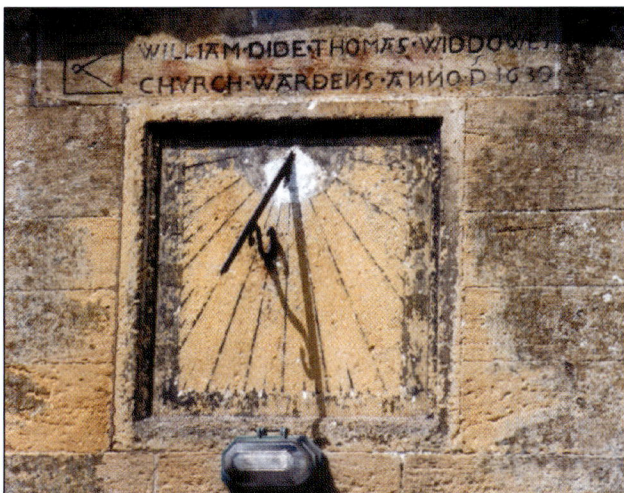

Vertical Dial at Blockley, Gloucestershire _{AW}

now so faded as to make casual reading difficult. The hour marking is down to quarter-hours and the gnomon is probably a replacement in the 370 year lifetime. The insetting of the dial face may cause problems early and late in the sundial's day. Above are recorded the names of the Church-wardens, even more important than the incumbent as they held the purse-strings and are frequently commemorated by inscriptions on building works and other additions to their churches. Dide and Widdowes must have paid the mason in 1630 and maybe held back a few pence over the reversed 'N's. The symbol to the left is most intriguing. So far the guesses include: black-smith's tongs, sheep shearing shears (we are in sheep country here) and even the Greek letter alpha with a missing omega to go on the right.

BOURTON ON THE WATER, Gloucestershire

A fine Cube Dial with magnificently curlicued gnomons sits on the entrance porch of the Dial House Hotel in the middle of this small Cotswold

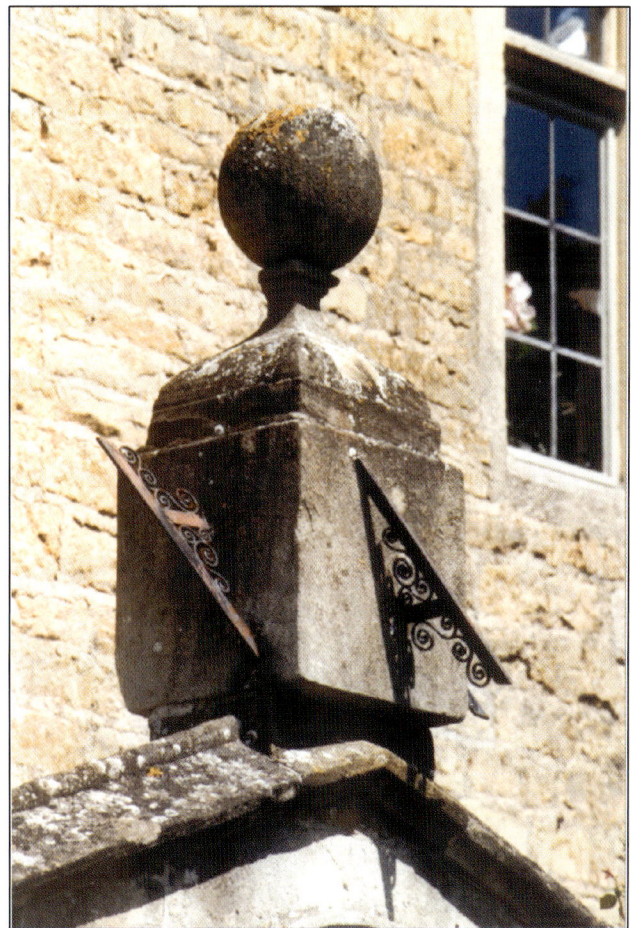

Cube Dial at Bourton on the Water _{AW}

town, now the magnet for thousands of summer tourists with its gentle river, a large area of grassy banks and low level bridges. Although a diallist's 'conceit' insofar as the faces are north-west, south-west and south-east, there is some justification in that the cube can be aligned with the doorway. The dial is a striking feature of the hotel but it is so old that all the markings have practically disappeared. The date is likely to be 17th century and the dial also features the ball finial typical of that era. Such a dial on a house is unusual and one wonders if its original position was on the church.

Bourton on the Water is well provided with sundials; there is a Horizontal Dial below that just described and a Vertical Dial on a shop in the square in front of the Hotel. The 'Old New Inn' has one as does the Tea Rooms opposite. The

Modern Vertical Dial at Bromsgrove

Circular Vertical Dial at Bradley in the Moors

model village attraction features the inn dial in miniature - perhaps the smallest fixed dial on public view?

BRADLEY IN THE MOORS, Staffordshire
The church of St Leonard in this small and secluded village was built in 1750 to replace an earlier one and, unusually, it remains architecturally unaltered. The yews in the churchyard are older than everything else except an ancient coffin lid. The visitor should not be misled by a tombstone in the churchyard bearing the date 1307. Even masons can make mistakes and the

3 should be an 8. The simple circular Vertical Dial on the tower is so much at one with the rest of the building that it must have been there from the beginning. No dial could be plainer, which argues that it was put there for everyday use rather than ornament, being the only timekeeper in the village for as much as a century.

At Croxden, not far from here are the impressive ruins of a Cistercian abbey.

BROMSGROVE, Worcestershire
This neat modern Vertical Dial can be seen on the wall of the Methodist Centre in Bromsgrove and it was put there in 1983. It is calibrated for British Summer Time, a sensible arrangement since the days are longer and sunshine more likely when this is in force. Technically it is a 'vertical west dial declining south' but all that this really means is that it is facing more west than south. It does not catch the sun until about an hour before noon but after that it can continue to tell the time well on into the evening. The designer has wisely decided to avoid any link with antiquity in a late 20th century dial and has provided a motto in English, which on older dials used to appear as:

Non Sine Lumine

BROSELEY, Shropshire
Visitors to the National Trust property Benthall Hall should certainly walk the short distance to look at the church. The present St Bartholomew's was built in 1667 to replace the old church, which was burnt down during the Civil War. The slate

The Vertical Dial at Benthall Hall

Slate Sundial at Burton Hastings, 1867

sundial with its mosaic eye at the root of the gnomon (which may be incorrectly fitted) is a later addition but it has an unusual story behind it. It carries the motto:

**OUT OF THE STRONG
CAME FORTH SWEETNESS**

which is familiar to anyone who has ever bought a tin of Lyle's Golden Syrup. It is, of course part of the riddle posed by Samson, which appears in Judges xiv, 14 concerning bees nesting in the carcass of a lion. What has this to do with sundials? Below the dial can be seen a carving of a lion's head whose open mouth was once the entrance to a bee-box in the gallery of the church. The churchwardens acted as beekeepers and sold the honey they produced to help the poor of the parish.

BURTON HASTINGS, Warwickshire
To the southeast of Nuneaton lies the village of Burton Hastings. The ashlar west tower of the church of St Botolph is of perpendicular style but inside, an earlier font can be seen. The Vertical

Dial on the church is of slate with a bronze gnomon and is dated 1867. The motto:

CARPE DIEM

literally means 'seize the day' and has been roughly translated in today's idiom as 'enjoy, enjoy'. Not surprisingly it can be seen on many sundials, especially those like this one from an age when Latin was more widely known. It comes from the Odes of Horace and those familiar with the original would have mentally completed the line - 'trusting little in tomorrow', or as we might say today, 'It is later than you think!' There is a Mass Dial on the south wall also.

Dial House, Chipping Campden

Cotswold House Hotel, Chipping Campden

CHIPPING CAMPDEN, Gloucestershire

The High Street of the small market town of Chipping Campden in the north Cotswolds has been described as the most beautiful village street now remaining in England. It certainly contains a remarkable number of Vertical Dials; seven in all. The earliest are from the 17th century and the most recent from the 19th. Over the Campden Bookshop is a dial dated 1690. Local tradition has it that it was used to set the Town Hall clock until the coming of the railway in 1854 when Railway Time replaced local solar time in the town. This dial is something of a puzzle. It is canted out from the wall so that it faces more south-westerly than the building. The moulding above is carried out and around it. The question is, was it designed for here and intended to show more afternoon hours or was it moved from elsewhere when the facade of the building was undergoing refacing?

An outbuilding of Cotswold House, now a hotel, built by a rich grocer in 1815 has a fine dial with a sunburst.

Dial House, once the premises of the Warners, a Gloucestershire clockmaking family, has a dial with Roman numerals which perhaps also doubled as an advertisement and shop sign.

COLEFORD, Gloucestershire

At Speech House in the Forest of Dean, near Coleford, is a truly vernacular Horizontal Dial. It is set at the entrance to the Cyril Hart Arboretum, next to the Speech House, ancient seat of democracy in the Forest of Dean. There are dials in stained glass; this dial has stained glass in it; a 'window' in the upper part of the gnomon has a decorative panel just visible in the photograph. The really chunky gnomon is around six feet high at the tip and the hour scale is a semi-circular bench seat in the sun. The hours are marked on copper plates fixed to the seat. The dial has a name; the 'Passage of Time Sundial', which is not unknown but unusual. It was created by the FORGE Visually impaired Woodcarving Group and a large name board with text in braille also is alongside. The whole Arboretum has similar provision for the visually impaired and other

Wooden Dial at the Cyril Hart Arboretum

Cube Dial now placed on the top of an old Preaching Cross at Coughton

Cube Dial at Coughton

DROITWICH, Worcestershire

War memorials which incorporate a sundial are less common than you might expect and it is a sad reflection on our times that this one in Victoria Square should have had its massive stone base vandalised. It was restored but that was not the end of its troubles since it was noticed that the gnomon was inclined at an angle of 30° to the horizontal which would have been fine if it had been in somewhere like Cairo but was not much use in Worcestershire. Whether it had always been like that or whether a new gnomon had been fitted following a previous episode of vandalism is not known. It was also found that the dial plate was misaligned by 11°. These faults were corrected in 1995 and the new gnomon is a robust

work by the Woodcarving Group is also featured there. The dial was opened in September 2002.

COUGHTON, Warwickshire

At one time many churchyards had tall preaching crosses that were no longer required after the Reformation. Some were removed entirely, but others were re-used to serve other purposes. One of these was perhaps that in the churchyard of St Peter's Church, Coughton, where in the 17th century the old cross base was given the addition of a column bearing a Cube Dial. This has a dial on each of its four faces. One wonders whether it was placed in position before or after 1605, when the wives of several of the Gunpowder plotters waited for news within Coughton Court nearby. The mansion is now a much-visited property of the National Trust.

Horizontal Dial on War Memorial at Droitwich

17th century Multiple Dial, a memorial to the Savage Family, at Elmley Castle ^{AW}

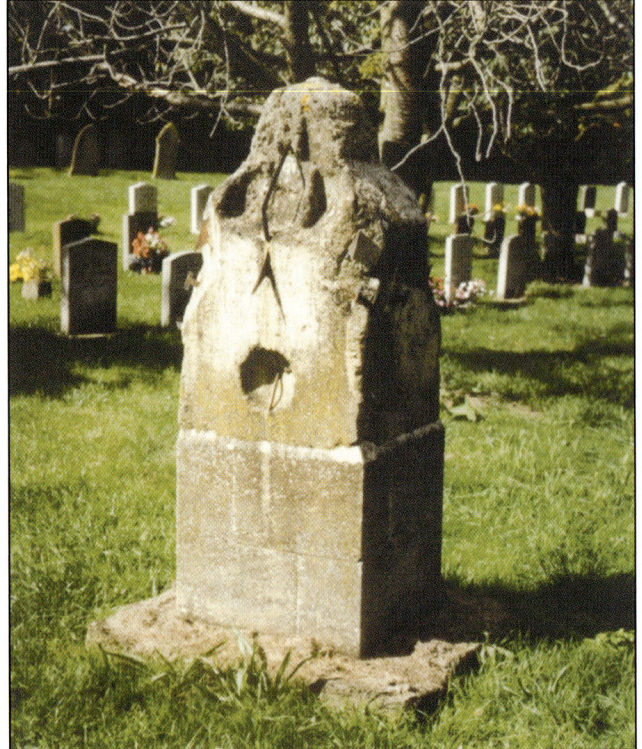

Second Multiple Dial at Elmley Castle ^{AW}

in good condition with multiple gnomons covering all four faces. The original gnomons had long corroded and were replaced by S H Grylls (Chief Engineer, Rolls-Royce Motors) in 1971. The remnants of the gnomons are on display in the church

one designed to resist further attack. The dial plate is uncluttered and the whole dial is a good workmanlike instrument worthy of the men it commemorates.

ELMLEY CASTLE, Worcestershire

To the left of the path to St Mary's Church stands a large Multiple Dial of a type that would look more at home in Scotland. It has a dozen or so small dials of various types, (Vertical, Scaphe, Polar and Reclining), on its faces and is topped by a small recent Equatorial. The dial, which dates from the 17th century, is actually a memorial to the Savage family and their ornate tomb is within the church. Essentially a 'conceit', the numerous dials are quite difficult to read or interpret. The dial has been raised somewhat from its original height and now stands about 18 inches higher on a matching stone block.

Further to the east in the churchyard stands another multiple dial, more elaborate, if that were possible, but still at its original height and looking

West face of Cube Dial at Great Malvern ^{AW}

108

Cube Dial at Great Malvern

are solid shapes; all in new copper. Above is the original stone capping and finial mount but the ball finial has now gone. The original pillar was probably an adapted 'preaching cross' and now stands to the north-west of our dial with a gabled cross atop.

HOCKLEY HEATH, Solihull

Hockley Heath straggles along the road from Birmingham to Stratford and there is little there to catch the eye of the traveller apart from the fine memorial to those lost in the First World War. It takes the form of a sundial on a tall column. The decision to construct this type of memorial may have been influenced by the Ash family who lived at nearby Packwood House and clearly were sundial enthusiasts. The three vertical faces are carved, unusually on a vertical equilateral prism, enabling passers by to tell the time from sunrise to sunset throughout the year. It was dedicated in 1921 when grief for lost loved ones was still fresh and continues to awaken

and the British Sundial Society is entrusted with a 'watching brief' over the dials.

The rear of the church faces south and on a quoin stone about four feet up is what appears to be noon marker stone in 'Scaphe' or hollowed form.

GREAT MALVERN, Worcestershire

The Priory Church of St Mary and St Michael is a large one as befits a former priory. It was originally a church of the Priory and was purchased by the villagers in 1539 for £20. The original monastery was founded in 1085 but the Priory was later and built on land belonging to Westminster Abbey. The church was completed by 1501. Its stained glass is particularly noteworthy. Somebody has made a considerable effort to restore, more probably replace, the Cube Dial, which once stood on a pillar in the churchyard. Now brought down to low level on the gentle hill to the north of the church, the cube is carefully marked out on all four faces: north, south, east and west. The east and west gnomons are wedge or spade shaped, the north and south gnomons

Three Sided Dial on War Memorial at Hockley Heath

**King's Norton Church with one dial over the porch
and the other above the south window**

thoughts of pity and sadness into the 21st century. It is a shame that its location is spoiled by a canopy of power lines and the proximity of a brash motor showroom.

KING'S NORTON, Birmingham

The town is now engulfed by the spread of Birmingham and yet the area round the church still manages to maintain an air of old-world tranquillity with the half-timbered 15th century Saracen's Head Inn to the south of the church and the equally old Grammar School to the north. There

are two sundials on the medieval church of St Nicholas but the visitor could well be excused if he failed to notice the second one. The big circular brass Vertical Dial above the porch commands attention and can easily dazzle the bystander when the sun's rays are reflected from it. It is no newcomer since it was put here in 1813 but the smaller, less conspicuous Vertical Dial, on a gable to the east of the porch is older and harmonises far better with the fabric of the building than its brash neighbour. This smaller dial is carved into one of the stones of the wall as were many early dials and there is enough of its rusting iron gnomon left to enable it to function. In addition to the two sundials there is a Mass Dial on the porch.

LAPWORTH, Warwickshire

Packwood House was built in Tudor times and was restored in the first half of the 20th century by the Ash family who later gave it to the National Trust. The interior reflects the original 16th century design and the gardens too, with their yews and borders, repay close attention. It is not only sundial enthusiasts who find delight in the various sundials. In particular the dramatic group of vertical dials on the brick stables, the latter probably erected by John Fetherston in the

Brass Vertical Dial over the porch at King's Norton

Carved Vertical Dial over the window at King's Norton

Vertical South Dial at Packwood House

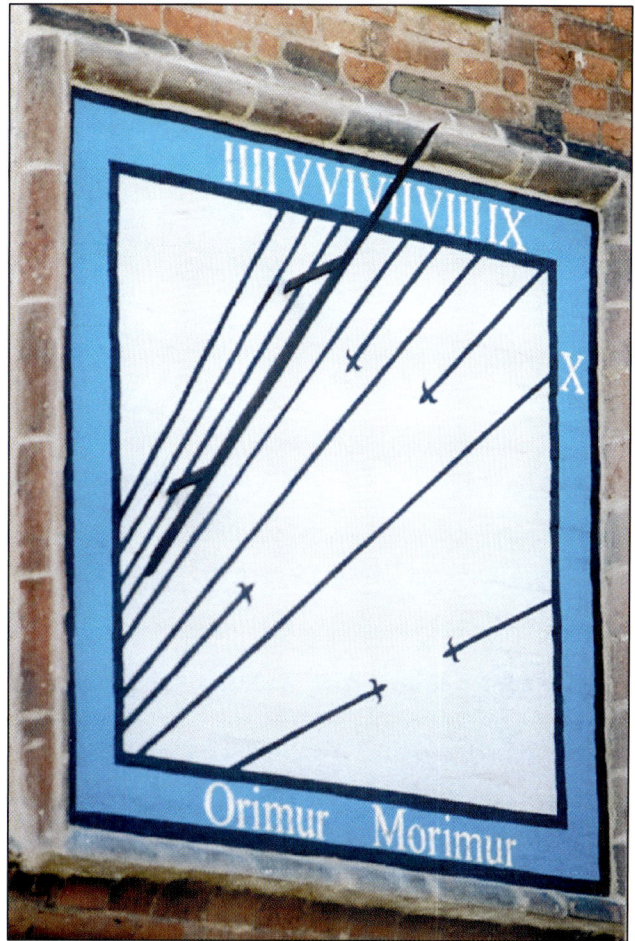

Vertical East Dial at Packwood House

1660s, are regularly the focus of visitors' cameras. Each is over a metre square, painted onto the brickwork. One shows the hours between 1pm and 8pm with the Latin remark:

Septem Sine Horis
(Without Seven Hours)

This rather strange motto refers to the fact that near to the longest day there are seven hours during which the dial will not function.

Another begins the day at 4am and continues until 10am saying:

Orimur Morimur
(We arise, We wither away)

whilst the third fills the gap covering the hours from 6am to 4pm, without a Latin comment. In the garden there is also a Cube Dial of 1667. The Ash family were obviously enthusiastic about sundials since their memorial in the churchyard at Packwood takes the form of another Cube Dial. Apart from this, the church has its own Vertical Dial on the tower and several Mass Dials.

LILLESHALL, Wrekin
(Formerly Shropshire). The ruins of the abbey founded in the 12th century are the main attraction for visitors to Lilleshall, but there are no dials here and we must look to the parish church to provide one. Vertical Dials are often painted directly on to masonry and their calibrations may become increasingly inaccurate with each

Packwood House with the large direct south Vertical Dial over the stables on the right

111

Painted Dial with Equation-of-Time plate at Lilleshall

working in London in the 18th and 19th centuries either on their own or in partnerships. Whoever it may have been, he has produced a fine instrument, which tells the time from **IIII** in the morning until **VIII** in the evening with divisions down to five and even one minute, a feat which is only possible on a large dial. The potential of this dial is to some extent wasted since it is positioned quite close to the south wall of the church and will inevitably be in shade for a significant part of its time range.

MORETON IN MARSH, Gloucestershire

Certainly one of the finest examples of a direct east Vertical Dial in the country is on the Market Hall (Redesdale Hall), well up and unnoticed by anyone driving through a busy town centre. It is probably 18th or 19th century, and its hours run from 3am to 11am. The dial face itself is offset to the left of the panel which is richly carved and features a puffing cherub with wings along the top. Unusually, the hour marking is confined to the dial face surround; most east facing dials have the lines complete, forming a parallel grid. The carving is very fine and the delineation is down

re-painting. Many have not been restored for years and are flaking badly like this one on the south wall of St Michael's church. A recent costly restoration of the church roof has left no money to spare for any work on the dial. The interest here lies in the bronze plaque at eye-level which is an Equation of Time table showing the difference between sun and clock time at dates throughout the year. It was put there in 1896 and bears the initials of the local sundial enthusiast Mr C C Walker FRSA who restored the dial and was the donor of the plaque. According to that great pioneer of sundial studies, Mrs Gatty, the dial once bore the motto:

Our days on earth are as a shadow

(1 Chron. xxix, 15) but there is no trace of it now. On top of a hill near the church is a 70ft column in memory of the last Duke of Sutherland.

LINDRIDGE, Worcestershire

Outside the church of St Laurence is an unusually fine Horizontal Dial. It is large - the circular dial plate is nearly two feet in diameter - beautifully engraved and very handsome. The maker's name appears on the dial plate as 'Watkins, London' but this is not sufficient to identify with certainty the manufacturer or to date the sundial since there was a dynasty of sundial makers of that name

Horizontal Dial by Watkins, London at Lindridge

Fine east facing Vertical Dial at Moreton in Marsh

cated. Good enough to set the clock by or even a watch later. When the timekeeping became good enough to keep regular hours it was found that the position of the spot at 12 o'clock varied throughout the year which could be noted (or even calculated in advance). The curve joining the successive 12 o'clock points is shaped like a thin figure '8' and is called an analemma. There are now no markings at all below the disc at Much Dewchurch but it is probable that only a simple vertical line was made on the stonework in some way. The device was more popular on the Continent perhaps and this example from the past may well be unique on an English church. There are other more recent ones in the country, perhaps the best being deep inside what used to be the Royal Aircraft Establishment at Farnborough.

to quarter-hours with the Arabic numerals skilfully fitted in the border. The gnomon is made from a copper rod, now green and contrasting nicely with its bronze supports. The Redesdales owned nearby Batsford House, whose grounds are now given over to Batsford Arboretum. The House itself has a dial which appears contemporary with the Hall's dial. It can be glimpsed from the Arboretum but is to the rear of the house from the front entrance and not accessible.

MUCH DEWCHURCH, Herefordshire

Heading towards the Welsh border, at the Church of St David there is a surprise in sundialling terms. In addition to a Horizontal Dial on a pedestal to the left of the porch we find a 'Spot Dial' consisting of a disc with a small central hole on the end of a rod supported over the porch door. There are frequent illustrations of such a device from the past, frequently with the disc looking like a sunflower. The spot of sunlight would fall on a marked line and in this way noon could be determined. Originally the line was a simple vertical line so sun time 'noon' was indi-

NAUNTON, Gloucestershire

St Andrew's Church in this little Cotswold village has managed to preserve a pair of Vertical Dials for two hundred and fifty years. They represent the end of the era of painted dials; painted, no doubt, because they were cheap. Unless maintained they will wash away in time and leave a lonely gnomon, which if noticed at all raises

Remains of 'Spot Dial' at Much Dewcurch

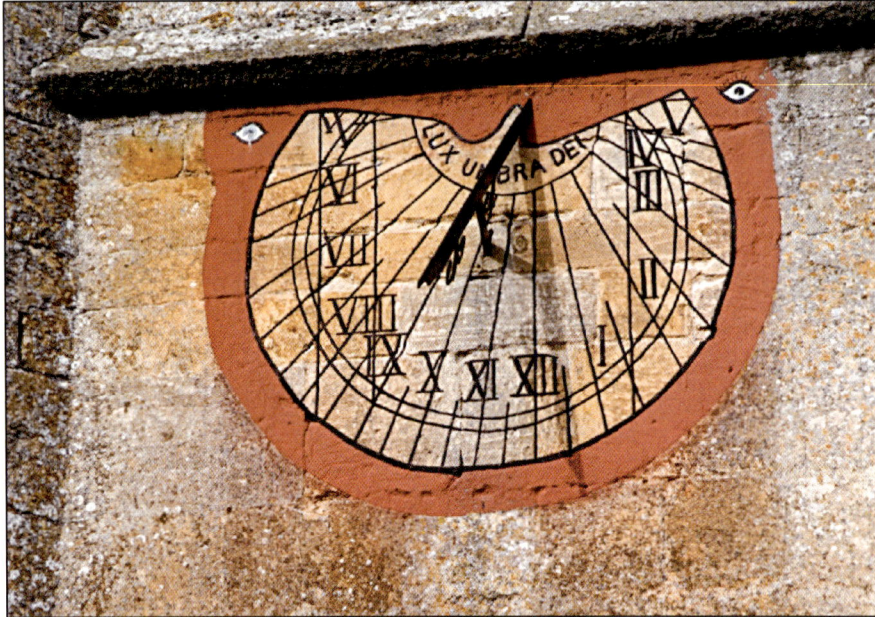

The two attractively painted Vertical Dials at Naunton

around noon and lengthen to a confusing quartet around 5pm. There seem to be two 'V's for 5am and the vertical lines are suspiciously like the borders of a square dial that has undergone some transformation in its lifetime. The motto:

LUX UMBRA DEI

(Light, the shadow of God) has been adopted from this dial for their Golden Jubilee dial at Ruardean, in the Forest of Dean.

Round the corner and bearing the date and the edge of an awning, the west facing dial features a red cloud. It looks like red oxide paint on both dials so may last a little longer than the original. The hours run from 11am to 8pm but the **XII** for noon does not have a vertical hour line as it should if correctly delineated. It isn't possible to say whether the line or the position of **XII** is wrong but the original is unlikely to have been incorrect. It is also unlikely that the style of the Roman numerals would have differed between the two dials.

intriguing questions as to the original design. Such has happened to the formerly painted dials at Leckhampton and Prestbury (both near Cheltenham) in Gloucestershire and other churches across the country. The wall has been cleaned and prepared but the mortar lines are still visible. The painting has presumably been carried out at intervals since 1748 and has probably resulted in changes to the original design on each occasion, maybe accidentally, maybe intentionally.

The south facing dial features a couple of eyes, an irregular top boundary and some improbable vertical lines. It declines east and the hours run from 5am to 5pm. The half-hour lines get shorter

The fleur-de-lis decoration on both gnomons is known on other examples in Gloucestershire; one of which is at Leckhampton, another painted dial.

OVERBURY, Worcestershire

Walking up Bredon Hill from Overbury (where there is a church with several Mass Dials) one sees on the skyline to the left an isolated building, a barn in fact, which has been restored and now features, quite high up on the wall, a splendid modern dial. It is beautifully carved from limestone, and is west declining. It is marked accordingly from 7 in the morning to 6 in the evening. The numerals stop at 6pm to allow space for the date **A.D. 1992**. The date and hour line in-lining

Modern Vertical Dial at Overbury

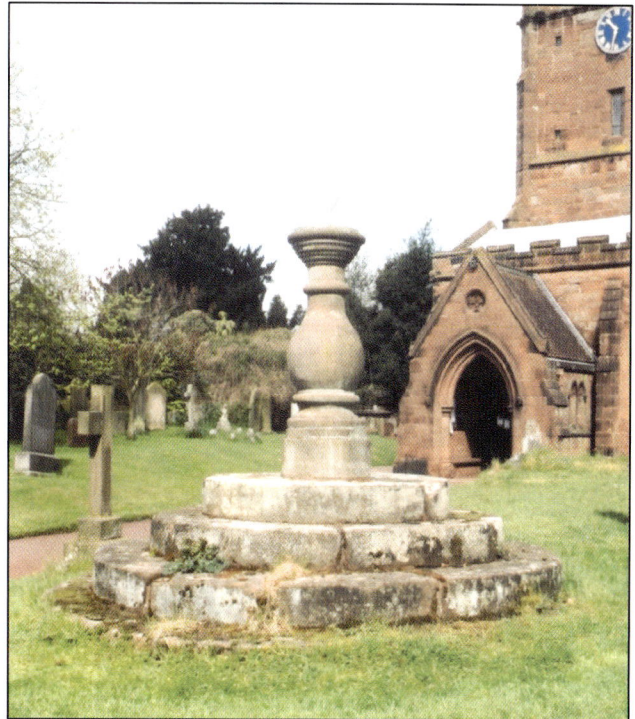
Baluster Sundial at Pattingham

is in gold with the motto:

NUNQUAM REDITURA

(I, Time, never will return)

and numerals in a rich reddish brown. The brown in-lining is to be much commended as preferable to black in this rural context. Its maker is Fred Heathfield, now in Cheshire but originally living locally and with other dials to his credit, notably another west decliner on a chimney in Chapel Lane, Kemerton. Coming across this dial whilst walking around the Cotswolds is an unexpected delight and must intrigue all who see it.

PATTINGHAM, Staffordshire

Pevsner, in his admirable series of books, 'The Buildings of England' seldom makes mention of sundials so that, in the entry for St Chad's church even the laconic '. . . *also a baluster sundial'* indicates that he has found something of unusual interest. He was probably more concerned with the fine pedestal than the dial itself though he might well have found it worth inspection. Its maker, Jno Baddeley had reason to be proud of his creation for as well as the usual time markings it has a circular equation of time scale labelled **Clock Slow** or **Clock Fast** at the appropriate points in the calendar. The church has Norman origins though much of it is 14th century and on its south wall are two mass dials described by Arthur Mee, (who always noticed sundials in his 'King's England' books), as *the oldest clocks in the village'.*

RANTON, Staffordshire

All that remains of the original Ranton Abbey is the 15th century tower next door to the shell of an 1820 brick mansion which took over the name.

Archaeologists hope that the mansion will fall down and allow them to excavate the abbey remains which lie beneath it. In the village nearby is the tiny church of All Saints' consisting only of a 13th century nave and a later brick chancel. The wall of the nave bears a sundial, which has some claim to be the most interesting dial in the county. It is carved on a stone which

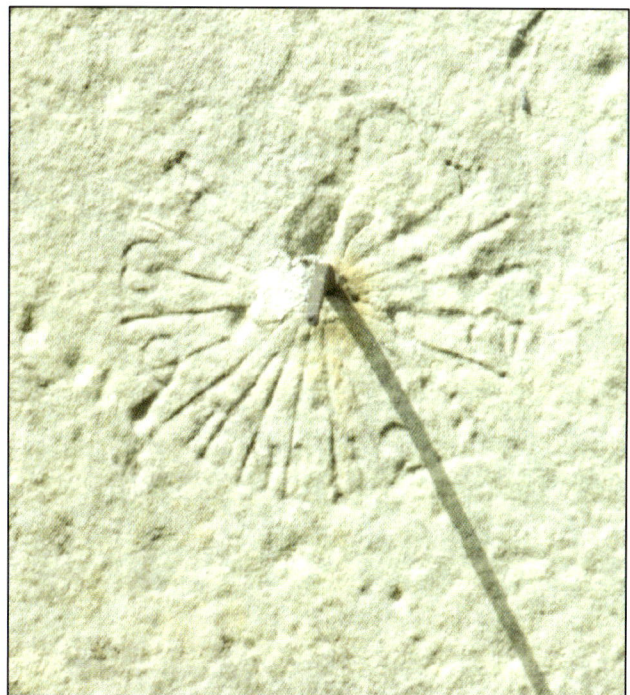
Large Mass Dial at Ranton

differs in colour from the rest of the wall surrounding it, suggesting that it may have been transferred from an earlier building. It is about a foot in diameter, which is larger than most Mass Dials, and it is placed much higher too. It has an iron gnomon, probably not original, which is bent downward as if to approximate to the polar-aligned gnomon of a 'scientific' sundial. There are some barely distinguishable Arabic numerals which may possibly be a later addition. What can it be? A Saxon origin seems unlikely but it could be a late Mass Dial or even a transitional form half way to becoming a sundial. As well as this possible hybrid there is a Mass Dial of the more usual type by the south door of the church.

SAINTBURY, Gloucestershire

At the roadside on the B4632 will be seen Vale Farm. Its well-known Cube Dial is also well worn. The thin triangular iron gnomons may be original but the west gnomon looks as though it is re-placed or re-fitted. The cross on top indicates some connection with the church but that is well

Saxon Dial at Saintbury

away up a hill on the other side of the road. The faces are aligned directly to the cardinal directions, all the gnomons are intact and the numerals are still readable on the south face at least. The dial is probably 17th century and worth being replaced on the top of a 'preaching cross'; but why it is out here on the highway is a mystery.

The church at Saintbury has a fine Saxon Dial. This is one of at least five or six Saxon Dials in Gloucestershire and is in quite good condition considering that it has been outside in the British weather for about eleven centuries. It is in relief as are many Saxon Dials and has only the noon line and two others in addition to the diameter marked. The gnomon would have been horizontal. The church is up a hill and the dial is on the south wall 'round the back'. It is fitted neatly into the wall stonework but must have come from an earlier building as the church itself is Norman.

SEZINCOTE, Gloucestershire

The 'classic' bronze Horizontal Dials of the 18th and 19th centuries have usually acquired a layer of patination, or worse, making the engraving difficult to decipher, let alone photograph in situ. Here we are able to read the maker's name, **Berge, LONDON** and see the numerals and 16 point compass star with its directions. Matthew Berge was a London instrument maker who worked from 1800 to 1819 from premises at 199 Piccadilly. The delineation is down to two minute intervals along the outermost scale. A small

Cube Dial at Saintbury

Horizontal Dial by Berge of London at Sezincote

now the only Moghul building in Western Europe. Its extensive gardens are open to the public.

STOCKTON, Shropshire

In St Chad's churchyard, mounted on a perilously thin sandstone column, there is an unassuming Horizontal Sundial which the casual visitor might dismiss as just a routine churchyard dial. This one is worth further inspection for its 16 sided dial plate is dated 1717 and has engraved on it the maker's name and address:

Daniel Blackman, The Farmyard, Brockton

Blackman, who came from a hamlet a few miles to the north, is not known to have made any other dials but he took a good deal of care over this one. Inside the chapter ring showing the hours, half-hours and quarters is a second ring of XIIs telling us when it is noon at other places across the world. Only two of these place names are decipherable now - Port Royal and Constantinople - and the former presents a problem until we realise that it was once the name of Kingston, Jamaica. Guesses can be made as to the longitude of the others from the time difference be-

curiosity is in the photograph; the gnomon shadow doesn't actually reach this scale at around 4pm in summer. The dial is Berge's only recorded example and, as mounted, forms an important part of the garden layout along the avenues from the House and orangery. A substantial octagonal stone garden seat provides the mounting; fortunately it is low enough to be able to read the time easily. Sezincote, near Moreton in Marsh, is a large stately home built from 1800 to 1805 by Charles Cockerell in the Indian manner and is

Horizontal Dial at Stockton

Horizontal Dial at Stretton Grandison AW

We miss the hand clasp, miss the loving smile
Our hearts are broken, but a little while
And we shall pass within the golden gate
God comfort us, God help us while we wait

The date on the inscription round the mount stone is 1928, within the era when sundials were chosen as a commemorative feature in churchyards. The dial plate is engraved:

FRANCIS BARKER & SON
DIALISTS [sic] **LONDON**

and as an honest dial is a fitting and well preserved memorial.

WALFORD, Herefordshire

(Sometime known as Walford-on-Wye.) The Cube Dial at the Church of St Michael and All Angels is a bit of a mystery as it is in exceptionally good condition but the delineation and numbering appear to be about 17th or 18th century. It may be a good copy or possibly modern recarving, with new gnomons. The cap and finial are original and heavily lichened. The gnomons on the east and west faces are wedge shaped

tween our noon and theirs; Rome, the Azores and Barbados are three likely candidates. Back in 1717 the easternmost island of the Azores provided the prime meridian and not Greenwich which is a good reason for its inclusion.

STRETTON GRANDISON, Herefordshire

At St Lawrence's Church we have what is everybody's idea of a Horizontal Dial, mounted on a baluster pedestal, again as expected. In fact it is a commemorative dial, dedicated unusually to a daughter who must have died young. The dial itself is by Francis Barker, a company that survived as dial makers into the late 1920s and produced relatively inexpensive but accurate and workaday instruments in a large variety of sizes and styles. Quite often in books about sundials their dials are overlooked in favour of the individually produced hand engraved 'classics' or more modern 'specials' but Francis Barker and Son must have produced hundreds of these dials and many will go unrecognised. Amongst the services offered was the engraving of mottoes and so we have, in addition to an eight-point compass star and the hour lines, the parents' memories:

Cube Dial at Walford AW

Horizontal Dial at Waterfall

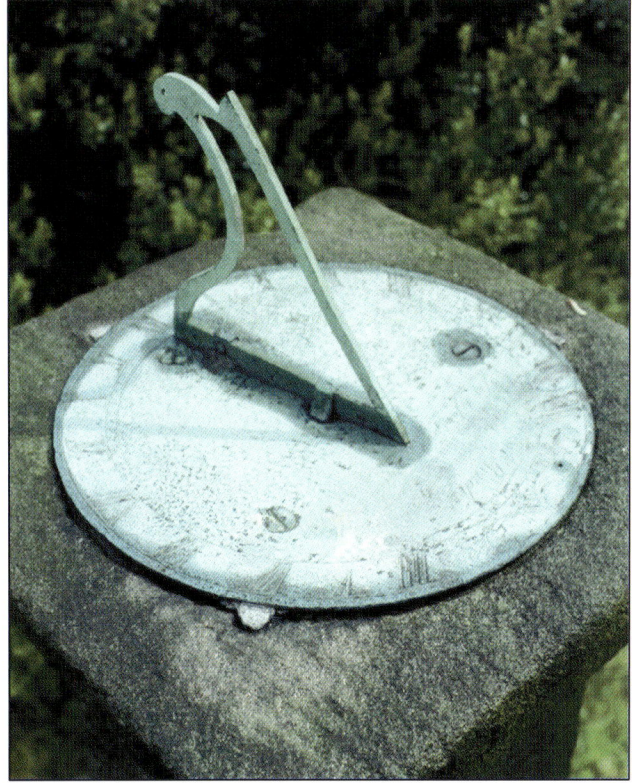
Dial plate at Waterfall

whilst the north and south ones are solid. In the time the dial has been in position since 'renewal' the south face has already become worn and pitted, the north face remaining sharp and clear. The numerals are distinctly medieval in character, copied or not. There is a cross for noon, the 10, 11, 1 and 2 follow the slope of the hour lines on the south face and the 8 has the flat top of that era. All faces are in Arabic numerals, (perhaps Roman would have been expected on the south face as happens on occasion). The 5pm on the north face has acquired an extra flourish on its top stroke; the 5's on the whole look awkward to our eyes anyway. The pedestal is possibly original and gives us a low mounted-dial, which one suspects was not a long lasting fashion as most church dials are mounted well up on a wall or column.

WATERFALL, Staffordshire

There is no waterfall here but the River Hamps and the River Manifold into which it flows vanish underground for part of their courses and this aptly named village is not far from their confluence. In the churchyard of St James & St Bartholomew is a Horizontal Dial set on a square section pedestal. In many ways it is unremarkable; there are many similar sundials in churchyards all over the country. What sets this one apart is the verse inscribed on the dial plate, now much more difficult to decipher than when the pioneering sundial scholar Mrs Gatty recorded it in the 19th century. It reads as follows:

> **Reader Use well the present moments**
> **as they fleet,**
> **Your life, however short, will be complete,**
> **If at its fatal ending you can say:**
> **I've lived and made the most of every day**

which is marginally more cheerful than a lot of sundial mottoes.

WEOBLEY, Herefordshire

The Vertical Dial at the Church of St Peter and St Paul, Weobley, appears to be made of slate and is beautifully carved with fine hour lines, half-hour lines and quarter-hour lines. We have a grey sunburst with gilded corona and the numerals and motto are finely gilded also. The motto is:

> **ONE DAY TELLETH ANOTHER. AND -**
> **ONE NIGHT CERTIFIETH ANOTHER**

which is a shade obscure but probably gloomy in intent. The hours go from 5.30am to 5pm (and no further although there is space) implying an east declination and this is confirmed by the gnomon offset. Close examination of the picture reveals a little mystery however. The stonework

Fine Vertical Dial at Weobley MN

Cube Sundial at Blakesley Hall JL

over the porch door is created specially for the dial and the top shadow shows it slightly inclined eastward from the wall line. The bronze rod gnomon is supported by a simple rod whose foot is now at 11.25am but which may have been originally at 11.45am as there is a small hole there. It looks as though the dial may have been turned more eastwards during its lifetime - why? The shadow of the left panel brickwork is already moving across the dial face and will probably move completely across before the delineation ends at 5pm. Is this why the hour line markings stop there? For all the above it is a fine dial and judging by the clock in the photograph it is quite accurate - allowing for the odd hour from GMT to BST.

YARDLEY, Birmingham
Blakesley Hall is a late 16th century half-timbered farmhouse which is now in the care of Birming-

ham City Council. Those who visit it should make a point of examining the sundial in the garden. This dial somehow acquired the reputation of being a Moondial and while it is quite possible for a sundial to tell the time by the light of the Moon it requires either a very complex dial-plate or the use of a table of corrections. Both methods demand a knowledge of the number of days elapsed since the previous New Moon. At Full Moon, for example, a sundial will be exactly twelve hours wrong indicating midday when it is midnight. This is not in fact a Moondial but a Multiple Dial. It is approximately a cube of stone which has dials on all four vertical faces and on the top. Some of the dials on the vertical faces are Scaphe Dials, which are hollows in the stone and the hour lines are marked in the hollows. The shadow may be cast by a metal gnomon or, in the case of a square hollow, by one of its edges. The dial was not always here but was moved from the village of Elford near Lichfield around the middle of the 20th century. It was in poor condition and to make matters worse it was positioned facing in completely the wrong direction. An extensive restoration of the hall was recently completed and it was decided at the same time to do some minor restoration to the sundial and turn it the right way round.

EAST MIDLANDS

Robert Ovens & Walter Wells

Derbyshire, Leicestershire, Lincolnshire, Northamptonshire, Nottinghamshire, Rutland

To many travellers on their way through the English Midlands the scenery may seem featureless and rather dull. No rugged cliffs or mountain ranges bound the observer's view. The ancient rocks of the region are for the most part concealed beneath a covering of low limestone escarpments and wide clay valleys draining gently to the sea. This is a countryside dedicated to prosperous farming and studded with small market towns and quiet villages.

Within living memory the area also supported some heavy industry concentrated in a few towns such as Derby and Corby, which profited from local deposits of coal and iron. Now that those resources are largely exhausted the activities they encouraged are no longer important although they have certainly left their mark.

Market Harborough Church with the Old Grammar School close to its south door

by the character of the stone in their local quarries, such as the grey limestones of Stamford or the warm ironstone of Braunston in Rutland.

For more than a thousand years local communities centred their lives on the parish churches, which regulated their daily affairs in more senses than one. The tapering spires and squat towers sent out their ringing message to a wide area and lives were paced according to the tolling of a bell. Much depended on the ability of the parish sexton to judge the passing hours. Still, today, we find primitive scratch dials carved on church buttresses and porches, where the sun's creeping shadow would mark the time.

These primitive dials must have been of rather limited use, even on the sunniest days. Fitted with a horizontal rod to cast the moving shadow, the so-called 'Mass Dial' divided the period from sunrise to sunset into twelve quite arbitrary and irregular 'hours'. These classical hours would have been longer both in summer and, on any day of the year, in the early morning and late afternoon and correspondingly shorter in the winter

Even so, the East Midlands region has a long history of prosperous activity and still preserves many treasures to please the discerning visitor. Public buildings and the elegant houses of wealthy landowners proclaim the skill of local stonemasons. Whole villages are distinguished

and nearer to mid-day. However mid-day itself would have been indicated correctly at the end of the sixth hour, if the horizontal rod was properly directed towards due south.

By the late 14th century the construction of weight-driven mechanical clocks began the more rational measurement of time. Even where these existed they were poorly regulated until the introduction of the pendulum about three centuries later. From there on we find modern equal-hour sundials, with their accurately tilted gnomons, put up near church clocks in order to check their time keeping. From their purely functional role at that time sundials came to be appreciated as a charming adornment for any building or pleasure garden.

For anyone with an eye for ornamental or skillfully constructed sundials the East Midlands region is richly endowed. Many of these instruments were created by local people, who were well informed about scientific matters through reading the reports from learned societies that were published extensively. However these early sundial-makers were faced with problems which are more easily solved today. Primitive maps, uncertainty of the orientation of the building and determining the exact moment of local noon were all difficulties which were not easily overcome. To set the angle of the gnomon they needed to measure the latitude by taking readings with a sextant. To have achieved so much, as indeed many certainly did, was truly remarkable in the circumstances. Often the maker has also incorporated some exhortation or an appropriate quotation as a motto on the dial.

The counties of the East Midlands vary in the history of their development. The hills of Derbyshire were, since Roman times, famed for the valuable veins of metallic ore and other minerals which they contained and here it was that the vigorous streams that drain the high ground were

Millennium Sundial at Barrow upon Soar

later harnessed by pioneers of the Industrial Revolution to power their factories.

Nottinghamshire had a historic role as the strategic centre of the Kingdom in troubled times. Later it provided coal which fueled many industries. Rural Lincolnshire, the most extensive of the six counties, owes much to the engineers from the Low Countries who drained its low-lying acres for agriculture.

Leicestershire, Northamptonshire and Rutland profited from their central position on the great roads linking more distant areas with the capital. Thriving towns developed as they provided staging posts and markets along the way, particularly those which were located at important crossroads.

The building materials of this large region range from the hard gritstones and limestones of the Peak District through successive layers of softer and more easily worked deposits of ironstone and limestone towards to the North Sea coast. Certain layers, like the Barnack limestone to the south, were highly prized by medieval builders for their provision of large regular blocks of 'freestone'.

Some of the interesting sundials to be found in this region are described and illustrated on the pages that follow. Many are erected on places of worship, public buildings and the stately homes that grace this area of England. Others were set up more recently in public open spaces to commemorate the new Millennium.

BARROW UPON SOAR, Leicestershire

At Barrow upon Soar near Loughborough a very large dial has been set out on the crest of a meadow in the Millennium Park. In a simple but sturdy design by Patrick Powers that incorporates a ring of pillars of the local granite, the shadow is cast by an inclined steel gnomon, 6.75 metres in length, and the hour-markers follow a circular paved walkway on the grass. The analogy with the stone circles executed by ancient observers of the seasonal rise and fall of the sun is a fitting tribute to the past at the beginning of a new millennium.

BASLOW, Derbyshire

In St Anne's churchyard at Baslow, north of Matlock, there is a sundial in the form of a recumbent cross on a pedestal. It is known as a Cruciform Dial and is a memorial to Lt Col Wrench who was an army surgeon in the Crimean War before he took over a medical practice at

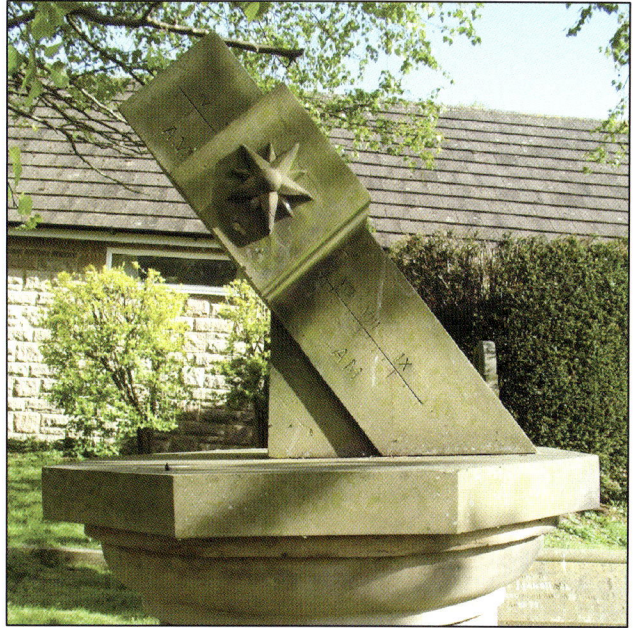
West View of Cruciform Dial at Baslow

Baslow in 1862. When a new church clock was installed to mark Queen Victoria's Diamond Jubilee, Dr Wrench designed the east dial, thus confirming his interest in recording time.

The sundial was designed to present a multiple set of straight edges and flat surfaces to the moving sun. Originally, the front of the cross was in the plane parallel to the Earth's equator. One set of edges at right angles to the front was therefore directed towards the pole. Unfortunately, the pedestal now leans away from the vertical, and an overhanging tree shades the dial for much of the day.

When set up correctly the hours around noon could be read from the top faces of the horizontal arms. At mid-morning and mid-afternoon the shadow of the arms moved across the sides of the lower vertical shaft.

It has the following inscription on the cross arm:

**IF O'ER THE DIAL GLIDES A SHADE
REDEEM THE TIME FOR, LO,
IT PASSES LIKE A DREAM
BUT IF TIS ALL A BLANK
THEN MARK THE LOSS
OF HOURS UNBLESSED
BY SHADOWS FROM THE CROSS**

BILLESDON, Leicestershire

At Billesdon, just off the A47 road from Leicester to Peterborough, the Old School has a significant

Cruciform Dial at Baslow

123

The North-West and South-West
Dials at Billesdon

The four dials at Billesdon,
NW, NE, SW & SE

set of four 17th century sundials, probably designed by Samuel Herrick, recently graduated from Isaac Newton's Trinity College, Cambridge. These simple dials trace the direction of the Sun from sunrise to sunset as it shines on each elevation of the building in succession, particularly on the longest days of summer.

The building was provided by William Sharpe of Rolleston in 1650 for use as a free school by those who could not afford to send their children to one of the local private schools. It remained in use until 1878 when the National School was built. In 1959 it returned to its former use when it became an overflow for the village school.

COLLYWESTON, Northamptonshire
Collyweston, near the Northamptonshire border with Rutland, is best known for its stone slate that has been mined there for centuries and used as an attractive roofing material all over the country. Less well known is that the village once had a

royal palace which was built by Sir William Porter early in the 15th century. It was later owned by Ralph, 4th Lord Cromwell, Treasurer of the Exchequer. When Cromwell died in 1456 it was sold

Dial at Collyweston

Detail of dial at Collyweston

124

to the Earl of Warwick for £1066 13s 4d. By 1640 it had fallen into disrepair and demolished. The materials were finally removed in 1780 when the Earl of Exeter sold them for £400. Some walls and buildings survived, however, including a magnificent sundial in the form of an elliptically headed alcove of grey limestone standing some four metres high by three wide. It is in excellent condition, having been restored with a new gnomon in recent years. It can be seen in a paddock just to the west of the church, near the junction of New Road and Back Lane.

COTTESBROOKE, Northamptonshire

Reputed to be the inspiration for Jane Austen's *'Mansfield Park'*, Cottesbrooke Hall is located ten miles north of Northampton. It represents a fine example of Queen Anne architecture in a stunning parkland setting. The building dates from 1702 and is home to a superb collection of sporting paintings.

The Hall and its grounds are open to visitors on certain days in the summer and a most unusual sight is the heraldic vertical sundial overlooking the Dutch garden, bearing the arms of Buchanan. The shield displays a lion rampant and the crest is in the form of a coronet. The motto reads:

GLORIOR HINC HONOS
(I take pride and from this comes Honour)

Dial at Cottesbrooke Hall overlooking the Dutch Garden

Below this is a fine Armillary Sphere.

Nearby, All Saints' Church contains a most extraordinary pew of the Langhams of Cottesbrooke Hall. The pew has tenants' seats in front, then a curving staircase up to a gallery which is like a box in the theatre. A fireplace ensured that the family could sit there in comfort.

DEENE PARK, Northamptonshire

Not far from Corby, off the A43 Kettering to Stamford road, is Deene Park, home of the Brudenell family since 1514. It is a very grand country house, with features that have developed over the centuries. When entering the central courtyard through the north door there is a set of three vertical dials to be discovered. These all date from the time of the 4th Earl of Cardigan in the 18th century. Facing south is a beautifully figured dial dated 1769 and painted on ashlar stonework below the parapet. The wall faces slightly to the east of south and the gnomon, which has a delicately scrolled support, is offset

South facing Vertical Dial at Deene Park

East Dial at Deene Park

from the vertical to compensate. Set high in the gable ends on the East and West sides of this open space are two sturdy gnomons, each presenting a straight edge that slopes upwards towards the Pole. Unfortunately the dial lines and numerals have long since weathered away.

EYAM, Derbyshire

There is a large and unusual sundial above the priest's door of St Lawrence's Church at Eyam in the Hope Valley. The village is well known for its experiences during the plague years of 1665 and 1666, the fleas carrying the plague apparently

The Sundial at Eyam, Derbyshire

arriving in a bundle of cloth from London. The villagers isolated themselves to prevent the disease from spreading, but unfortunately three quarters perished. There is a museum to the plague in the church.

The sundial was erected for Churchwardens William Lee and Thomas Froggat in 1775, and tradition maintains that it was designed by a Mr Duffin, Clerk to Mr Simpson of Stoke Hall, and executed by William Shore, a local stonemason. It is considered more likely, however, to have been made by the eminent clockmaker John Whitehurst FRS at the request of the then Rector, Canon Seward.

The sundial was originally located above the south door and was moved to the present position when the south aisle was widened and a porch built at the restorations of 1868 and 1882.

The dial carries an abundance of geographical and astronomical information and its motto is:

INDUCE ANIMUM SAPIENTEM
(Cultivate Wisdom) or
(Take to thyself a wide or enquiring mind)

Around the root of the gnomon, the dial shows the times when it is noon at 13 different towns of the world from 'Calicut' (Calcutta) to 'Mexico', and its nodus traces out the appropriate month of the year between the two Tropics. Even the corbels that support the heavy dial remind us:

UT UMBRA SIC VITA
(Life is no more than a passing shadow)

A very useful leaflet, which describes the sundial in some detail, is available in the church.

FOLKINGHAM, Lincolnshire

Folkingham is 8 miles north of Bourne on the A15 towards Lincoln. Until the late 17th century, the present wide and attractive market place was by no means delightful because here there were stacks of timber, a horse pond, a dismal town

126

The Greyhound Inn, Folkingham

Father Time supporting a Horizontal Dial at Belton House

hall, a butchery and a foul-smelling open well. In 1788 the Lord of the Manor was in financial difficulties and he sold the estate to Sir Gilbert Heathcote whose great-grandfather was Governor of the Bank of England. He did much to transform Folkingham and his changes included clearing the market place and equipping it to cater for the stage coaches using the main London to Lincoln road. The Greyhound, a substantial coaching inn, dominates the former market square, which is now partly grassed over.

The Inn boasts a large painted sundial above the main entrance. It is furnished with zodiacal lines for the summer and winter solstices and the equinoxes of spring and autumn. Unfortunately, the dial has been furnished with a replacement gnomon which is set at the wrong angle for accurate time indication. A small nodus fitted to the original gnomon enabled it to indicate the passing of the seasons as well as the hours of the day.

GRANTHAM, Lincolnshire

Belton House near Grantham was built in 1685-88 and is considered by many to be the perfect English Country House. The interior is magnificent and it has fine silver and furniture collections. In the gardens there is a fine Horizontal Dial which is supported by a pedestal standing between a pair of figures that symbolise the passage of time. In Father Time and the cherub we are reminded that the same progression that leads to maturity and eventual old age is also fostering youth and renewal. The pedestal was beautifully sculptured by C. G. Cibber.

GREAT GLEN, Leicestershire

Stoneygate School at Great Glen has a delightful example of 20th century pargeting where the twelve signs of the Zodiac support an attractive vertical sundial. It is located on an outbuilding on the south side of the school, looking out over tennis lawns. This independent school occupies the house which was originally built in a country location for an eminent magistrate. Arrangements should be made with the school to visit this sundial. (Photograph on page 128).

HAMBLETON, Rutland

Since the flooding of the twin valleys of the River Gwash in 1976, the village of Hambleton now stands on a peninsular stretching out into the centre of Rutland Water. Hambleton Hall is now well known as an exclusive restaurant and hotel, but when built in 1881 by Walter Gore Marshall, a wealthy brewer, it was a hunting lodge. Mr Marshall had chosen a location in Rutland to enjoy the fox hunting, and in particular the intensive social activities that went with it.

127

Stoneygate School, Great Glen ^{WW}

village having provided a new school, post office and several cottages before he died in 1899. He even erected a clock on the front of the post office in the style of his sundial and stable clock.

KEDLESTON HALL, Derbyshire

All Saints' Church at Kedleston, formerly the private chapel to Kedleston Hall, was taken over by the Churches Conservation Trust in 1983. At the same time the Hall, built between 1759 and 1765, and its park, became part of the National Trust, ending nearly 850 years of ownership of Kedleston by the Curzon family.

The church, which is immediately adjacent to the west side of the Hall, has a magnificent east-facing

On the front of the Hall he erected a beautiful sundial in the 'Arts and Crafts' style which has Latin and French inscriptions:

NUNC HORA BIBENDI
(It's drinking time now)

and

le·temps·passe·l'amité·reste·
c'est·l'heure·de·bien·faire

(Time passes, friendship remains,
it is the time to do good)

He also placed a turret clock above the adjacent stables that has four dials in the same style commemorating the sixtieth year of the reign of Queen Victoria. He was a great benefactor of the

Vertical Dial at Hambleton Hall ^{RO}

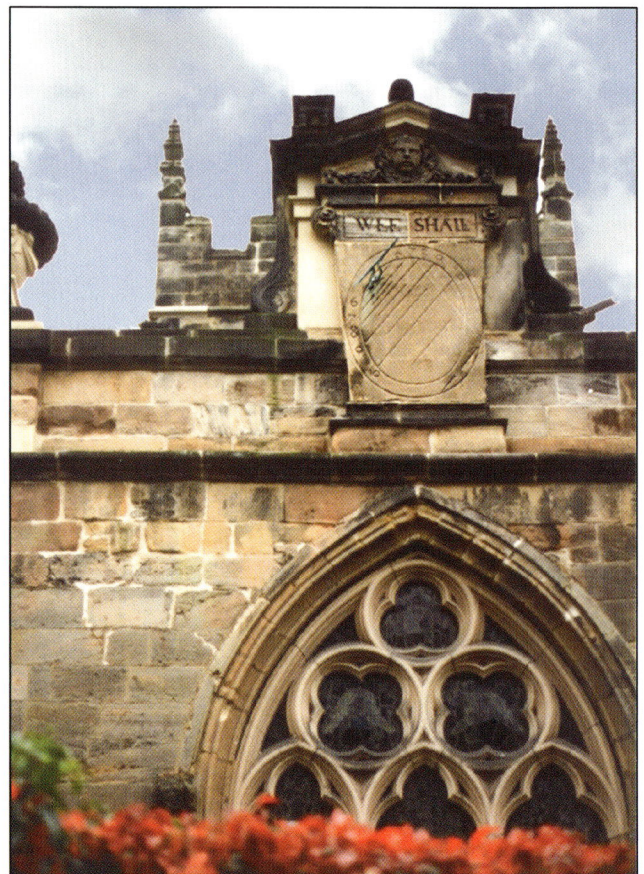

East facing Vertical Dial at Kedleston Hall ^{WW}

pedimented Vertical Dial suspended above the Gothic east window. Above the dial are two hourglasses and a skull. The rather sobering and cryptic motto should read:

WEE SHAIL - DIE ALL

(- di-al!), but the lower part is now missing.

The mansion has a typical Palladian frontage linked by colonnades to matching pavilions on either side. The entrance leads through a very grand hall to a series of rooms in the same style. It is strange, therefore, to be confronted through an upstairs window by the prominent sundial on the east end of the church. This church is all that remains of Kedleston village following its clearance by Nathaniel Curzon, the first Lord Scarsdale.

LEICESTER

Cathedral. On the south elevation of the south aisle of Leicester Cathedral is a carved and gilded slate dial. It was erected to mark Queen Victoria's Diamond Jubilee in 1897. The face shows a large Greek cross and displays the symbols of the four Evangelists between its arms. The motto is:

CŒLI·ENARRANT·GLORIAM·DEI

(The Heavens declare the glory of God)

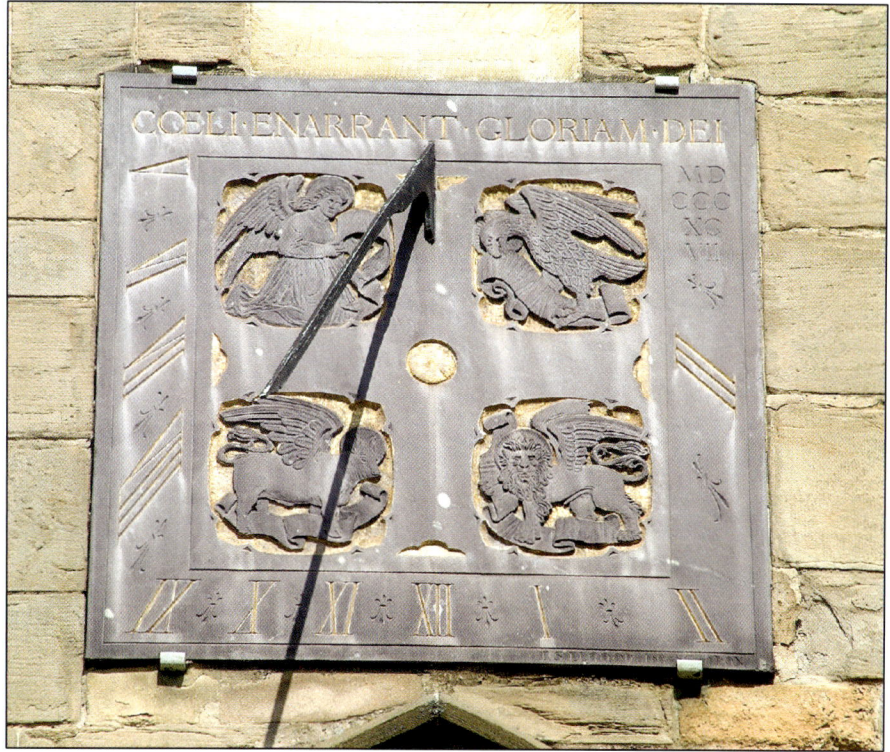

Vertical Dial on Leicester Cathedral

which is the first line of Psalm 19, used by Haydn in his oratorio, 'The Creation'.

University Campus. The campus lies between University Road and Victoria Park and is open to public access. A visit here will be rewarded with three interesting sundials and a magnificent astronomical clock, all of which have adjacent plaques describing their significance.

Above the entrance to the Bennett Building are a pair of Welsh slate dials designed by Allan Mills and carved by Michael Fisher. They were erected in 1994.

Slate Dials by Allan Mills at Leicester University

129

The Isaac Newton Dial at Leicester University ^{WW}

The first dial indicates 'antique hours', the earliest timekeeping system. Its gnomon is simply a horizontal south-pointing peg, the shadow of its tip being the time marker. In this system the daylight part of the day is split into 12 equal 'hours': the summer 'hour' is therefore longer than the winter 'hour', and midday is at the end of the sixth 'hour'.

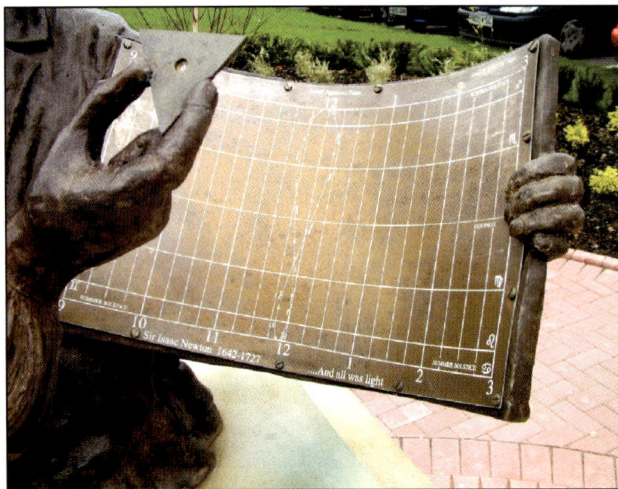

Isaac Newton Dial, detail ^{RO}

The second dial indicates equal hours, where each hour is 1/24 of the full day, which includes daylight hours and night hours. The gnomon is parallel to the Earth's axis and points to the pole star. The shadow of its entire upper edge indicates the time of day. Midday is now 12 noon.

These dials therefore serve to demonstrate the essential difference between the medieval mass dial and the later scientific sundial.

The gilded curves mark the paths travelled by the tips of the gnomons at the summer and winter solstices, and the ammonite on the first dial and Man on the modern dial symbolize geological time and evolution. The quotation:

Time measures every thing
but is to Nature endless

is from James Hutton's 'System of the Earth', 1785.

Near the entrance to the next building is the great Astronomical Clock designed and constructed by Allan Mills and Ralph Jefferson. It indicates Greenwich Mean Time, Solar Time and Sidereal Time, all of which is explained on the nearby plaque.

A new sundial in the garden beside the Centre for Space Research pays a fitting tribute to two aspects of the science of Sir Isaac Newton (1642 – 1727) who had a lifelong interest in sundials. His own earliest sundials were constructed when he was a grammar school boy and before he had knowledge of geometry or trigonometry. In fact he made '...dyals of divers forms and constructions every where about the house, in his own chamber, in the entrys and rooms wherever the sun came in' - (William Stukeley). One of his dials can be seen behind the organ in Colsterworth Church, Lincolnshire.

The University sundial is a sculpture by Vanessa Stollery supporting a design by John Davis of Flowton Dials in which a small aperture in a prism-like object trains a narrow beam of sunlight upon the engraved dial-plate. This device celebrates the seminal work of Newton, both on the nature of light and on the motion of the Earth as it rotates about the Sun.

The recently restored Vertical Dials high up on Lincoln Cathedral

The Direct South Dial at Lincoln Cathedral

LINCOLN

Around the year 48AD the Romans came to what we now know as Lincoln, its position at the top of a steep hill being an ideal place to build a camp. As a result the city is now rich in Roman history with many preserved artefacts and standing proudly on the fortified hilltop is the splendid medieval cathedral.

Two elegantly simple sundials on a south buttress of the cathedral carry sobering thoughts drawn from ancient poets. Facing south is the motto:

CITO PRÆTERIT ÆTAS
(Life is passing swiftly)

adapted from verses by Ovid, and facing east:

PEREUNT ET IMPUTANTUR
(Our finest days are slipping by
and will not come again)

as the hard-working lawyer, Martial, complained to his friend. The east-facing dial carries a simple

sloping gnomon and its hour lines are all set at exactly the same angle across the face. These two dials have recently been repainted and restored to their former splendor.

The Roman camp was known by its Celtic name of Lindun of which Lindum was the version used later by the Romans. They built stone walls and gates around their camp. Today the north gateway still survives and is known as Newport Arch. In 1068, William the Conqueror ordered a castle to be built. Close to the castle and joining to the lower town is an unusual street known as Steep Hill. Two of the more interesting houses on Steep Hill are the Jew's House and Aaron's House.

Just below the Cathedral is the Usher Gallery which boasts a fine collection of clocks, coins and memorabilia of the Lincolnshire poet Alfred Lord Tennyson.

MARKET DEEPING, Lincolnshire

The small town of Market Deeping lies at the southern extremity of Lincolnshire, a few miles north-west of Peterborough. Here there is a

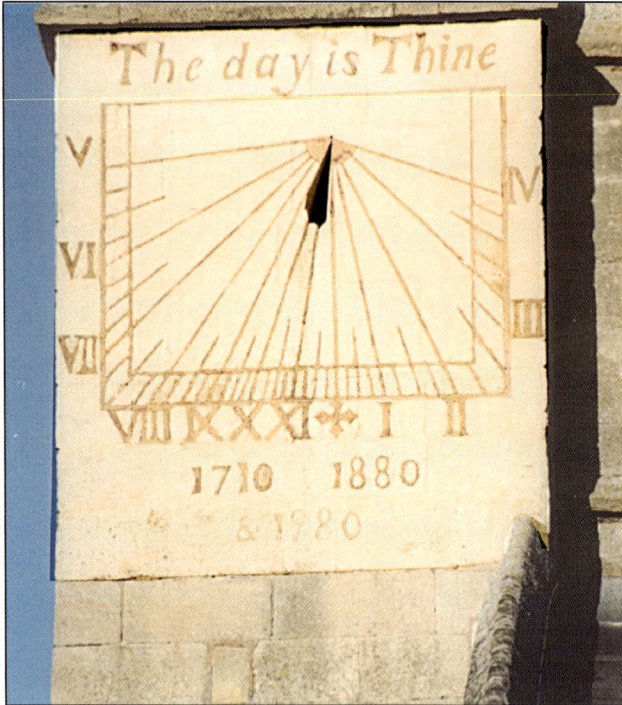

The 'daytime' Dial at Market Deeping ^{WW}

large painted dial. A particular feature is iron-work supporting the gnomon which has been beautifully wrought into delicate scrolls by the

The magnificent church tower at Market ^{MC} Harborough

delightful pair of 18th century Vertical Dials on the tower of St Guthlac's Church. The pious mottoes proclaim, on the south-facing dial:

The day is Thine

and, on the reverse side of the tower, where sunlight falls on summer evenings:

The Night cometh

The 'evening dial' can also be used for a brief period on summer mornings.

MARKET HARBOROUGH, Leicestershire

St Dionysius' Church. On the south side of the lofty tower of St Dionysius' Church is a very

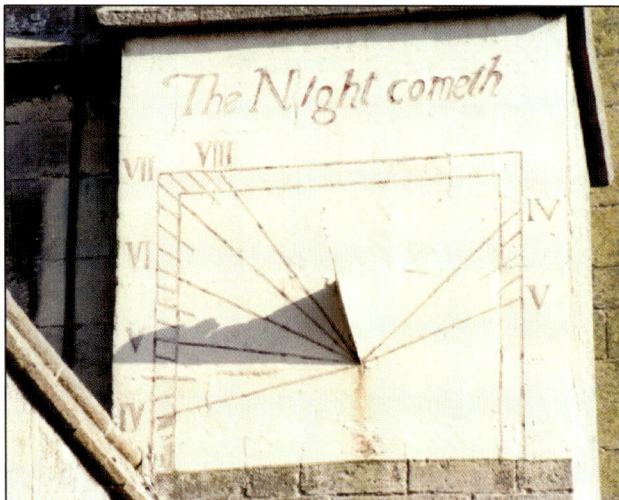

The North Dial at Market Deeping ^{WW}

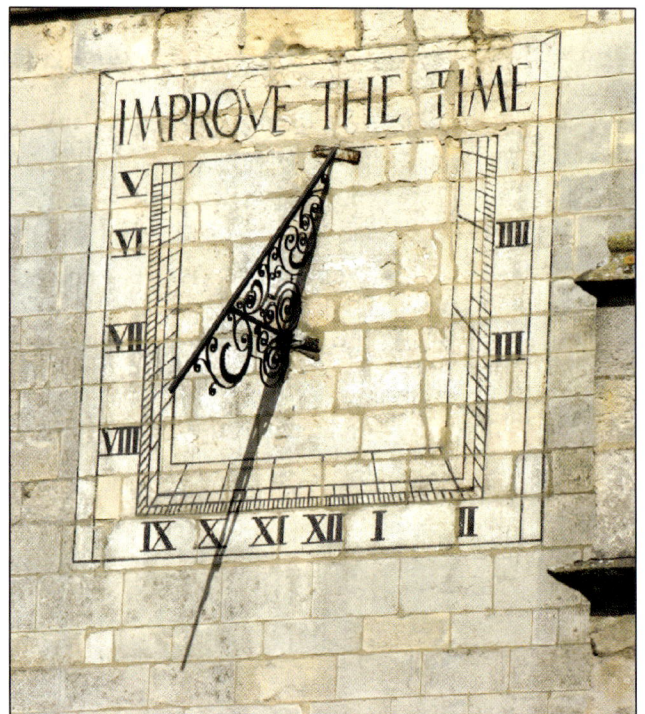

Ornate gnomon at Market Harborough ^{MC}

local smith. The motto is:

IMPROVE THE TIME
apparently directed towards the scholars of the Old Grammar School adjacent to the south door of the church. The Grammar School building itself is of great interest being built in 1614 on wooden posts, the area beneath it being used as a butter market. St Dionysius' dial was designed and given in 1771 by Samuel Rouse, a local shopkeeper devoted to the latest developments in astronomy. For his studies he constructed a large brass and timber sextant with which in 1737 he established the exact latitude of the town. For his dial he made a series of measurements of the orientation of the church tower, which he found to decline from the south towards the east by 16° 45'.

Old Union Canal Basin. Just north of the town centre a fine sculptured dial was erected near the Old Union Canal Basin development in 2000. It is in the form of a 'Man with a Plank'. A nearby plaque states that *This Community Sundial Sculpture is dedicated to the canal builders and the long line of Company servants, boat people and wharf employees who made their living from the Leicestershire and Northamptonshire Union and Old Grand Union Canals'.*

It was commissioned by Harborough District Council and the design is the work of Marjorie and Michael Clements. The large plank casts its shadow on the seat board and further to a surrounding border of carved slate plaques. An additional feature is the small gnomons that show

'Man with a Plank' Dial at the Grand Union Canal Basin in Market Harborough

the hours on either side at the top of the plank. The equation of time is also shown on the west side of the plank. The slate plaques were designed by artist Liz Minichiello from ideas submitted by the local community. They were carved by stonemason Alf Herbert.

The town has also acquired more recent sundials that are worth visiting. An Armillary Dial in the new shopping area centred on St Mary's Place celebrates the town's former sheep and cattle

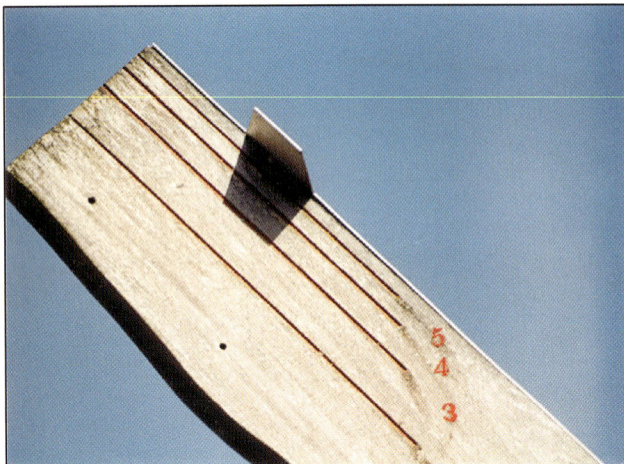

Market Harborough canal dial showing the gnomon tip with its additional hour scale WW

Rotary International Dial at Melton Mowbray AM

market in cutaway silhouettes on the hour-ring. A more recent addition is a Polar Dial installed in the Park with an Equation of Time graph carved on its back.

Further canal connections may be found at nearby Foxton Locks. Here there is an impressive flight of locks, the remains of a Victorian inclined plane and a small canal museum which has two modern sundials on adjacent corners.

MARSTON, Lincolnshire

St. Mary's Church, Marston, which is near the Great North Road just north of Grantham, was the place of worship for the Thorolds, a great Lincolnshire family. They arrived in Marston in the 14th century. In the south chapel is the tomb of the most famous member of the family, Sir Anthony Thorold, who was High Sheriff for Lincolnshire during the reign of Elizabeth I.

In the churchyard, near the south porch, a tall stone column carries west-, south- and east-facing sundials.

MELTON MOWBRAY, Leicestershire

In Egerton Park, just off the Leicester Road a large Horizontal Sundial celebrates 75 years of public service by the local branch of Rotary International. The dial plate is massively constructed in stainless steel and mounted on a circular pillar

Multiple dial at Marston RO

Dial by Tony Moss at Melton Mowbray RO

Modern Dial with conical gnomon at Newark

Lady Ossington Hotel, Newark

topped with granite. The gnomon is supported by part of a gearwheel which is representative of the Rotary International logo. Eight stone pedestals set into the paved area surrounding the sundial mark out the cardinal points of the compass. Adjacent to the dial, made by Tony Moss of Lindisfarne Sundials, there is a tablet explaining how to convert sun time to clock time.

NEWARK, Nottinghamshire

A most unusual Horizontal Dial to commemorate the new Millennium is situated in the park by the River Trent. The large conical gnomon casts its shadow across the grass and the hour numerals set into it. The time is read by estimating the centre of the triangular shadow.

Winding round the cone is a symbolic 'river of time' (the nearby Trent) alongside which are marked 13 dates of events in Newark's history. Each of the 13 hour markers is engraved with a simple cartoon of one of these events. As a history lesson, local school children are asked to pair up each event with its correct date.

The dial was designed by the Free Form Arts Trust and was delineated by John Moir.

Not far away a more conventional sundial with the motto:

MAKE HASTE - TIME FLIES

greets the visitors to the Lady Ossington Hotel opposite the Castle. Lady Ossington was the widow of a former Speaker of the House of Commons. She was a strong supporter of the Temperance Movement and the 'Lady Ossington Coffee Palace' was provided as an alternative place for non-alcoholic refreshment.

OAKHAM, Rutland

The county town of Rutland is perhaps best known for its 'Castle' which houses a magnificent collection of ornate horseshoes donated by visiting members of the Royal Family. Nearby is the

Old Vertical Dial at the Hospital of St John & St Anne, Oakham

Dial on the former Matkin's Building, Oakham

old schoolroom where Robert Johnson founded Oakham School in 1584. He also founded the Hospital of St John and St Anne in 1590 on the western edge of the town, now adjacent to the railway line. The chapel, which has recently been renovated, has a 17th century direct south limestone sundial set into its south-west corner.

In the Market Place is the late 16th century Buttercross, above which is a Cube Dial; a block of limestone with dials facing the four cardinal points of the compass. Sadly, all the gnomons are now missing, but photographs taken in the 1920s show the dials to be in working order.

The vertical sundial with the motto:

TEMPUS FUGIT
(Time flies)

located on the former Matkin's building at 13 High Street, is a local landmark. Charles Matkin was a well known local printer.

PINCHBECK, Lincolnshire

Close to Spalding lies the village of Pinchbeck with its large church of St Mary. The Vertical Sundial, boldly positioned over the south porch dates from 1799. It has recently been restored under the guidance of architects Marshall Sisson, working from old photographs and notes. These also confirm that the dial was restored in 1892 and 1934. Its motto is:

Pereunt et Imputantur
(They [the hours] pass away
and are given account of)

The dial also carries the names of two Churchwardens, R.Clark and T.Plowright who were pre-

The recently restored dial at Pinchbeck

sumably in office at the time of its original installation. Note that the equinox line is not straight, an error which has probably been copied from an earlier restoration. St Mary's Church itself dates back to Norman times but the building seen today is a mixture of styles between c.1150 and 1500. Inside is an impressive hammer-beam roof with gilded angels on the brackets.

POTTERSPURY, Northamptonshire

Just off the old Watling Street, now the A5, between Milton Keynes and Towcester, is the village of Potterspury. St Nicholas' Church has a fine mass dial high up beside a window. Its calibrations are rather puzzling, apparently having two sets of lines that were carved at different

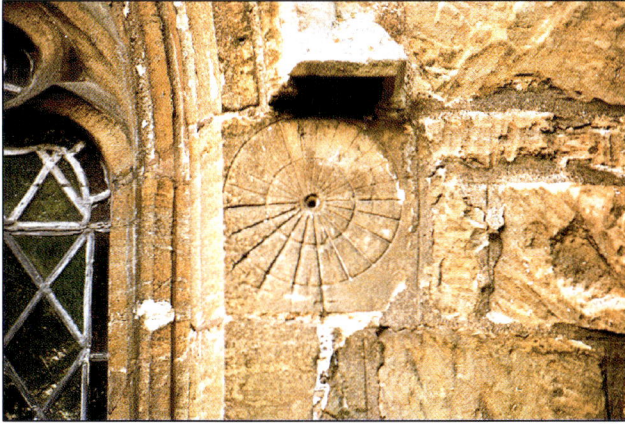
Early Mass Dial at Potterspury ^{MC}

Dial at Repton School through 'The Arch' ^{WW}

Vertical Dial at Repton Scool ^{WW}

times. One set divides the circle into 13 spaces and the other 10. Four of these last lines project beyond the outer circle and are marked with a cross. This dial was recorded by Eden & Lloyd in 1900 and shows that it has only slightly deteriorated over the last 105 years.

REPTON, Derbyshire

Repton School was founded in 1577 by Sir John Post who had taken over the buildings of a 12th century priory in the Derbyshire village of Repton. The school is entered via the Priory Gateway, known as The Arch, through which may be seen an engraved stone sundial over the main door to a building which is still known as The Priory.

The neatly rhyming motto in two lines across the top of the dial is directed at those who enter:

**FUGIT HORA
ORA LABORA**

(Our lease of time is short.
So we must work and we
must pray)

RUSHTON, Northamptonshire

An early sundial maker was Sir Thomas Tresham, an independent character who maintained his Catholic faith in difficult times. He found a Latin affinity between the family name of Tresham and the number 3 and conceived a way of celebrating that connection in an elaborate monument that he built in stone. We can still stand in wonder at his Triangular Lodge, built for the warrener, or estate rabbit-keeper, near his house at Rushton.

Triangular Lodge near Rushton with its strong Catholic links ^{WW}

One of the three dials at Triangular Lodge ^{MC}

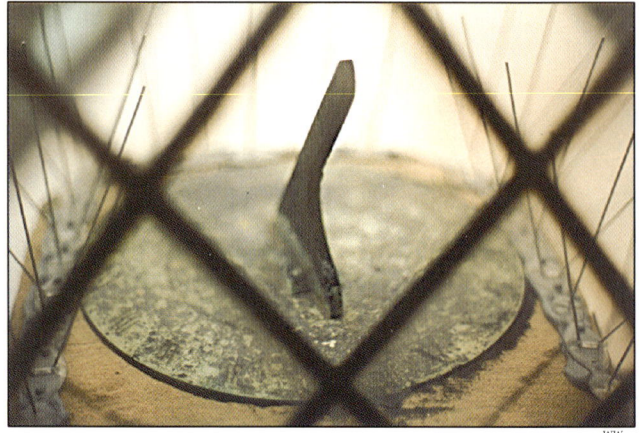

Dial outside the clock chamber at Southwell Minster ^{WW}

The building is in the form of an equilateral triangle with three storeys and three windows on each side of each floor. Each side has three gables rising to three tapering pinnacles. A single three-sided chimney rises from the meeting point of the nine gabled roofs. Each side carries the date 1595. Below the gables is a continuous frieze each side being 33 feet long with 33 letters. Its interior is simple with a large hexagonal room on each floor and three small triangular spaces in each of the corners.

The windows and gables are all designed in triangles and, not surprisingly, there are three sundials mounted high up on the three outside walls, which face roughly to the south-east, north and south-west. Their hour line markings have long since disappeared and the north dial has had its gnomon inverted by some later restorer. The building bears the dates of 1593 and 1595 and a wealth of religious symbols and Latin inscriptions, including a direct reference to the Holy Trinity placed above the entrance door. The words used are:

TRES TESTIMONIVM DANT

(There are three to bear witness)
a direct quotation from John's First Epistle in the Vulgate New Testament.

SOUTHWELL, Nottinghamshire

Southwell Minster, the Cathedral and Parish Church of the Blessed Virgin Mary became a cathedral church in 1884 when the Diocese of Southwell was formed. The original church on this site was founded by Paulinus, the missionary Archbishop of York, in 1627. Set in a quiet market town, the Minster, with its majestic Norman nave and glorious 13th century chapter house, is one of the least known jewels in the crown of Nottinghamshire. Here, a small Horizontal Dial has been set up in a curious location. It is on a precarious shelf outside a lattice window near to the clock room. It was presumably provided for the verger to check the time of day

Horizontal Dial at Sulgrave Manor ^{MC}

Dial Plate at Sulgrave Manor

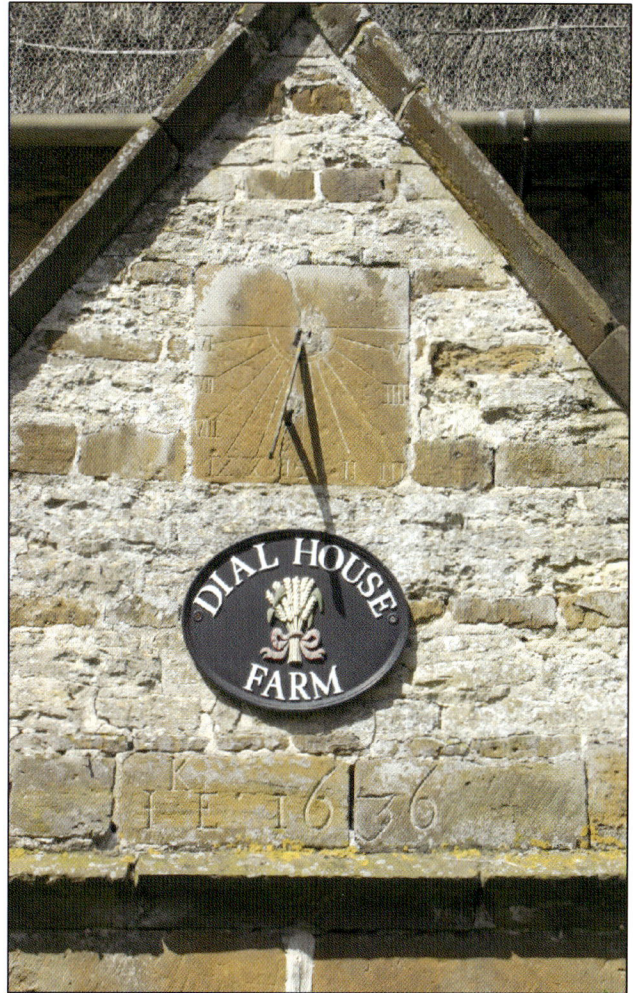
Dial House Farm, Sulgrave

so that he could regulate the church clock, a common use for sundials before standard time was introduced.

SULGRAVE, Northamptonshire

This attractive village nestles in the valley of the River Tove. Probably the most famous building in the village is Sulgrave Manor, once owned by the Washington family, creating strong links with the USA. It has two sundials; a simple Vertical Dial over the porch and an attractively situated Horizontal Dial in its gardens. This square dial is dated 1579 and carries the maker's initials 'G.N.' Its numerals have unusually been applied by punching, a technique seldom used for Horizontal Dials. The dial itself has been mounted more recently on a larger plate with the following inscription around its edge:

> PRESENTED·TO·SULGRAVE·MANOR·
> BY·HIS·WIFE·AND·DAUGHTER·IN·
> MEMORY·OF·GILMER·S·ADAMS·
> PRESIDENT·OF·THE·NATIONAL·
> SOCIETY·OF·COLONIAL·WARS·IN·
> THE·COMMONWEALTH·OF·KENTUCKY·
> 1925·

There are four other sundials in the village but perhaps the most attractive is at Dial House Farm. The house is of local grey stone with brown Eydon stone window dressings and its Vertical Dial, dated 1636, is over its rather low porch.

SUTTON IN ASHFIELD, Nottinghamshire

Situated on the western edge of Sherwood Forest, the town of Sutton in Ashfield occupies a prominent position overlooking Derbyshire to the west. It is a sprawling, heavily industrialized town with a modern shopping centre. Portland Square in the town centre is believed to be the site of the

Modern Dial in Portland Square, Sutton-in-Ashfield

Dial at Pury Hill near Towcester

The fine but neglected dial at the former Unicorn Inn, Uppingham

first village green which was known as Swine Green when the central pond was used to water livestock. Later it became known as Engine Green when a building to house the town's manual fire engine was erected on the site. The current name of the square was adopted in honour of the Duke of Portland's family, who owned most of the land. Now a pedestrianised area, it is dominated by a sundial which has a gigantic gnomon elevated on pillars and reaching some 10m into the sky. One of the tallest sundials in Europe, it attracts visitors from far and wide.

TOWCESTER, Northamptonshire

At Pury Hill Farm, Alderton, near Towcester, a new business park has been created. On one new building, in the form of a dovecote, and which is visible from the public highway, there is a large Vertical Dial of limestone which carries the date it was installed and the motto:

Post Umbram Redit lux - Y
(Shadows pass and the light returns)

Note that the larger initial letters of the Latin slogan spell 'PURY'. This dial was designed by Walter Wells and executed by Robert Ovens in 2003.

UPPINGHAM, Rutland

The second town in Rutland is the home of Uppingham School, another school founded by Robert Johnson in 1584. Like Oakham, the original school building survives, and until 1990 it had a magnificent sundial over the south wall. It was removed because its weight was causing the wall to bow.

Uppingham has a number of interesting sundials to see, three of which are located on the north side of High Street East. Walking in an easterly direction from the Market Place, the first is on the south wall of the former Unicorn Inn. This was one of 28 public houses and inns in Uppingham, but parish records show that the Unicorn was the venue chosen by the Churchwardens for their meetings throughout the 18th century. The sundial is but a shadow of its former self, all details having been painted out for many years. It is best viewed in the late afternoon sunshine when the oblique light throws the former dial markings into relief, although an adjacent downpipe casts an unhelpful shadow. It is then possible to see hour lines and Arabic hour numerals

Artist Kitty Wigmore's watercolour
impression of the 'Unicorn Dial'

Sundial House, Uppingham

Dial at the Crown Hotel, Uppingham

and some of the motto can also be picked out. The gnomon has fortunately survived. This dial was recorded by Eden & Lloyd which confirms that the motto was:

IMPROVE THE TIME

and that it was dated 1765. A local artist, Kitty Wigmore, has recreated this dial in watercolour to give an impression of its former glory.

Continuing along High Street East, the next sundial can be seen high under the eaves of the Crown Hotel. This dial is also recorded by Eden & Lloyd as:

NON REGO NISI REGAR

(I rule not if I be not ruled) The dial is square – black and gilt – and the motto acknowledges submission to the sun. It also illustrates the profound truth that – as

Modern dial plate used outside at Sundial House, Uppingham

Millenium Sundial at Welford

Sundial and Toposcope at Welford

Thomas à Kempis expresses it – 'No man ruleth safely, but he that is willing to be ruled'. (He was a 15th century Augustinian monk and writer, born near Kempen in Germany.)

Today, the dial is in good condition, but part of the gnomon is missing.

A little further down High Street East, just before the Town Hall, is Sundial House with its direct south Vertical Dial in the gable of the dormer window. Above is a fire insurance plaque and attached to the wall above the house name, to the left of the door, is a dial plate from a modern sundial.

Another sundial, thought to be the earliest dated sundial in Rutland, is preserved in the south wall of a row of Victorian cottages in Pleasant Terrace. It is dated 1661 and has the initials RPA in a circle about the gnomon root. The gnomon is missing.

WELFORD, Northamptonshire

At Welford, near the northern boundary of the county, a Millennium sundial in the parish churchyard is combined with a toposcope indicating the direction to neighbouring village churches. It was presented to the parish by the Welford Women's Institute to mark the year 2000 and has the following inscription around the centre of the dial plate:

All time is no time when time is past.
Make time, save time, while time lasts.

142

NORTH WEST

Alan Smith & Robert Sylvester

Cheshire, Cumbria, Isle of Man, Lancashire

The North West area of England, with its current population approximately 10% of that of the whole country, is probably best known for its tremendous industrial development from the 18th century onwards. At first water-power, then coal, steam engines, ship building, the development of heavy engineering, textiles, chemicals, to be found in the rest of the country, but it is its industrial past which tends to dominate outsiders' perceptions of the region.

In terms of communication the North West has scored several 'firsts'. The first navigable canal cut in England was the Sankey Canal in 1757,

Modern man looking back over the historic Bewcastle landscape. A mural by Kate Norris ^{RS}

paper making, glass making and other industries – all contributed to the densely populated region embracing South Lancashire, Greater Manchester, Merseyside and parts of north Cheshire. The North West contains rural areas dominated by agriculture, and stretches of wild moorland too, all comparable for beauty and tranquillity to any

followed shortly by the Duke of Bridgewater's canal to carry coal from his Worsley mines to Manchester (with underground canals to the coal faces) in 1759. Later, in the early 1890s, the Manchester Ship Canal was built to bring ocean-going liners thirty miles inland to the Manchester Docks from the Mersey estuary. The first public

143

railway, in a modern sense, was the Manchester Liverpool Railway, opened in 1830, and the first stretch of modern motorway, built over fifty years ago, was made to by-pass the Lancashire town of Preston.

Amongst the many smaller industrial towns such as St Helens, Wigan, Bolton, Burnley, Accrington, Preston, Oldham, Rochdale, Stockport and many others, one would not automatically think of seeing sundials, for these would seem to be somewhat anachronistic in an industrial area and more fitting in the rural landscape. However several modern dials from the North West are illustrated, all created during recent years.

Further north is Cumbria, perhaps not the most accessible part of Britain. It contrasts the sparsely populated central area with peripheral towns. What we call Cumbria consists of the former counties of Cumberland and Westmorland as well as what was previously known as 'Lancashire-over-the-Sands' (stretching as far north as Hawkshead) and just a tiny piece of what was Yorkshire. Everyone knows about the beauties of the Lake District but what an opportunity

The Saxon Dial on the Bewcastle Cross

a visit is to combine the enjoyment of them with a spot of dial hunting!

Most people fail to realise that there is far more to Cumbria than the National Park. The Furness and the Cartmel Peninsulas lie in the south and their relative isolation give the impression of a little bit of England which time has passed by. The west coastal area has its own charms, with views to the Isle of Man and occasionally, as far as Ireland. These give way as we move northwards to the Solway Plain and vistas of southern Scotland. A little explored area is the Eden valley with its charming red sandstone villages and a wealth of folklore.

Cumbria can boast Britain's earliest dial at Bewcastle as well as modern ones such as the slate dial at Holker Hall, proving that the county can hold its own in both the ancient and the modern. Although not outstanding in architectural gems, the visitor to Cumbria can discover sundials in unlikely places and therein lies the charm of the subject.

BEWCASTLE, Cumbria

A superbly carved Anglian Cross lies in the churchyard at Bewcastle, an isolated hamlet 25 miles north-east of Carlisle and within a few

The Bewcastle Cross, with
Britain's earliest known sundial

miles of the Scottish border. This very remarkable freestone monument shows a distinctly Mediterranean influence in the carving of 'rich fruit'.

Each of the four sides of the monument is profusely sculptured but the west side is the most interesting to the archaeologist, as it contains a long inscription in runic characters, the interpretation of which reveals the origin of the column. It is thus read (substituting Roman letters for the Runic characters):

+ THISSIG BEACN THUN SETTON HWAETRED WALTHGAR ALWF- WOLTHU AFT ALCFRITHU LAN KYNIING EAC OSWIUING. + GEBID HEO SINNA SAWHULA

(This slender pillar Hwætred, Wæthgar,
and Alwfwold set up in memory of
Alcfrid, a king and son of Oswy.
Pray for them, their sins, their souls)
It is therefore considered that the Cross was erected by three men in memory of a fourth.

On the south side between the compartments are traces of four inscriptions, which read thus (again substituting Roman letters for the Runic characters):

FRUMAN GEAR ECGFRITHU RICES THÆS KYN- INGES

(In the first year of the reign of
Egfrid, king of this kingdom)
(the kingdom being Northumbria). If this reading is correct it indicates most probably the year of the erection of the monument. Egfrid was a son of Oswy and brother of Alcfrid, mentioned in the inscription on the west side, and ascended the throne of Northumbria in the year 670.

At the time of its construction, masons from Mediterranean lands were working in Jarrow and Monkwearmouth and Bede had yet to write his histories at Jarrow. During the Middle Ages there were reports of the Cross having a greasy feeling and as late as 1685, Bishop Nicholson described it as being *'washed over....with white, oily cement'*. It is now considered that this was the remains of an early undercoat. When constructed it would likely as not have been painted, as formerly there was not the aversion to painting statuary and monuments that there is today. The cross head is missing and was said to have been appropriated by the Elizabethan historian Camden for study in London. It has never been seen since - vandalism in the name of scholarship! Although the gnomon on this Saxon Dial has gone missing with the passage of time, it would probably have been a horizontal peg. The dial itself is about ten feet from the ground and it shows the

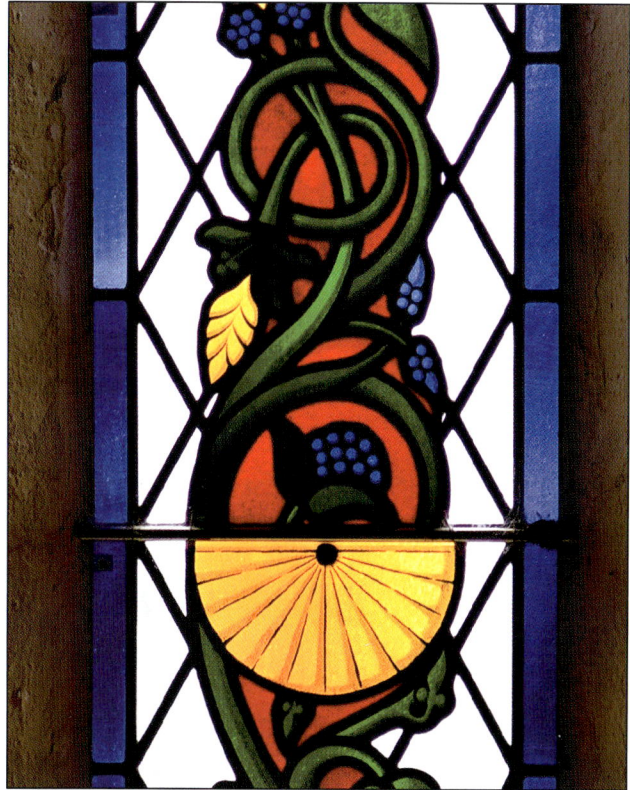

Inside Bewcastle Church the Cross is detailed in a modern stained glass window

old tidal system of time keeping. A memory of this persists in words still in use such as 'noontide' and 'eventide'. In well over 1,000 years of exposure, it is now a shadow of its former self but is acclaimed as being the oldest sundial in Britain.

Unfortunately it is in a poor condition having been defaced by the addition of spurious lines and water erosion.

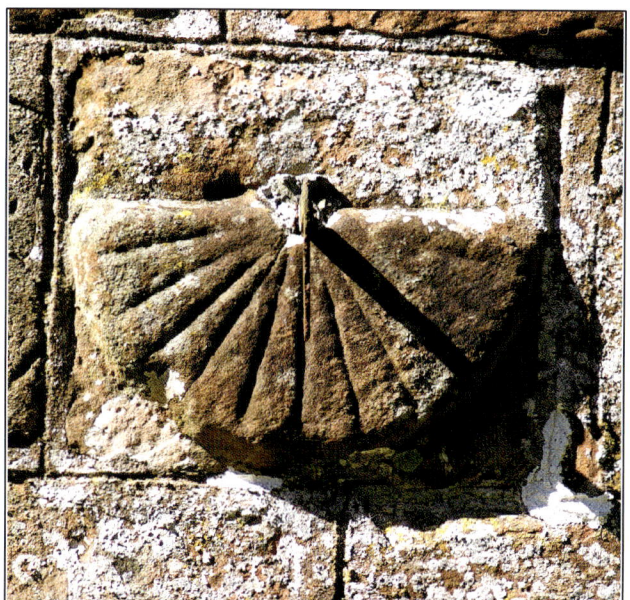

Early Dial at Newbiggin Church near Temple Sowerby (details on page 146)

145

Interestingly, a dial with an 'L' shaped gnomon but probably medieval is at the church at Newbiggin near Temple Sowerby, just east of Penrith. It is high on a buttress dated to the 12th century. The Revd W S Calverley writing in the 19th century comments upon the similarity in shape, placement and size to that at Bewcastle. There is also a conventional Horizontal Dial nearby, possibly by the Porthouse family of Penrith.

BOLLINGTON, Cheshire

Polar Dials are not very common but a fine example is situated at the Arts Centre Garden at Bollington in Cheshire. It was designed by Christopher Daniel, who conceived it as a memorial to the British Sundial Society's founder and first Chairman Dr Andrew Somerville.

Andrew was a valued supporter of the Centre, singing in the Bollington Festival Choir, a founder member of Bollington Chamber Concerts and the dial was commisioned by Dr John Coope on

The Dial in Memory of Richard Towneley[AS]

Polar Dial in the Arts Centre Garden at Bollington, Cheshire

behalf of the Arts Centre. Andrew's widow Anne continued to support and work for the British Sundial Society and Bollington Arts Centre.

BURNLEY, Lancashire

Towneley Hall, now an art gallery and museum, was the ancestral home of the Towneley family which produced three men of great distinction. The first, Christopher Towneley (1604-1674) had a fine scientific library and was patron to Jeremiah Horrocks (see Much Hoole) and other early English astronomers; the second was Richard Towneley (1629-1707) Christopher's nephew, who was a meteorologist and astronomer and who, amongst other things, designed the escapements for the pair of astronomical clocks made by Thomas Tompion for the newly built Royal Observatory at Greenwich in 1676; the third was Charles Towneley (1737-1805) a classical scholar whose collection of ancient Greek and Roman sculptures is now in the British Museum. The Vertical Dial at Towneley Hall was designed and made in 2001 by Alan Smith as a memorial to the outstanding life of Richard. On it can be seen a

The Market Cross at Carlisle

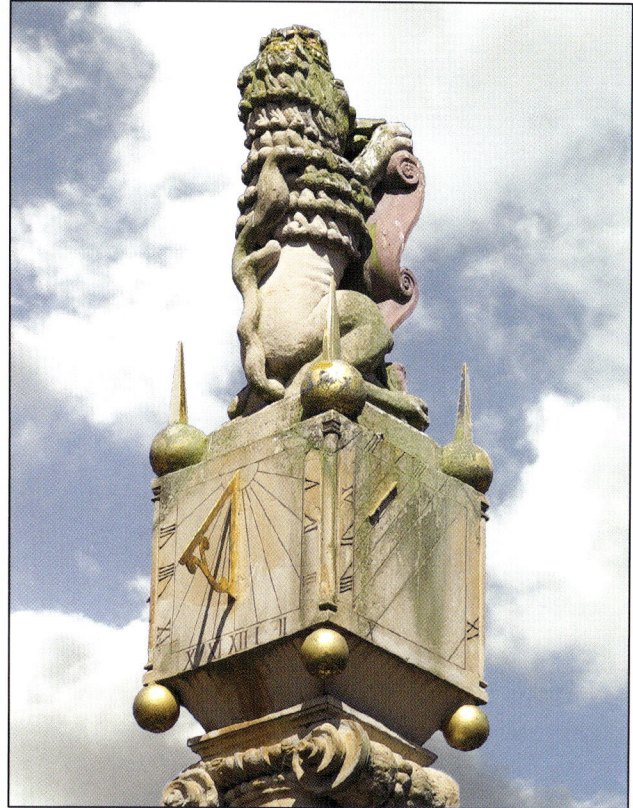

Cube Dial on Carlisle Market Cross

representation of the clock escapement mentioned above (top left), the Towneley coat of arms (centre), **MM** – for the Millennium (top right) and the noon line elongated into a cross, symbolising the Catholic faith to which the family belonged. The clocks at Greenwich were used by the first Astronomer Royal, John Flamsteed (1646-1719) for measuring the daily rotation of the Earth, for determining the Equation of Time and to assist in the creation of his astronomical atlas.

CARLISLE, Cumbria

The Carlisle market cross features a four sided sundial surmounted by a lion holding the old coat of arms of Carlisle, and sitting upon a book. The exact date of the erection of the cross is not known for certain, but an inscription on one side reading 'Joseph Reed, Mayor, 1682' seems most likely.

This structure, which probably replaced an older cross, stands in the centre of Carlisle and marks the site of the traditional market place. In an age before clocks or watches were common, the sundial (Cumbrian weather permitting) would have shown the market opening time - an important

function when opening a stall early (known as forestalling) was against city by-laws.

The market cross faces the old town hall of Carlisle, itself rebuilt in the 1680s on the site of a medieval predecessor. In both cases, the rebuilding represents the council investing in the city after centuries of border warfare and lawlessness. In 1717 the town hall was enlarged, including a clock tower to augment the market cross sundial.

As well as marking the rebuilding of Carlisle, the cross is also interesting as a political statement. At the time of its construction, King Charles II was making great efforts to exert royal control over what he saw as dangerously independent towns such as Carlisle. Two new charters were issued to the city, giving the King considerable powers over the council. With this in mind, it would seem that supporters of the King had a hand in designing the new market cross and included very provocative political symbols. The lion on the cross is not just decoration, but represents royal authority. It is shown very deliberately facing the town hall, its paws holding the city coat of arms and sitting upon the Carlisle Dormont Book - which contained the city's by-laws. These carefully chosen symbols represent not so much royal authority invested in the city as royal control *over* it.

147

Colne with its dial high above the porch ^{AS}

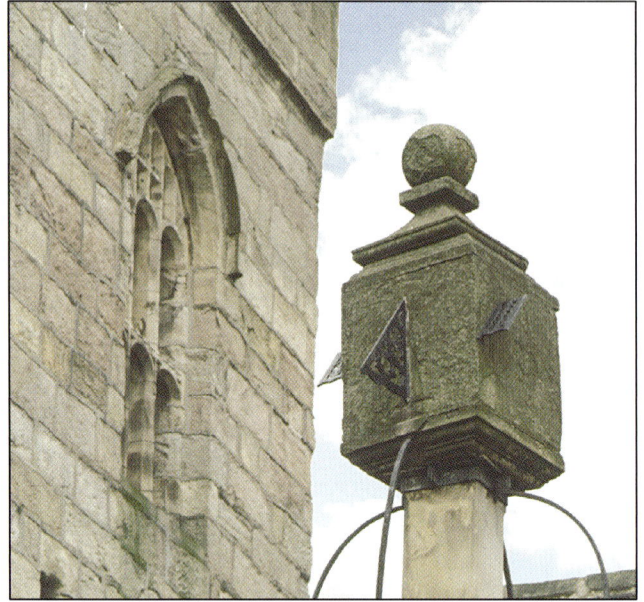

Colne Cube Dial ^{AS}

COLNE, Lancashire

The parish church of St Bartholomew is mostly 15th century in date, standing in the long main street of the town, a place formerly devoted to textiles. The south porch itself is probably of 14th century date but its cubic sundial on the apex is likely to be an 18th century addition. It has gnomons facing south, east and west, but the hour lines have now weathered away. Its considerable weight is supported on a square plinth, strengthened against the wind with four graceful brackets, and above is a moulded canopy and ball, also much weathered.

CULCHETH, Warrington

The district of Culcheth has set up a dial to commemorate the year 2000 in front of a row of shops and car park, aptly named 'The Millennium Garden'. Perhaps its unique feature is a clock set towards the top of its 15 foot high gnomon. The whole gnomon is made of cast iron and steel as are the hour markers spaced round the outer edge of the pavement. Much of the relief decoration on these features is based on wheels and ratchets to be found in clockwork, reminding us that Culcheth is close to Warrington where the famous Peter Stubs set up his factory for making clock tools and parts in the late 18th century. The whole area of SW Lancashire in fact has its roots in clock and watch-making going back as far as the 17th century. The

The Millennium Garden Dial at Culcheth ^{AS}

The hour markers at Culcheth ^{AS}

dial was erected by the Council of Culcheth and Glazebury in the County of Merseyside.

DOUGLAS, Isle of Man

Now situated in the Manx Museum at Douglas, this elaborate multiple faced sundial of Poolivaish limestone was made at Ballafreer, Marown by John Kewley in 1774 for Sir George Moore's estate at Ballamoore. The locations marked are for Peking in China, Boston in America, Port Royal in Jamaica and also Jerusalem as well as for the Isle of Man. Sir George had trading interests in Jamaica which is probably why Port Royal is

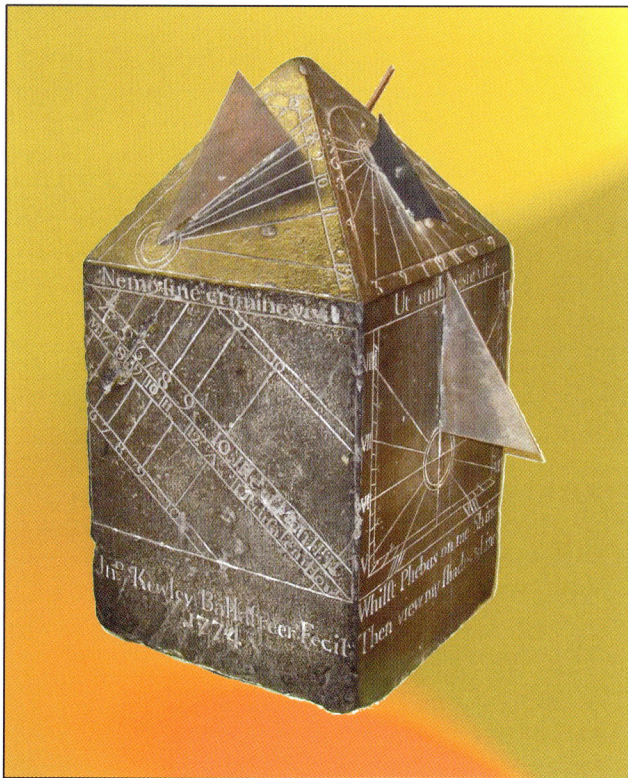

Polyhedral Sundial from Ballafreer in the Manx Museum, Douglas ^{CMC}

shown. The mottoes are in Manx, Latin and English.

EDGWORTH, Blackburn with Darwen

Edgworth is a small community dating back to the 13th century, about five miles north of Bolton. It stands in cultivated countryside, on the edge of moorland, and is characterised by having several 'Folds', the 17th century name for farmsteads in Lancashire. In 1909 Sir Thomas Barlow, a native of Edgworth, with his brothers and sisters financed the building of the Barlow Memorial Hall in memory of their parents. Sir Thomas, who became physician to Queen Victoria and Kings Edward VII and George V, died at the age of 99, and the hall bearing his name still serves today as a meeting place for the flourishing village community. On the south side of the building is a wooden Vertical Dial which probably dates from about 1910, cleated to the wall and carved and painted, but now in considerable need of restoration. Whoever designed it clearly understood dialling because its small declination to the west has been recognised and the latitude 53° 33' 39" and longitude 2° 23' 33" have been accurately inserted.

Edgworth's Barlow Memorial Hall Dial ^{AS}

NORTH WEST

149

Gosforth, Cumbria

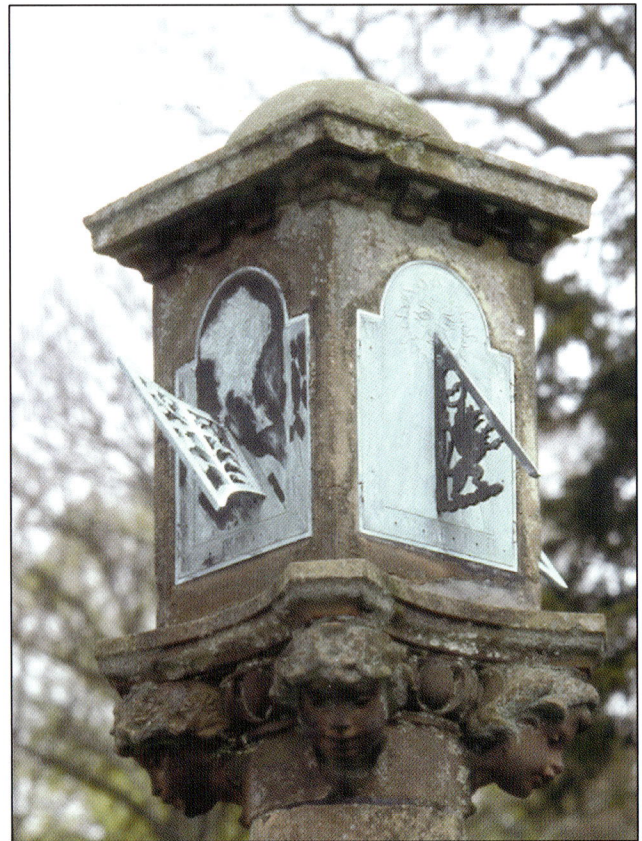

Cube Dial at Graythwaite Hall

GOSFORTH, Cumbria

Motorists who travel along the busy A595 up Cumbria's west coast will pass Sally Hill farmhouse that sits adjacent to the road, just north of Gosforth at New Mill. The east declining Vertical Dial is above a door and is a delicate piece of work dated **1783**.

The dial bears the Latin motto:

Atas Rapuit Diem
(Time will hurry away the day)

The dial declines east of south, so it will pick up the morning hours from about 4.30am in summer until nearly 4.00pm. It has the hours, halves and quarters marked and the half hours are indicated by fleur-de-lis motifs on the end of the half hour lines.

There is another east declining Vertical Dial in Cumbria, at Ratten Row near Dalston.

GRAYTHWAITE, Cumbria

Graythwaite Hall has parts that date from the 16th century. It is the residence of the Sandys family and lies to the west of Windermere. The Dutch Garden has the pillar dial as a central feature. At Graythwaite, only the gardens are open to the public and they still retain much of the workmanship of Thomas Mawson from the early 1890s. Dan Gibson, who also made the handsome iron gates, made these dials at the beginning of the 20th century. To a quick glance, the three dials on bronze plates appear to be facing east, south and west, but on closer inspection, the dials decline from cardinal points of the compass. The obvious reason is that the pillar has been made to sit squarely in the garden and the Dutch Garden does not align exactly north-south. To make these dials, Dan Gibson has shown a sound understanding of the science of dialling.

There are also interesting aesthetic features such as a grimacing detailed sun at the gnomon root of the south facing dial. There is similar ornamentation on the east and west dials but it is tilted to the line slope. The north face is not marked as a dial but bears Sandys family Arms above the date in Roman numerals. The gnomons are elaborate. A lion rampant (as on the Arms) is cut out of the south gnomon and the east and west gnomons have cross with foliage sprouts, one above the crossbar on each.

In the adjacent Rose Garden at Graythwaite sits a fine Horizontal Dial.

GREAT CROSTHWAITE, Cumbria

The Church of St Kentigern, just north-west of Keswick, is the final resting place of the poet Robert Southey (1774 – 1843) as well as Canon Rawnsley (1851 – 1920) vicar, reformer and one of the founders of the National Trust.

The south facing east declining Vertical Dial at this church is situated high above the south clerestory. It is inscribed with the date **1602** and bears the letters **TIH** in monogram form. One suspects that this is the name of the maker but

150

Vertical Slate Dial at Great Crosthwaite, dated 1602

that **TIH** means something more profound. In Greek, these letters symbolise the crucifixion: 'T' has become the cross and '**I H**' the name of Christ. We may be wrong, but it is certainly intriguing and the study of sundials leads us up many roads.

While at the church, look at the south-west buttress in glancing light. Now almost effaced is what is said to be a scratch dial bearing Roman numerals and a cross patty for noon. This large dial, scored into the stonework, is not a scratch dial by the normal understanding of the term, but it is certainly an unusual time indicator of a bygone age.

HAWKSHEAD, Cumbria

The sundial above the front door of the old Grammar School at Hawkshead (now a museum) is of slate and would originally have been easier to see before a canopy was added to the frontage during restoration in the 1890s. Although the inscription was badly weathered by the 1990s,

care must always be taken in assuming anything from what is written on a sundial as churchwardens or donors sometimes have their names carved. In this case, it is tempting to speculate

only a small part of the Equation of Time table was indecipherable and this was reconstructed with information supplied by The Royal Greenwich Observatory. The sundial dates from 1845 but may replace an earlier one. It is highly possible that it was constructed with some input by the boys at the school as the Grammar School had a long tradition of mathematics, astronomy and the study of globes. The astronomer William Pearson attended Hawkshead Grammar School and some of the pupils' calculations are still in the school and are available for inspection upon request.

Regarding the declination of the dial that is accurately stated at its head, it would seem to have been made to encompass a specific hour range. When the young William Wordsworth attended this school in the early 1780s, he had to be at his school desk at 6am. This practice of early starting continued for some time into the

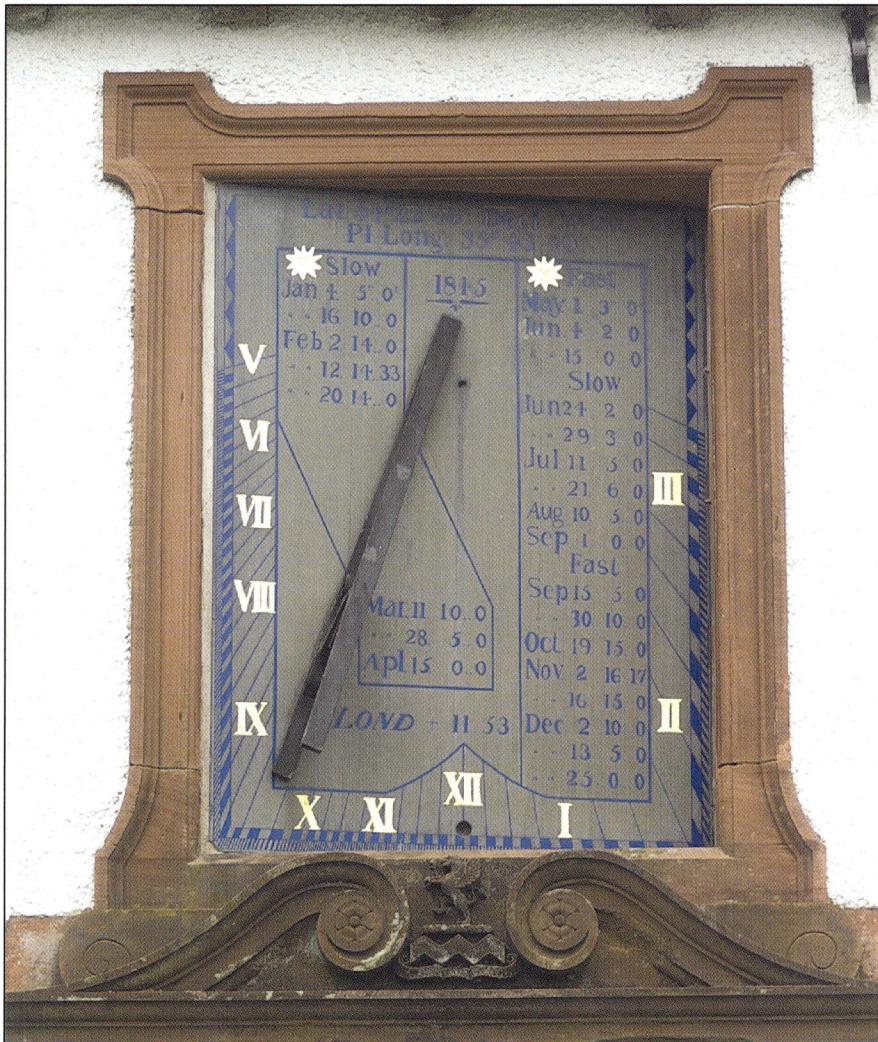

Vertical Dial at Hawkshead Grammar School

The door to Hawkshead Grammar School RS

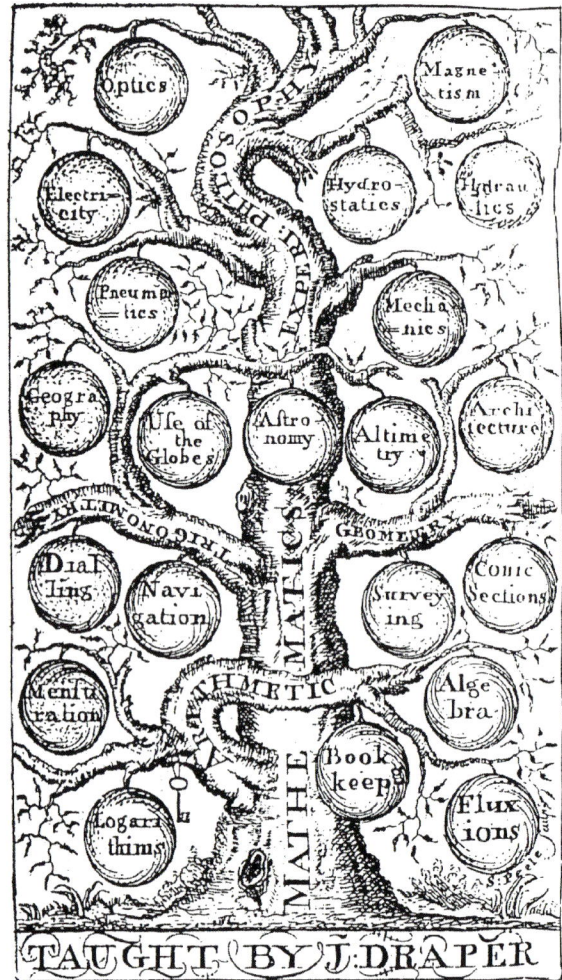

The 'Mathematical Tree' by John Draper snr. of
Whitehaven, the sort of text that includes Dialling
and would have been in use at Hawkshead.
From: *'The Young Students' pocket companion
or Arithmetic Geometry, Trigonometry
and Mensuration'* of 1772.

19th century. By declining the dial some 30° from the line of the building frontage (which faces not quite due south) maybe the intention was for the dial, at least in the summer half of the year, to catch the hours most suitable to fit in with the school day. There is a mystery regarding the second line from the top which reads:

PI Long 35° 43' 40"

Reading eastwards on a globe this could be Jerusalem because if taken towards the west it is in the Atlantic Ocean! It could be speculated that westwards would indicate the Azores and that eastwards would indicate Pulkovo near St Petersburg, both regarded at the time as significant meridians, but the co-ordinates do not quite correspond.

The sundial was restored in 1997.

HOLKER, Cumbria

Sitting in a paddock near Holker Hall is one of the largest single piece slate sundials in the world. Designed by British Sundial Society member Sir Mark Lennox-Boyd, it is a Hemispherium based upon the work of the astronomer Berossus in the 3rd century BC. It was produced at Grizebeck at the Burlington Slate Quarries that belong to Lord

Hugh Cavendish, owner of the Hall. The sundial is laid out on Greek principles and has a vertical gnomon of phosphor bronze, which traces out throughout the year the shadow of its tip on the hour lines. Running across the hour lines are

Large Scaphe Dial at Holker Hall RS

152

Overhead view of the Holker Dial

lines indicating which sign of the zodiac the sun is occupying at the time, giving an indication of the solar declination throughout the year.

In essence, the dial is a development of the Horizontal Dial. Instead of a polar aligned gnomon, the gnomon nodus is held aloft on a vertical rod. Were it not for the curvature of the bowl, it would not be possible to indicate the time before 6am and after 6pm as the glancing rays of the sun would project the shadow of the gnomon tip clean off the face of the dial.

The dial was formed from a block of blue-grey slate and the dimensions of the dial are 1,550mm diameter and 380mm deep. The task of carving the detail, including the elaborate Equation of Time table, was the responsibility of Burlington Slate's Master Stone Mason, David Allonby.

ISEL, Cumbria

The isolated Church of St Michael lies to the north east of Cockermouth and is situated beside the river Derwent. Here is a set of Scratch or Mass Dials dating from the Mediaeval period and acting as event indicators. They are crude and

Pin gnomon in Holker Hall Hemispherium

Mass Dials at Isel Church

The Isel Mass Dials as sketched by Revd W S Calverley

153

were re-inscribed by successive priests, giving not time as we know it but the times for the people to attend mass.

Mrs Gatty states that Scratch Dials appear to have been cut and then set in place. Current thinking goes against this, the feeling being that they have been cut in situ. The south chancel window dials at Isel probably have been cut in isolation as there is a definite asymmetry about the jamb of the south chancel window where the dials are cut, as opposed to the right hand side upright of the window.

A fourth Scratch Dial at Isel is next to the church door and is now enclosed by the later-built porch. These dials, as well as others in Cumbria were studied in the 19th century by Revd William S Calverley who served as the incumbent at Dearham and Aspatria. He sketched them before the depredations of acid rain made them a shadow of their former selves. Other Scratch Dials in Cumbria can be seen by the perceptive eye at Beetham, Caldbeck, Cliburn, Dearham, Great Salkeld, Kirkoswald, Milburn, Newton Arlosh, Urswick and Warcop.

Tabley House, Cheshire

KNUTSFORD, Cheshire

Tabley House is one of the finest Georgian houses in Cheshire, designed by John Carr of York and built between 1761 and 1767 to replace an older house dating from the medieval period. On the south side, in the middle of the spacious lawn, is a Horizontal Dial on a plinth which is built from the top layers of stone from a Greek Doric column. With its fluted shaft, grooved fillets, curved echinus and square abacus on the top, it is a perfect example of the Doric order of architecture and the whole is mounted on two circular steps. This pedestal is the real centre of interest rather than the dial itself, for this appears very small in proportion to its support. It would appear likely

The new Liverpool Museum Dial

that this pedestal was originally part of a complete Doric column, dismantled from another building, but where it came from is not known.

LIVERPOOL

A new dial has been made for the south-facing frontage of Liverpool Museum this year, 2005. Because it is sited in a recess in the museum wall a special framework has been constructed to bring it level with the surrounding wall surface, so that it appears to 'float' in the recess aperture. The dial is made of Welsh slate, designed, carved and gilded by Alan Smith as a memorial to a past Curator of Physical Sciences at the museum. The fret design round the edge is derived from the border pattern of a 7th century Anglo-Saxon brooch which was excavated in Kent in the 18th century and known as the Kingston Brooch. This is one of the most treasured historical artefacts in the museum's collection. The gnomon is made of stainless steel and because the dial does not face due south but declines 3.5° to the west there is a slight 'skew' in the layout of the hour lines. The dial also has solstice and equinox lines added and in front of the museum is a viewing area for the use of school parties and other observers when the sun is shining.

MANCHESTER

UMIST. The University of Manchester Institute of Science and Technology began life originally as the Mechanics' Institute, later the Municipal College of Technology. It was founded by (amongst others) the famous engineer Sir Joseph Whitworth (1812-1887) who was also a great benefactor of institutions in the city. Later still the college combined with the Victoria University of Manchester to form what is today one of the largest university campuses in the country. The Vertical Dial we see here replaced an older one

Dial at UMIST, Manchester AS

Horizontal Dial at Manchester Cathedral AS

which was made to commemorate the 150th anniversary of the college foundation, and the dates can be seen on the dial **MDCCCXXIV – MCM-LXXIV** (1824-1974) along with the university coat of arms. The present dial is on the modern Renold building and features lines of declination (the height of the Sun throughout the year). Each zodiacal band is coloured differently (cold colours in the winter and warm ones in summer) and because the wall declines to the east the dial appears skewed. Unfortunately the dial has had to be mounted rather high to avoid shadows from other buildings making it difficult to read.

About the time that the UMIST dial was made (1997) another and more modern dial was designed to fit on the wall of Grosvenor House, one of the students' hostels. It is 'modern' in the sense that it is made of unconventional dial materials (tubular steel) against a plain brick wall surface and unconventional too in that its form is simplified into that of a 'Suntime Bird', which is its title on the descriptive panel below. Designed by Christopher Rose-Innes and calibrated by Alan

Smith the lower edge of the tail and upper edge of the wing show, by the tip of the gnomon's shadow, the date of the summer and winter solstices respectively, while at the equinoxes when the sun is overhead at the equator the shadow follows the straight line forming the lower edge of the wing.

Manchester Cathedral. A rather fine Horizontal Dial is to be found on a pedestal on the south side of the Cathedral. Unfortunately, since the 18th century when it was first installed, other buildings have been erected which put the dial almost permanently in shadow. During the summer months it will get some sunlight in the evening, but not for the rest of the day. The baluster shape of the stone pedestal is very satisfying, but the dial plate and gnomon are now in a sad state of decay, and it is not possible to decipher the name and markings on it. The decorative scrolls of the gnomon are mostly broken, but the base has two unusual semi-circular flanges to give it stability. Close to the dial, above the door of the Choir School, is a sculpture of St Mary, St Denis and St George by the famous sculptor Eric Gill, carved in 1933.

Liverpool Road Station. Now incorporated into the Manchester National Museum of Science and Industry, it is thought to be the oldest surviving passenger railway station in the world. It first

Grosvenor House Students Hostel AS

155

NORTH WEST

came into use at the grand opening on 15 September 1830 as the Manchester terminus of the Liverpool-Manchester Railway, which was engineered by George Stephenson at a cost of £800,000. On that day eight trains left Liverpool for Manchester carrying about 800 people, and the day was auspicious for two reasons. It was the day when William Huskisson (1790-1830) British statesman and pioneer of free trade, was killed on the line while dismounting, against all advice, to meet the Duke of Wellington with whom he had been having some disagreement, and was hit by a train on the opposite line. This sad event is chronicled in great detail in Fanny Kemble's 'Record of a Girlhood', and naturally it cast a serious damper on the day's celebrations. Also during the formal opening ceremony in Manchester, the Duke of Wellington, a distinguished guest, remained in his carriage during the luncheon, while outside the poverty-stricken mechanics and artisans of the city staged a demonstration with hisses and shouts of 'no corn laws' and 'vote by ballot'. Today the station remains virtually unaltered and has been restored after many years of decay and neglect. Outside the first-class waiting room window, on a pedestal visible from the road, is the sundial, available for the use of passengers looking through the window while waiting for their train. The original dial and its gnomon may be seen in the museum inside the building, but the one outside and the pedestal are modern reproductions. It is interesting that just as the distribution of a time standard was to come into being with the invention of the electric telegraph, that a sundial should still have been used by the newly-emerging railway system. It was primarily with the coming of railways that a need was desperately felt for the national use of 'mean time', since 'local time' as provided by sundials made efficient time-tabling between cities and towns virtually impossible because of the necessary corrections for longitude. Such, however, was the intransigence of many places throughout the country that mean time was not universally adopted until the 1870s.

The Railway Dial at Manchester's Liverpool Road Station

MARTHALL, Cheshire

The sundial on an old farmhouse at Marthall, about three miles from Knutsford on the A537, came about in a rather curious manner. While renovating the old building the present owners came across a large, thick, rectangular piece of slate which had formerly been a working surface in the dairy and they got the idea that it would make a good base for a sundial for the slightly west declining wall of the house. Graham Aldred of the British Sundial Society was consulted, and after many visits to determine the exact declination of the old brick wall he calibrated and drew out the hour lines and lines of declination of the sun and made a suitable gnomon to be bolted on.

Vertical Dial on an old farmhouse at Marthall

156

At this point he invited Alan Smith to carve the lines, numerals and other details on the slate surface, and then to gild it. The great weight of the dial is supported on brick corbels, and it can be seen from the main road but only obliquely and not straight on. As well as the solar information the dial has its date, **1999**, and owners' initials at the top, and the designer's and maker's initials below.

MUCH HOOLE, Lancashire

For two reasons the dial on the parish church at Much Hoole is included in this survey. The first is because of its unusual size, being nearly 16 feet in diameter, and the second because of its association with a church where England's first great astronomer, Jeremiah Horrocks (1619-1641) officiated. At one time it was thought that he was a curate, but whatever his role it was from Carr House, in the village, that he witnessed the transit of Venus across the sun in 1639, the first man ever to do so. The astonishing fact is that Horrocks, a poor man with limited scientific equipment of his own making, was able to predict that this astronomical event would occur. He died at the age of 22 having established principles of astronomical observation which formed the bedrock of the work of later astronomers and he collaborated with William Crabtree (1610-1644) of Broughton, near Salford (now part of Greater Manchester). Eulogies to Horrocks later came from Sir Isaac Newton and other leading astronomers of his day. In the church at Much Hoole a memorial plaque and two stained glass windows are dedicated to him and also a plaque in Westminster Abbey. The sundial on the church tower is dated 1875, but Eden & Lloyd date it as 1815, so perhaps it was repainted.

The large south-facing dial has the motto:

SINE SOLE SILEO
(Without the sun I keep silence)

and it appears to be painted on a cement or plaster rendering on the sandstone tower.

NEAR SAWREY, Cumbria

On a window ledge inside Hill Top, the home of Beatrix Potter, children's writer and illustrator, is a small Inclining Dial. The plate is hinged and

Large Vertical Dial on the tower at Much Hoole
AS

can be inclined to compensate for operation at different latitudes. Its maker is Gabriel Davis of Leeds who made it about 1850. The dial plate is of brass, mounted on a turned and polished wooden plinth that is attached to a square base. It is in excellent condition apart from the missing gnomon. The former gnomon's position is indicated by the two alignment holes where the gnomon origin would have been. These occur as is correct at the mid-point of the 6am to 6pm line. The thickness of the gnomon is indicated by the gap at the noon position on the plate. At this time, the sun would strike the gnomon directly instead of glancingly as at other times and the shadows of both edges of the gnomon would indicate noon.

The misleading feature when first seen is the curved graduated tongue that appears to be a wrongly shaped gnomon. It is engraved in individual degrees and marked in Arabic-style numerals at 10° intervals. It covers a latitude range from 60° north to 20° north, from Gothenburg to Timbuktu. The missing gnomon would have had a curved slit running deep into it to accommodate this graduated tongue. A window or a pointer would have been present on the gnomon to allow setting of the scale. The spacing of the hour lines is for a horizontally set plate of 60°. The hour range marked is adequate for its use in higher latitudes in mid-summer but the back-hours (those before 6am and after 6pm) will become superfluous at lower latitudes where day lengths vary little throughout the year, regardless of the season.

Inclining Dial at Near Sawrey RS

This is an unusual sundial. It does not conform to a classical portable sundial design as it has a plinth and no inbuilt magnetic compass for alignment. It is best regarded as a fascinating conversation piece.

NELSON, Lancashire

The Polyhedral Dial (many sided) in Marsden Park is a most unusual type of dial for this part of the country, being more commonly associated with Scottish dial designs. It has twenty separate dials, each with its own gnomon, and several show the time in different cities. It will be noticed that all the gnomons have their shadow edges (the style) aligned in exactly the same direction (i.e. pointing to the celestial North Pole) and those on the north side will only register shadows during the summer months. With the surrounding vegetation it is unlikely that the dials underneath would now ever receive sunlight. Several of the dial markings became difficult to read because of corrosion, and by the 1970s the dial was in a poor condition. However it was restored

Polyhedral Sundial at Marsden Park, Nelson PS

by Roger Macaulay Lord in 1986 and the gnomons appear to be replacements of that time. The whole dial is mounted on four elegantly shaped stone brackets on a circular stone base of two steps, with a central iron bar to give extra support. Marsden Park contains many interesting architectural features created in the 19th century by R T Walton, a descendant of the Walton family that owned Marsden Hall from the 16th century. The town of Nelson was formerly Great and Little Marsden and was re-named after Lord Nelson

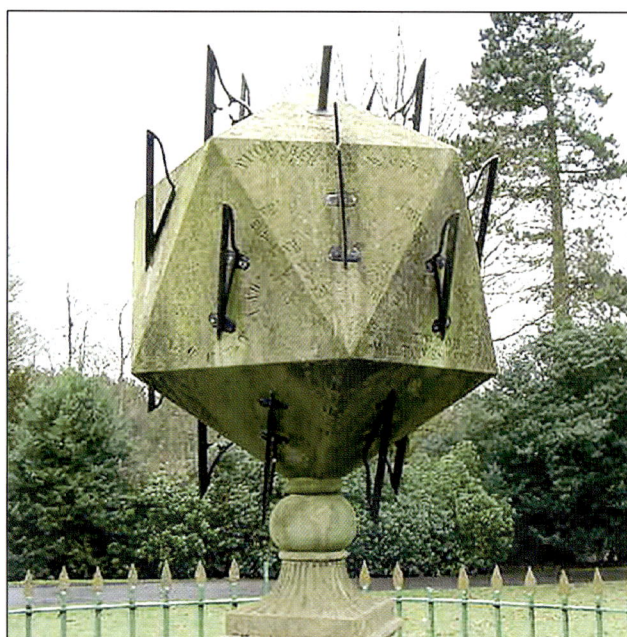

Polyhedral Sundial at Nelson AS

early in the 19th century. The park is now the property of Pendle Borough Council.

PAPCASTLE, Cumbria

At the Grove and the adjacent Grove Barn in this village outside Cockermouth are two beautiful 18th century gold leaf on slate Vertical Dials. The one at the Grove is a great decliner, so much so that it is almost direct west facing.

It is signed **J · FALDER F**[T] (Falder is a fairly common Cumbrian name) about whom nothing is known.

In a great decliner, the gnomon appears to be lifted away from the face of the dial. This is because only a small part of the possible dial has been used and the gnomon origin and the convergence of the hour lines could be several metres away from the building.

The listed sundial at the Grove Barn is a south facing Vertical Dial. It has been re-erected adjacent to an archway in recent years. Obviously by the same hand, it is puzzling. It declines slightly, evidenced by the asymmetry of the hour lines, yet the close parallel lines inscribed to accommodate the gnomon width run vertically close to the noon position. It is difficult to imagine anyone with the skill to delineate the first sundial finding difficulty in making the gnomonically much simpler second one.

Both dials need re-alignment if they are to function correctly.

PENRITH, Cumbria

Sundials can not only be tellers of life's passage, they can also be set up as memorials or function as focal points in a community. An example of the latter was the old sundial at Cartmel Fell where the Parish Clerk would read out news and announcements. An example of the former is the striking pillar dial known as the Countess' Pillar, set beside the busy A66 near Brougham, just east of Penrith.

This monument was set up to commemorate the last parting with her mother in 1616 of Lady Anne Clifford, Countess of Pembroke (1590 – 1676). On her arrival in the north, Anne started at once to rebuild or repair six of her ancient castles: Appleby, Brougham, Brough and Pendragon in Westmorland; Skipton Castle and the tower of Barden in Yorkshire. 'Her passion,' writes the author of her biography in the National Dictionary, 'for bricks and mortar was immense.' She restored no less than seven churches or chapels -

The two slate Dials at Papcastle

The Countess Pillar near Penrith

the churches of Skipton, Appleby and Bongate and the chapels of Brougham, Nine Kirks, Maller-stang and Barden. She founded the almshouses that we can see today at Appleby, and restored the one which had been built and endowed by her mother at Bethmesley. It was her custom to reside at fixed times at each of her six castles. '*She lived in vast hospitality,*' writes Pennant, '*at all her castles by turns, on the motive of dispensing her charity in rotation among the poor of her estates.*' Sedgwick records that she continued '*a year or two in Yorkshire and a year or two in Westmorland, to the great benefit of both counties, expending the rents that she had in these counties.*' Her journeys from one place to another were like royal progresses; she travelled in a horse-litter, and often took new and bad roads from castle to castle in order to find a reason for spending money among the indigent by employing them in the repairs.

The Countess' Pillar stands about 14 feet high on a hillock and is an octagonal shaft with square facings above; the sides of the square portion bear east, south and west facing sundials and two shields of arms (Clifford impaling Vipont, and Clifford impaling Russell) and the following inscription:

This pillar was erected Anno 1656 by the Hon. Anne, Countess Dowager of Pembroke, and daughter and sole heir of the Rt. Hon. George, Earl of Cumberland, and for a memorial of her last parting in this place with her good and pious mother, the Rt. Hon. Margaret, Countess Dowager of Cumberland, the 2nd. of April, 1616. In memory whereof she also left an annuity of four pounds to be distributed to the poor within the parish of Brougham every 2nd. day of April for ever, upon the stone table here hard by.

ST HELENS

There is evidence that a Vertical Dial was originally planned for the Hardshaw Friends' Meeting House as early as 1691, and that a stone slab was provided for it in March, 1692. The designer was Alexander Chorley and he may have actually calibrated the dial at that time. The present date 1753, however, appears on the dial and this was probably added during one of the re-paintings during the first sixty years of its life. By 1995 it had degenerated to a miserable condition and was restored by Alan Smith.

The restoration of the dial provoked some controversy at the time. When he started to remove no less than five layers of over-painting the original hour lines and other markers were discovered on the bare stone surface. By modern computation

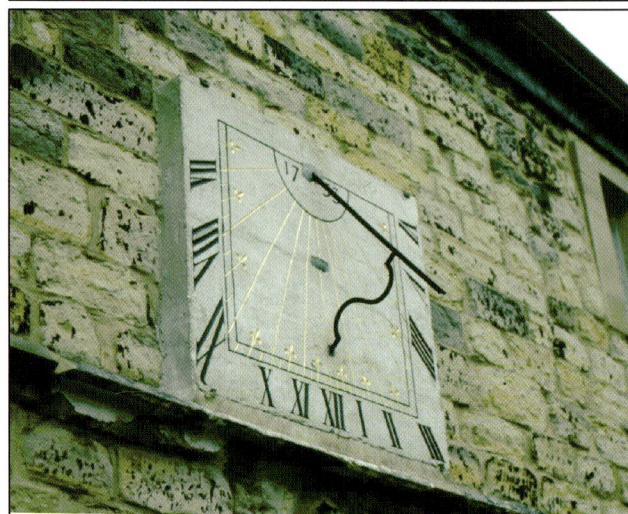

Hardshaw Friends' Meeting House, St Helens [AS]

these were found to be remarkably accurate, having taken into account that there was a small 2° declination of the wall to the west. Also the gnomon was not original and a second base for an original support was discovered beneath the debris. The stone slab itself was badly decayed, especially at the edges, and needed consolidation. What was the best way forward in restoring this ancient dial? After much soul-searching and consultation with the authorities representing the Society of Friends', to make the original lines and numerals *permanent* it was decided to carve them into the stone, with subsequent gilding, and to design a new gnomon to be made in non-corrosive materials. To do it this way would preserve the exact layout of the dial during any future restoration, because past re-paintings had produced errors of up to two or three centimetres compared with Mr Chorley's original design.

Conservationists argued that this would destroy the artefact, but it was felt that the action taken *preserved* the design for all time. In the event the restored dial was found to marry happily with the building's stone facade, and the Quaker members of the Meeting House and the Trustees of the Hardshaw Estates were very pleased with the work and are proud of their dial above the front door.

SEASCALE, Cumbria

Most sundials have a gnomon that slopes at an angle that is the same as the latitude at which it is situated. Any other angle would mean that the passage of the shadow between the hour lines would vary as the seasons progressed. There are some exceptions, notably the Analemmatic Sundial. The gnomon is vertical and the seasonal variation is allowed for by moving the foot of the gnomon along a scale.

An interesting development of this is the Human Analemmatic Sundial where the shadow of the observer acts as the time indicator. This lends itself to an unusual dial where the person using the dial becomes actively involved. The months are marked out on a paving stone and the observer merely places his toes on

Human Analemmatic Dial at Seascale

the appropriate month line. Many Human Analemmatic Sundials are featured in school playgrounds as they teach the pupils the basics of the Sun's daily motion.

This attractive example is photographed from the Fort on the seafront at Seascale and was set up as a Millennium project by Thom States, using plans supplied by a member of the British Sundial Society and the services of local stonemason Mike Gibson. Buried deep under the stones used to mark the hours are stainless steel time capsules filled with memorabilia representative of all aspects of village life in the year 1999.

Note that as well as the normal hours marked in the outer arc, the inner arc is similarly marked but with one hour displacement to accommodate the variation for British Summer Time.

STANDISH, Wigan

The old parish church of Standish, near Wigan, has a vertical stone sundial on its porch which is difficult to date with certainty. To judge from the form of the gnomon and lettering one might consider a date of about 1910. It is very plain with the hour scale arranged in a 'U' form with a long motto beneath, of which the lettering is very much decayed, especially on the right. The motto itself reads:

LET NOT PASSING CLOUD
OF BITTERNESS
THINE ACCUSTOMED SERENITY
OERSHADOW

This is a curious phrase, never before encountered on a sundial.

Vertical Dial at Standish near Wigan

161

Cube Dial at Ormathwaite Farm near Keswick ^{RS}

UNDER SKIDDAW, Cumbria

This Cube Dial is situated on a barn at Ormathwaite Farm just north of Keswick at Under Skiddaw close to Applethwaite. It dates from 1769 and it is in fine condition. This sundial is situated at the apex of a barn roof making the dial very difficult to read. All four faces have sundials delineated. There is a Greek and Latin inscription around the base that reads:

ΡΑΙ ΟΔΕ ΚΑΙΡΟΣ ΟΞΥΣ

CARPE DIEM

ΒΙΟΣ ΒΡΑΧΥΣ Η·····

(Shatter Swift Time)

(Seize the Day)

(Life is Short)

The south dial tells the time from 6am to 6pm while the north dial tells the 'back hours', that is earlier and later times which are only available during the summer half of the year. The east and west dials are complementary dials and give the hours from sunrise to noon and from noon to sunset respectively.

Other Cube Dials in Cumbria are situated at Egremont Castle and in Farlam churchyard near Brampton.

WARRINGTON

Searchers for this dial at Centre Park, once they have negotiated the convoluted roads and heavy traffic of Warrington, will find one that is, to say the least, eccentric. Central Park is situated just off the A49 on the south side of the town, most easily identified by its entrance over a steel bridge painted bright blue. Made of concrete, stainless steel and partially clad with copper sheeting, the dial stands 18m high and more closely resembles the bowsprit of a sailing ship than a sundial. Passers-by might be forgiven for not recognising it as a dial at all! Its enormously long shadow outstrips, at any time of the year, the hour batons set in the brick pavement, and because of the lack of hour numerals the corroded bronze plaque explaining the Equation of Time is virtually meaningless. As a piece of 'modern sculpture' it has some merit in its setting of young trees, garden plants and fountain but it has little horological significance. The structure was made by the engineering company Birse Group Plc and was designed, presumably, by the Operations Research/Management Services, and erected in 1990. If after successfully finding it the visitor needs rest and refreshment, the Village Hotel & Leisure Club can provide it nearby!

Centre Park, Warrington ^{AS}

WHALLEY Lancashire

Now a Jesuit private school, Stonyhurst College, near Whalley, moved to the present site from the Continent in 1794. In the magnificent countryside of the Ribble Valley the college is approached along a tree-bordered avenue and the south side overlooks the majestic Pendle Hill that forms a barrier between the industrial south of Lancashire and the rural north. The fine group of buildings date back to the late 14th century, and is one of the architectural glories of the county. In the garden stand two 18th century pavilions or gazebos and the Observatory built in 1838, reminding us that Stonyhurst was, for many years, an important centre for astronomical research in the north of England. It was kept alive by distinguished astronomers selected by the British government, notably for surveying solar eclipses, recording sun-spots, and in the 20th century for work on magnetic variation of the North Pole. Although this work declined after the Second World War there are signs of a revival today.

In view of this association with solar observation it is not surprising, therefore, to find sundials in the garden, and two can be seen in the photograph - one a Spherical Dial and the other a Horizontal Dial on a square plinth, both close to the ornamental lake. In the background is the college chapel built 1832-35. Restoration of the gardens and Observatory are currently in hand and can be visited during July and August and the whole college complex is more than worthy of a visit by anyone interested in architecture and the college's renowned art collections.

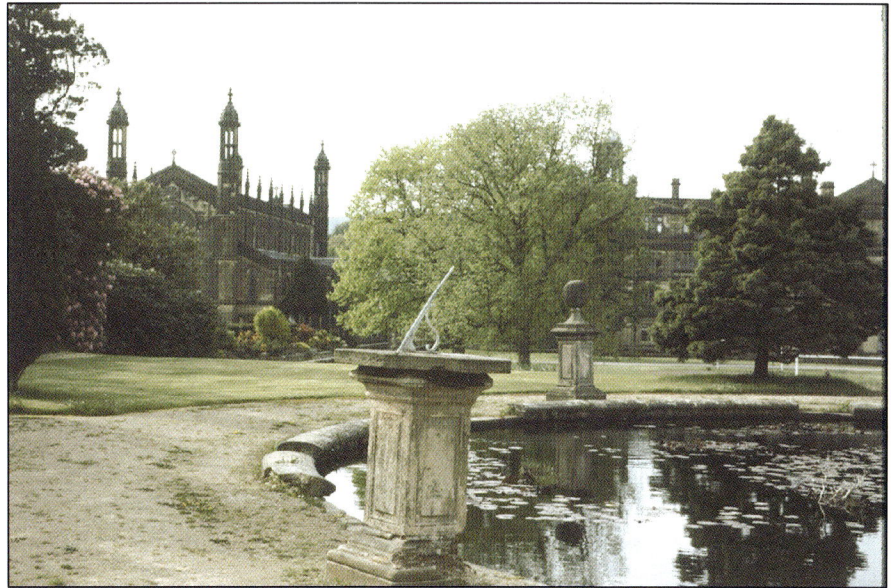

Two Dials at Stonyhurst College, Whalley AS

WINDERMERE, Cumbria

At Holehird Gardens, a property of the Lakeland Horticultural Society, there are three fine sundials.

The Heliochronometer is about a century old and was designed by Pilkington & Gibbs of Preston. Probably the most sophisticated type of sundial, it adjusts for the difference between sundial time and clock time and is accurate to about one minute. Heliochronometers were sold worldwide and are occasionally seen at country houses. The main disc is rotated until the sunlight passes through a perforation on one of the uprights and projects a point of light onto a line on the opposite upright. The date is entered on the top disc and Mean Time is read on the edge of the instrument. Their demise occurred after World War I with the introduction of free broadcast time signals (the 'pips').

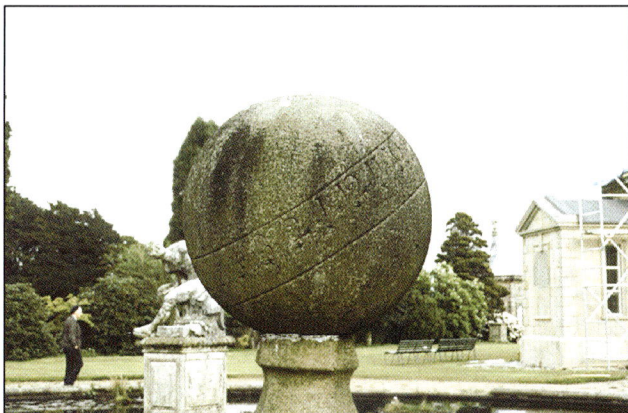

Spherical Dial at Stonyhurst College AS

Heliochronometer at Holehird Gardens RS

163

Equatorial Dial at Holehird Gardens RS

In 1908 the Groves family, who had commissioned Thomas Mawson to design an orchid house for the walled gardens, purchased the Holehird Heliochronometer. This Heliochronometer was protected for most of its life by a glass bell jar near the Octagonal Pool on the Lower Terrace. After falling into disuse, it came into the care of the Lakeland Horticultural Society. It was restored to working condition by British Sundial Society member Graham Aldred and repositioned on a pedestal in the Walled Garden. It is now placed outside on sunny days but is kept inside in the Winter and when wet to reduce deterioration by the elements.

Central within the Walled Garden is a modern Equatorial Dial. This is basically an Armillary Sundial stripped of its superfluous bands and the gnomon is the arrowed rod that represents the projecting polar axis of the Earth and casts a shadow onto the numbered, curved rim that represents the celestial equator. It is easy to lay out as the line spacings are equal, being 15° for each hour as measured from the gnomon.

This simple but striking model was a prize winning entry in a competition for a memorial for Professor H W Miles of the Lakeland Horticultural Society. It was designed by Sidney Brown of Kirkby-in-Furness and made by the apprentices of Barrow-in-Furness shipyard Skill Centre.

There is a Horizontal Dial with a thistle design on the plate on a sandstone pedestal by the northern end of the Lower Terrace near the Octagonal Pool.

'Gentleman at a Sundial' by Arthur Devis

GENTLEMAN AT A SUNDIAL by ARTHUR DEVIS c 1755

Arthur Devis (1712-1787) was born in Preston, Lancashire, but he spent most of his working life in London. He became a highly successful painter of small portraits and conversation pieces. His patrons were mostly wealthy middle class merchants and squires and he portrayed them with great charm, simplicity and naivety. This picture of a gentleman in his country park, checking his watch by a sundial, is most apt in the present context, for it reminds us that in the 18th century the only means by which a man could check his watch in his country estate, away from a public clock, was by a sundial, always remembering that public clocks had to be adjusted by using a sundial too! The paintings of Arthur Devis may be seen in many art galleries and museums.

YORKSHIRE

Frank Evans, Robert Sylvester & John Wall
North Yorkshire, East Riding of Yorkshire

Yorkshire is England's largest county, and many would say (especially the natives) that it is also, on account of its rich diversity of scenery, the most beautiful. It is also unrivalled in the number of its castles, abbeys and stately homes. Yorkshire is bounded on the north by the River Tees, on the east by the North Sea, on the south by the Humber and its tributary the Don, and on the west by the Pennine hills. More precisely, although much of Yorkshire lies to the west of the Pennine watershed, most of the rivers that rise in these hills flow west to east, forming the celebrated 'Yorkshire Dales'. Further east, across the widespread Vale of York, lies the upland region of the North York Moors. The scarp of this anticline faces north (overlooking industrial Teesside and an out-lier, the Cleveland Hills) so the rivers here flow north to south, then through the flat lands of Ryedale and the Vale of Pickering. To the south these give way to the Howardian Hills and the limestone Yorkshire Wolds.

The county boasts an equally varied coastline. First, from Teesmouth, wide sandy beaches bordered by low hills of boulder clay and sand-dunes. Then from Saltburn to Bridlington, some of the highest cliffs in the kingdom, broken only by the estuaries of rushing streams, where shelter picturesque fishing villages. Then follow the flat lands of Holderness terminating in Spurn Head, jutting out into the Humber. By far the greater part of this vast tract is landscape of the highest quality, untainted by industry. The scene is far different in Yorkshire's south-west extension: here the steep-sided Pennine valleys were home to industries of every kind, from Leeds in the north to Doncaster and Sheffield in the south. Yet even here it is possible to escape the grime with ease into the surrounding hills and moors.

Because of its hilly nature there has always been an abundance of quarried stone, chiefly millstone grit and limestone. Vernacular buildings and sundials alike reflect in their place the dark hues of the one and the bright tones of the other. Durable stone was only one of Yorkshire's riches that attracted waves of invaders, colonisers and settlers. The Romans established a legionary fortress at York, *Eboracum*, where the Minster now stands. They were followed by the Saxons who by 600AD had driven the native Celts of Iron Age Britain to Elmet in the west, where most of Yorkshire's Celtic place names survive. By 800 the Anglo-Saxons occupied the whole of the Vale of York and its exten-

York Minster with its nearby Pillar Dial

YORKSHIRE

sions, where in Ryedale we encounter the most numerous group of Saxon sundials-with-inscriptions in the kingdom. Vikings, (more accurately Danes), sailed up the Humber and the River Ouse in their longships as far as York, where they established an independent kingdom, and far beyond. Here Viking place-names outnumber Saxon. However, we do well to remember that Harold the Saxon defeated a later wave of Danish invaders at Stamford Bridge near York, before his epic march south to defeat at the hands of William the Conqueror at Hastings in 1066. Yorkshire was one of the few areas where the Norman advance was briefly impeded, by the resident Danes. Thereafter England became truly a United Kingdom (apart from such disturbances as the Wars of the Roses, of York and Lancaster), Norman architecture gave way to the flowering of the Gothic - and the rest, as we say, is history.

ALDBROUGH, East Riding of Yorkshire

Aldbrough is situated in that part of south-east Yorkshire bounded by the North Sea called Holderness, formerly part of the East Riding. Inside the parish church of St Bartholomew, and built into the south side of the south Transitional nave arcade, there is a most unusual Saxon Dial. It is carved on a circular stone and its face projects from the wall in which it is set. It consists of a central hole where a gnomon (now missing) was fixed, and eight evenly spaced lines like the spokes of a wheel, which divide the day and night into eight three-hour 'tides'. The circumference is formed of two concentric circles which contain an inscription which, being translated, reads:

Ulfo ordered the church to be built
for himself and Gunwara's soul

The stone has clearly been ignorantly re-located, and what should be the perpendicular noon-line misplaced. When correctly positioned, the beginning of the inscription is at the lowest part just beneath the noon-line. A cross-mark known as a 'daeg-mel' occurs in its correct position marking the canonical hour of prayer, 7.30am (as on the sundial at Kirkdale). The inscription is of added importance since it is written in a type of 11th century Danish-English - 'The only short specimen of Old English (that is Saxon) written by a Dane'. He was probably that Ulphus whose famous horn is still preserved at York Minster. Gunwara is mentioned in the Doomsday Book. She was therefore still living at the beginning of

Aldbrough Saxon Dial, now inverted

1066 and since, according to the inscription, she had died by the time the sundial was made, this date is the earliest possible for both the sundial and the church.

BIELBY, East Riding of Yorkshire

John Smith, the son of a farmer, was born in the village of Bielby, near Pocklington in the East Riding of Yorkshire, in 1807. He was a self-taught diallist of remarkable erudition and ingenuity. His masterpiece of a sundial is in the Albert Park,

John Smith's first known dial at Bielby Wesleyan Chapel

Middlesbrough. At the age of 18 he made a wooden sundial which indicated the time at Bielby and New York. Shortly after he made another sundial, which stood on a pillar in his father's garden, and showed the time either by Sun, Moon, or stars - termed 'the astronomical clock'. This was followed by an astronomical almanac for his own private use, and when he was only 22, by a spherical or terrestrial globe dial. This curious construction had drawn and painted on it a map of the world, and incorporated a device 'which being adjusted to the true latitude of the place, gave the correct solar time without the need of a gnomon'. It appears to have been a type of armillary sphere.

In 1837, before leaving Bielby to farm on his own account in 1842, he made a south declining Vertical Dial inscribed on stone for a Wesleyan chapel newly erected in the village. (The sundial is still *in situ* although the building no longer serves as a Methodist chapel.) John Smith was himself a Wesleyan Methodist local preacher much respected in the district. The inscriptions are of style that became his hallmark:

SOLI DEO OMNIS GLORIA!
(To God Alone be the Glory)
BOAST not Thyself of Tomorrow
(Proverbs 27:1)
For on Thine Eyelids is the Shadow of Death
(John 16:16)

Other dials by John Smith at Spaldington, Middlesbrough and Grantley Hall are mentioned later in this chapter.

BULMER, North Yorkshire

Bulmer lies two miles south of Castle Howard and for a long time it formed part of its extensive estates. The parish church which serves the village and the mansion alike is dedicated to St Martin. Much of the present building is Saxo-Norman overlap of the 11th century, that is the nave, the tall narrow blocked north doorway, two small south windows, and the lower part of the chancel north wall. Of this date also is the head of a 12th century wheel cross, and a sundial. The nave south wall presents a confusing picture of a variety of window shapes.

There are no less than three sundials, situated in relation to these windows, as follows. First a 'modern' sundial just below the moulded corbel-table, to the right of a 14th/15th century window

Saxon Dial at Bulmer

in an upper 'clerestory' stage. Second, a Mass Dial on the left-hand jamb of a single-light window slightly lower to the east. The third, a Saxon Dial occurs inscribed on two blocks of limestone which form part of the left-hand jamb of a 15th century window immediately to the right of the easternmost 11th century window. It is therefore clearly not in its original position. The dial is a half-circle, about 12 inches in diameter, of the duodecimal 'twelve hour day' type, bordered by two concentric semicircles. The hour lines are clearly inscribed and the outer semicircle is accented on both sides of the meridian line to produce a crossbar.

CULLINGWORTH, Bradford

At the village of Cullingworth near Bradford, there is a well-inscribed declining Vertical Dial.

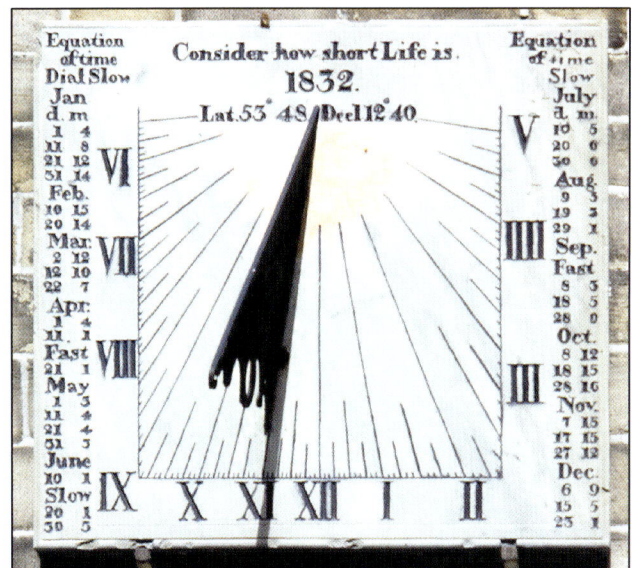
Dial at Cullingworth near Bradford

It is affixed to the old Wesleyan Methodist Church. The sundial is inscribed with an Equation of Time table as well as a suitably mournful motto, much favoured by the Victorians:

Consider how short Life is

The Equation of Time table correlates sundial time with clock time and although it varies over the years, will hold true for 1832, the year in which the sundial was made. The Grade II listed church to which it was attached had been sold in 1989 to a local building firm, A C Developments of Harden who wanted to convert it into flats. The sundial was moved from its 18 feet high position to make way for scaffolding when work began in January 1991. Nobody knew what had happened to the dial and the villagers were most concerned by its disappearance but A C Developments boss Adrian Curtis was given the task by Bradford Council's Planning Department of locating or replacing it. It was finally found in the hands of a dealer in Thirsk who had bought it in good faith.

The latitude of its site is marked as well as the declination of the dial from due south, so giving an indication of where it was intended to be used. It was retrieved in September 1991 and Adrian Curtis, helped by his 12 year old son Gregory removed paint from its face. It required a three and a half ton mobile crane and a lot of patience to secure the half-ton dial back into its rightful place.

GILLAMOOR, North Yorkshire

In this village two miles north of Kirkbymoorside there is one of the most remarkable sundials in Britain. It is situated by the roadside just outside 'Dial House'. It was formerly in the front garden

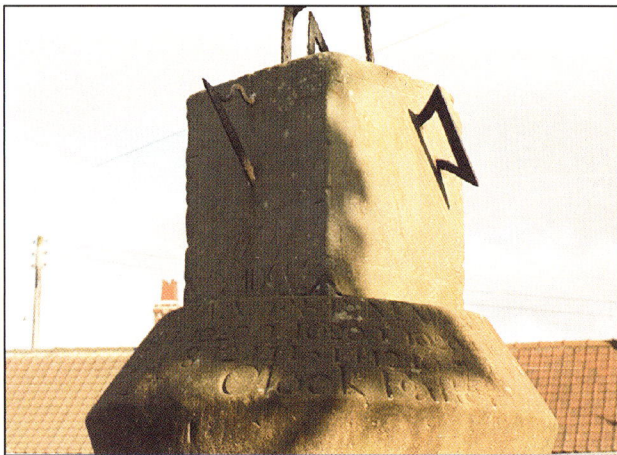
The Dial at Gillamoor on its pillar

Cube Dial at Gillamoor

until the wall was moved to make it more accessible to the public. There is a Vertical Dial on all four sides of a stone cube plus a Horizontal Dial on top of the cube surmounted by a stone sphere to represent the Earth. However, what is so remarkable about this sundial is that the Equation of Time values, in minutes for every day of the year, are carved in stone on the circular bevel below the cube.

It was designed by John Russell, a local schoolmaster, in 1902.

GREAT EDSTONE, North Yorkshire

There are three significant Saxon-sundials-with-inscriptions in Ryedale, of which the largest and most celebrated is at St Gregory's Minster in Kirkdale. Great Edstone is just over two miles from Kirkdale, as the crow flies to the south-east and its parish church sits on a hilltop site. The nave of the church is of the 13th century but unlike St Gregory's Minster there is little in the fabric to betray the presence of a Saxon church

The Saxon Dial at Great Edstone

on the site. The exception is the striking 11th century Saxon Dial over the south door, the position it would have no doubt occupied in the original church before its rebuilding.

A Saxon inscription; which (in contrast to that at St Gregory's Minster is very short) occurs immediately to the left of the sundial and being translated reads:

Lothan wrought me a(nd)..

There are three features to note. First, short as it is, the wording and the lettering are almost identical to a similar formula at St Gregory's. This suggests that they came from the same hand and that the inscribed stones originated in the same workshop. The second feature is the apparently incomplete nature of the inscription. It has been suggested that the inscription breaks off in the fourth word 'as if the writer had fallen from the scaffold on which he was working', or as if urgent business had called him elsewhere.

Perhaps the most interesting feature of the sundial, however, is an inscription above the horizontal equinoctial line. Unusually for a Saxon Dial it is in Latin and, being translated, reads:

Wayfarer's Day-Mark
or
Wayfarer's Clock

Like the inscription in a similar position over the sundial at Kirkdale, this one also provides a definition of its function. It also indicates by the use of the word 'wayfarer' (for whose benefit it was created) that a highway passed close to the church from which it could clearly be seen.

HALIFAX, Calderdale

In 1842, aged 35, John Smith removed to Spaldington village, near Howden in the East Riding, to farm on his own account. He immediately began devising a vertical south dial for a National

School that was being erected there by the landlord, Sir Henry Vavasour. Because the estate passed into other hands shortly after, the sundial was not placed on the school. Some years later it was erected in the People's Park, Halifax, in the West Riding, having been bought and presented by an alderman of that town. The dial shows the time both in Halifax and in New York, for which city John Smith had a strong attachment. There are four literary inscriptions, in three languages:

English:

Boast not thyself of to-morrow...
TIME BY MOMENTS STEALS AWAY:
FIRST THE HOUR AND THEN THE DAY

Latin:

TEMPUS EDAX RERUM
(Time is a Consumer of all Things)

John Smith's Sundial at Halifax

169

Old Father Time at Duncombe Park

Greek:

(Being translated reads..)
Redeem the Time because the Days are Evil
(Ephesians 5:16)

HELMSLEY, North Yorkshire

Duncombe Park is the seat of the Earls of Feversham. It was built for Thomas Brown Duncombe to the designs of William Wakefield between 1713-18. It is likely that he was advised by Vanbrugh who was working at nearby Castle Howard at the time. At the rear of the house is an extensive lawn that gives way to a grassy terrace overlooking the river Rye. In the centre of the boundary between these two features is a stone statue of Old Father Time with his scythe but without his hourglass which is replaced by a brass Horizontal Dial. The statue was commissioned from John Nost (1686-1729) in 1715. The sundial plate is 13½ inches in diameter and is inscribed T[homas] **Heath fecit**, (floreat 1714, died 1773). The furniture includes a compass rose of 32 points, alternately shaded and scrolled. Every sector of 10° is marked, as is every ten minutes

The simple Saxon Dial at Kirkby Moorside

on the hour ring. Two other Father Time statue/sundials by John Nost are recorded, one at Anglesey Abbey in Cambridgeshire and another at Welburn Hall in North Yorkshire.

KIRKBY MOORSIDE, North Yorkshire

There are two sundials at the parish church of All Saints, Kirkbymoorside - one inside (Saxon) and one outside. An inside sundial is no use at all, unless it catches the sunlight as this one does, built into a window jamb: except that it faces the wrong way. Although the hour-lines are very indistinct and partial, it was probably a duodecimal type.

Outside the church, a Horizontal Dial on a pedestal sits by the side of the path leading to the church porch, protected by iron railings. It is a very fine engraved brass dial. The furniture includes a compass rose and Equation of Time ring.

KIRKDALE, North Yorkshire

Without doubt the finest Saxon Dial in Britain is at St Gregory's Minster, Kirkdale, near Kirkby-Moorside. Providentially it is shielded from the ravages of the weather by a later porch, which accounts for its survival in such a clear condition. It is remarkable because it carries the longest Saxon inscription anywhere in England, from which we can calculate the history and the date of the church and the dial. The left hand panel being translated reads:

Orm, Gamal's son, bought St Gregory's Minster
when it was all utterly broken and
(continued on the right hand panel)
fallen, and he it let make new from the ground to

Dial at St Gregory's Minster, Kirkdale

**Christ and Holy Gregory in King Edward's
day and in Earl Tosti's day**

The inscription is completed below the sundial in the middle section, which reads:

And Howard and Brand the Provost made me

From that historical information we can date the church and its sundial to c 1064, but that is not all. The inscription on and above the sundial itself is an apt description of the function of a Saxon sun-dial:

**This is the day's sun-marker
at every tide**

Another Saxon Dial at Weaverthorpe has also been protected from the weather by a later porch. As at Kirkdale the long inscription tells us a great deal about the history of the church and the sun-dial.

LEEDS

At the Church of St John, off the Headrow and at the junction of Merrion Street and Mark Lane, is a fine south-facing Vertical Dial. This has a Sun motif bearing the usual depiction of a human face. Tradition decrees that the Sun's face is masculine and depictions of the moon with its gentler light show a feminine visage. Most people do not realise the relative brightness of these heavenly bodies and make comments such as of a winter Full Moon that the night is a bright as day. Moonlight is, at its brightest, almost half a million times fainter than sunlight, which might explain (together with its variable phases and complicated motion) why the moon is not a very practical way of telling the time by the shadow it

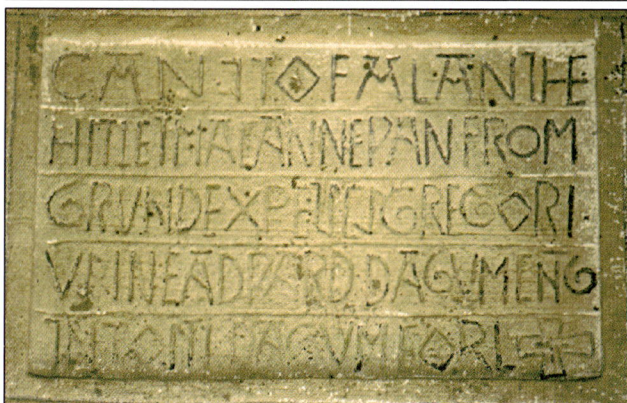

**Inscriptions left and right of the Kirkdale
Saxon Dial**

SO TEACH US TO NUMBER OUR

DAYS THAT WE MAY APPLY OUR

HEARTS UNTO WISDOM ✠

Vertical Dial at St John's Church, Leeds

casts. This church dial is a good clean example of a direct south facing Vertical Dial. The gnomon of a vertical south-facing sundial is angled to the face at the complementary angle of the latitude of the dial's site. As Leeds is at a latitude of 53°, the complementary angle is 90° minus 53° = 37°. In this example, unfortunately, the gnomon appears to be greater than 37°. This will have the effect of rendering the dial inaccurate and the shadow will vary in the time it takes to pass between each hour line as the seasons progress. The motto:

**SO TEACH US TO NUMBER OUR
DAYS THAT WE MAY APPLY OUR
HEARTS UNTO WISDOM +**

is from Psalm 90:12.

MIDDLESBROUGH

Another of John Smith's great dials is at the entrance to Albert Park, Middlesbrough. After the Spaldington dial, now located in Halifax, John Smith's creative talents seem to have lain fallow for 15 years. In 1865, aged 68, he retired to South Stockton on Tees, the neighbouring borough to

The pattern used by John Smith when
making his Albert Park Dial

John Smith's Dial at Albert Park,
Middlesbrough, recently restored

Middlesbrough in the North Riding. Now, for the first time, he enjoyed sufficient leisure to perfect the design of a masterpiece of a sundial that he had been contemplating for some years. It incorporates many features of his earlier dials, especially the multiplicity of time-related literary quotations. The sundial was commissioned by Henry Bolckow, a wealthy and prominent iron-master and MP for Middlesbrough. Bolckow was a generous benefactor to his adoptive town, to which he had given Albert Park, in which the sundial now stands, in 1867.

This dial, dated 1876, shows at the top a diagram of the Earth and the inner planets rotating around the sun and the various zodiac signs and outside this the planets of Neptune, Saturn, Jupiter and Uranus. The dial shows the time for Albert Park, Melbourne, Australia and New York. Its various mottoes in English, Latin, Greek and Hebrew are similar to those seen on most of his other dials. Below the sundial on its support there is a brass plaque on which is inscribed the Equation of

Time. We are fortunate to have a newly restored copy of the Albert Park sundial on wood. This formerly hung on the south wall of the stable block at Castle Howard and is now on display in the Dorman Museum near the west gate of Albert Park. It was probably John Smith's own working model for the Albert Park sundial. It reproduces all the features of that masterpiece with perfect clarity for our benefit. One difference between the Albert Park sundial, also recently restored, and the Dorman Museum model concerns the curious globe at the foot of each.

OLD BYLAND, North Yorkshire

As its name implies, Old Byland was one of a number of sites temporarily occupied from 1143 by a group of Cistercian monks until they finally settled at Byland in Ryedale, between Ampleforth and Coxwold, in 1147. All were Norman foundations and yet the sundial at Old Byland parish church is Saxon - evidence for a prior Saxon foundation in the vicinity. There was certainly an

Decimal Sundial at Old Byland Abbey ^{JW}

early Norman church here, but the sundial cannot have belonged to that. It is not now in its most probable original position over the south door but is most inappropriately relocated upside down on a quoin, facing east, on the south-east corner of the tower which was built onto the church in the 18th century. The mason responsible had no knowledge of, or scant regard for, its true purpose.

When functioning properly the sundial was almost unique in Britain in adopting a decimal system of time measurement, as the remaining hour lines indicate. The only other decimal dial

is in the vestry of Middleton St George, another Saxon church just over the border in County Durham. Two concentric lines mark out the circumference which contains an elaborate Greek key pattern, or meander. A Saxon inscription runs across the top which being translated reads either:

For Sumarlethi Huscarl made me
(Huscarl [as a proper name] made me), or
(**Sumarlethis Huscarl** [as a title] **made me**)

In either case the sundial can be dated to c1065, contemporary with the Kirkdale sundial.

OTLEY, Leeds

A highly unusual Double Polar Mean Time Sundial of green slate sits on Wharfe Meadows in Otley. It was designed by Christopher St J H Daniel in memory of entrepreneur Samuel Chippindale. He in collaboration with Arnold Hagenbach introduced the enclosed shopping mall to Britain and these became known as the Arndale Centres founded under the Arndale Property

Double Polar Mean Time Dial at Otley ^{RS}

Sundial by John Smith at Grantley Hall JW

RIPON, North Yorkshire

Close to Ripon, at Grantley Hall, is another quite elaborate dial by John Smith. It sits just outside the east gate in front of the Lodge on a gracefully sculpted column inscribed:

Jn Smith fecit Stockton on Tees

In many respects it appears to be a horizontal copy of his Albert Park sundial. The three hour rings are identical, although reversed. The inner one is labelled **N.York Time,** the middle one is labelled **Melb. Time** and the outer unlabelled one appears to represent Albert Park Time! However, there are differences between the two dials. Whilst four of the Albert Park quotations are repeated, five are not, and one new uncharacteristic quotation is added:

What is Deity? What is Eternity?

The Grantley Hall sundial carries no date, and the terrestrial globe and the astronomical diagram at Albert Park are omitted. Finally, the table of the Equation of Time is replaced by an Equation of Time ring on the dial inscribed:

Clock Fast: Clock Slow: Equal

Trust in 1950. The sundial has echoes of the instruments in the observatory at Jaipur in India. An unusual memorial was called for and the sundial functions by having the dial face aligned with the Earth's polar axis. The wavy lines graphically indicate the variation between sundial time and Mean Time. The Sam Chippindale Foundation has backed causes having wide ranging and long term benefits in the Otley and Bradford area encompassing medical support, sports facilities and important art acquisitions. The sundial was dedicated by Roger Suddards at a ceremony held in July 1993 in the presence of Sam's son and daughter and the Mayor of Otley.

Polar dials are not very common but a fine example is situated at the Arts Centre Garden at Bollington near Prestbury in Cheshire, also designed by Christopher Daniel.

SEATON ROSS, East Riding of Yorkshire

There is no limit to how large a sundial can be made but it will not necessarily be more accurate, as the sundial is limited to telling the time to only about the nearest minute. This is because the Sun is not a point source but has a measurable sized disk and therefore cannot cast a critically sharp shadow. A large sundial, however, does make for easier reading.

Dial Hall Farm, Seaton Ross with its large sundial RS

Siting a sundial at a farm would seem an excellent idea as the workers could easily glance at it and see how much longer their labours should continue (although other factors would indicate how much longer they had to work!).

Sundial making attracts a wide variety of people and occasionally one finds that an individual has gone to great efforts to add these old timepieces to the landscape. One such person was William Watson (1784 - 1857) who left his mark in this way in the East Riding of Yorkshire.

This fine example of a direct south facing Vertical Dial is near the village of Seaton Ross and is painted onto the brickwork of the farmhouse.

Another Vertical Dial of similar dimensions is at nearby Sundial Cottage and yet another is on the south wall of the church, all by the same maker. His tombstone there has the lines:

At this church I so often with pleasure did call,
That I made a sundial upon the church wall

WAKEFIELD

At Walton Hall is an unusual Icosahedral Sundial by George Boulby. He was a stonemason who lived at Crofton near Wakefield. It is said that he was untutored but had an insatiable thirst for knowledge. From his meagre earnings he bought a book that described mathematical figures and learned about platonic solids. There are five such regular solids, the cube, the tetrahedron, the octahedron, the dodecahedron and the icosahedron. These solids have unusual properties being the only ones which will permit the inclusion of a sphere which will contact the centre of each face and also permit the figure being enclosed by a sphere where the corner of each figure will touch the inside of the sphere. Accordingly, in 1813 George Boulby decided to make a sundial in the shape of a regular icosahedron and inscribe a dial upon each face, indicating the time locally as well as in 19 different parts of the world. So striking was this sundial that the renowned naturalist and eccentric Charles Waterton from nearby Walton Hall saw it as he rode past Boulby's home and bought it from him. It tells the time (amongst other places) at Demerara, Philadelphia and Rome, all places associated with Waterton's travels. The sundials are essentially declining, reclining, proclining and inclining. It says much for George Boulby's skill that he achieved such an unusual sundial. Ten of the twenty faces are

Icosahedral Dial at Walton Hall, Wakefield

functioning at any one time, the rest being in shadow. Although there are other Icosahedral Sundials in existence, this one is a particularly splendid specimen. The story of this sundial is recounted in Richard Hobson's biography 'Charles Waterton: His Home, Habits and Handiwork' published in 1866. Two other Yorkshire sundials by Boulby are in existence, being at churches and both are well-made direct south-facing Vertical Dials. They are at nearby Wragby and at Aberford.

WELLBURN, North Yorkshire

Wellburn Hall is now a residential home for handicapped children and lies five miles east of Duncombe Park. The original Jacobean manor house survives as one wing of the present much-enlarged mansion. John Nost's Father Time

Old Father Time at Wellburn Hall JW

Dial over the door at York Minster FE

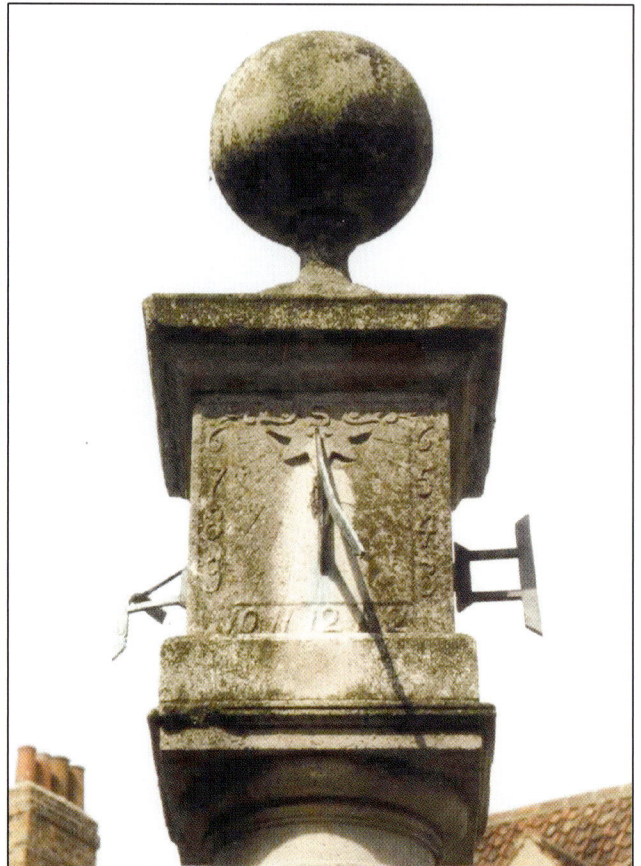

Cube Dial close to York Minster FE

statue overlooks the lake to the south of the house. In all its essential features it is identical to its counterpart at Duncombe Park. The dial plate is similar, but is of smaller diameter. Although the maker's name is not inscribed, it is possible that it was also made by Thomas Heath by association with nearby Duncombe Park and on stylistic grounds.

YORK

The Minster at York has two sundials. One is a simple Vertical Dial, installed in 1889, and is incised in the stone above the south doorway. Its motto is:

LVCEM DEMONSTRAT VMBRA

(The shadow shows the light)

Due to the fact that it is slightly recessed it shows only the hours from **IX** to **III**.

The other, a Cube Dial (also illustrated on page 165) is atop a fluted pillar at the east end of the Minster, near to the Treasurer's House. Its four dial faces carved in relief on the cube show the hours throughout the day.

NORTH EAST

Frank Evans

Durham, Northumberland

The Northumbrian region extends from the River Tweed in the north to the River Tees in the south. This land constitutes the core of the Anglo-Saxon kingdom of Northumbria, which for a time before the 9th century was a foremost centre of Christian learning, learning which included dialling. More recently the area was represented in the two ancient counties of Northumberland and Durham before being recently further cut into what are probably transient bits. In general terms the county of Northumberland lay between the

sundials have been made. This material is largely sandstone, and as a result weathering of this soft stone has been extensive and destructive. There are, nevertheless, dials surviving which are well over 1000 years old as will be noted, while many more recent ones have been eroded or lost.

The history of the two counties is substantially different from that of the rest of England. The Romans, of course, were prominent around Hadrian's Wall for 400 years but in the present

**The Gateshead Millennium Bridge over the River Tyne,
Northumberland to the left and County Durham to the right**

Tweed and the Tyne, and County Durham lay between the Tyne and the Tees. Geologically much of the area is carboniferous with extensive coal measures although towards the south-east more recent magnesian limestone deposits over-lie them.

The geological character of the terrain has dictated the material from which most of the region's

context, although they left this vast monument behind they left no sundials. They were followed by the Angles (there appear to have been no Saxons in north east England) who in due course became Christian. The Angles encountered Celtic people who had earlier been Christianised by Scottish missionaries, that is to say, by monks from Ireland (the people of Scotland were at that time not Scots but Picts). Two monastic centres of learning arose, one at Lindisfarne, or as it is now known, Holy Island, a half tide island south of Berwick which at first followed the Irish tradition, and the other at the twin monasteries of Jarrow and Wearmouth, which followed the Roman tradition. The principal difference between the two was the dating of Easter which was finally resolved in 663 at Whitby in favour of Rome. Unfortunately only the Roman discipline has left us sundials although Lindisfarne has bequeathed us the Lindisfarne Gospels and subsequently Durham Cathedral to house their principal saint, Cuthbert, who died in 687, while Jarrow-Wearmouth in the Roman discipline gave us Bede (673-735) and the 'Ecclesiastic History of the English Nation' as well as a couple of 8th century local dials.

Bede was plainly aware of the shape of the Earth and gave a method for discovering the time of day by measuring the ratio of the height of a vertical pole to its shadow length. He had a clear understanding of equal hours and indeed corrected Pliny's value of the tidal interval by just half a minute in a day even while sundials showing seasonal or unequal hours (Escomb, Dalton-le-Dale) were being created around him.

Horizontal Dial at Amble

Northumbria was never settled by the Norsemen (place-names are all Anglo-Saxon) but Viking raiding destroyed monasticism in the 9th century and no dials survive from this time. Later Anglo-Saxon dials, which reappeared in more peaceful times, while being carefully made, are less elaborate than the early ones. From around the time of the Conquest these in turn disappeared and from then on we have only rather crude Mass Dials of the 'scratch dial' type. There then was a long fallow interval during which, for centuries, border wars and lawlessness ravaged Northumbria. Dials slowly reappeared after the union of the crowns and most of our surviving modern dials are from no earlier than about 1700.

Modern dials continue to appear in Northumbria to this day, arising from the hands of skilled dialists, some working locally. In earlier centuries a dial was the workaday timepiece for the village. (They were never called 'sundials' until the advent of clocks, and Shakespeare speaks only of 'dials'.) Many Vertical Dials were on churches, visible to all. In Northumberland approaching a quarter of the parish churches, about 30 of the 130 churches of sufficient age, still bear dials. Horizontal Dials on a pedestal, on the other hand, tend to be privately owned.

AMBLE, Northumberland
Recently two very large horizontal dials have been erected, one at Shiremoor and the other at Amble. The one at Amble consists of a large spar gnomon of stainless steel tubing with three metal supports rising from the dial plate floor in curves recalling plunging waves, although a further metal strut had later to be inserted for extra strength. The dial is handsomely situated in the town square among houses of local stone. At the time of its erection it was said to be the largest dial in Europe. The time lines were calculated by the skilled diallist, Patrick Powers.

BERWICK-UPON-TWEED, Northumberland
The dial on Berwick parish church may be taken as an example of the early modern period in the region. It is on a church newly emerging from the centuries of Border strife and is one of the very few churches consecrated in the time of the Protectorate; it was finished in 1652. The dial, of stone with an iron gnomon, is clearly contemporary. It is symmetrically placed on the battlements surmounting the south aisle. It is a plain

Vertical Dial at Berwick-upon-Tweed

Sundial on bridge at Berwick on Tweed

south-facing dial with Arabic numbers and without a motto. Puritanical in aspect, its simple purpose was to tell the time. In this way it differed from some other local dials of the period, such as Hartburn, which are of the Scottish type.

A second dial in Berwick is also unusual for it is one of only about half a dozen dials in Britain sited on a bridge. It is a neat metal dial with hours cast in relief, to be found on the parapet on the south side of the old bridge crossing the Tweed. It is a modern replacement of an earlier dial and a tablet below states that it was restored at the behest of Ruth Lister of Berwick in 1995. It replaces a dial of unknown age which, it was reported, was lost in 1953 when a salmon poacher tied his net to it, wrenching it into the river.

CULLERCOATS, North Tyneside
The Cullercoats dial is a purely amateur effort, made by Frank Evans. It was on his retirement that he installed it on his place of work, the Dove Marine Laboratory in Cullercoats, a marine biological station of Newcastle University. It is dated 1994. Lacking metalworking skills he made it of marine ply with plentiful coats of paint and a final coat of varnish to protect it from the withering attacks of wind, sand and spray rising from the seashore. To find the deviation of the wall on which it is sited he followed the Sun's shadow until it lined up with the wall and calculated the sun's azimuth at that moment. The dial records the date of the foundation of the laboratory, 1908. It has been repainted once but as with other painted dials, e.g. that of the Walker (Newcastle) dial, will require further attention every so often.

DALTON-LE-DALE, Durham
The Anglo-Saxon Dial here was only recently defined and is set over the porch ridge of the parish church in a nave wall dated to the early 13th century, so it has clearly been moved at some time. It is more badly eroded than the Escomb dial. The dial plate is in the form described for Escomb, the semicircular plate, cut in relief, being again about 55cm in diameter. Above the semi-

Dial at Dove Marine Laboratory, Cullercoats

179

Remains of Anglo Saxon Dial at Dalton-le-Dale over church porch ^{FE}

Aperture for Durham Meridian ^{FE}

circle the stone is much eroded and fragmented, but there may at one time have been ornamentation here in the style of Escomb. At the lower edge of the dial plate there is further very badly eroded relief work but examination reveals it to be the remains of a double stranded barley twist surrounding the plate, again in the style of Escomb. This is clearly another early Anglo-Saxon Dial, closely related to Escomb. However, there is an important difference. Both dials have been carefully conserved and the gentle cleaning at Dalton has now revealed the time lines to be hour lines rather than the tide lines of Escomb.

These two dials belong to the sophisticated flowering of a civilisation later to be ravaged by invading Norsemen.

DURHAM
The purpose of a sundial is normally to tell the time over as much of the day as the sun allows. This wide purpose may be contracted to indicating merely one particular moment. This indication is sufficient for regulating a clock and the reduced dial takes the form of a Meridian or Noon Line, telling the time only at twelve o'clock. Such a line is to be found in the north cloister of the quadrangle of Durham Cathedral. The line, dated 1829, predates the arrival at the University Observatory of a transit circle which performed the same task, by some fifteen years. The time line in the cloister was the work of John Carr, the headmaster of Durham School and William Lloyd Wharton, the High Sheriff of the County. A small circle of sunlight is thrown onto the back wall or floor from an aperture in the fretted stone on the other side of the walkway. As the sun moves, the

spot of light traverses the Noon Line at local midday. A provision is made to catch the time five and ten minutes later in case of passing cloud. The noon line extends from a mark at one end on the floor which indicates midsummer, across the floor and up the wall to a second termination at midwinter. By observing the sun's image crossing the noon line the Cathedral clocks could be reset every few days.

A second, more conventional Vertical Dial appears on the south wall of what was once part of the monastery, in the green known as the College, attached to the Cathedral.

ESCOMB, Durham
All the Anglo-Saxon Dials known from Northumbria are in County Durham. The most famous is at Escomb. It is situated some 6 metres above

Durham Cathedral Meridian ^{FE}

Vertical Dial at Durham Cathedral College ^{FE}

Anglo Saxon Dial at Escomb ^{FE}

the ground on the nave wall of an Anglo-Saxon church which has been dated to about 700. The church is a national treasure and has been little altered. This is perhaps the earliest sundial in its original place in the UK.

The dial is trapezoidal in shape and about 55cm across. Although now very weathered, the whole dial appears to have been finely carved in relief, with a serpent or fish-like creature curled around the upper edge. The head is seen in plan and the triangular tail encloses a worn triquetra (triple-knot). The body seems to have been ornamented with 'V' shaped incisions. The lower, curved edge of the dial plate is surrounded by a simple two-stranded stone barley twist. It is difficult to tell whether the dripstone above the dial, which is probably also an animal head, is part of the scheme or a later insertion.

The time lines are not hour lines but 'tide' lines of three-hourly interval, an early form of division. Like in all early dials these time intervals are unequal and vary through the year.

The stone twist in relief surrounding the semicircle of the dial plate is a reference point to a related Anglo-Saxon dial at Dalton-le-Dale.

We here mention a curious dial of a much later period within the church at Dalton. This consists simply of a series of Roman numbers from seven to twelve carved in relief on stone blocks and fixed horizontally at intervals along the length of the north wall of the nave. Mrs Gatty in 'The Book of Sun-Dials' writes that they are said to

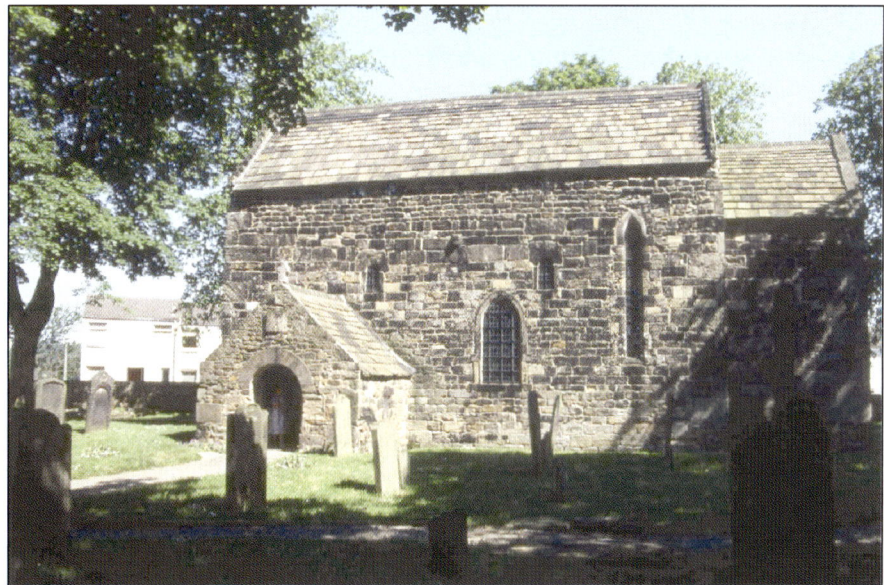

Escomb Church ^{FE}

conceal a set of older figures. They must have indicated time during the morning from a mark on a south window but no mark can now be discerned.

See also the dials at Hart and Pittington.

FERRYHILL, Durham

Pedestal dials are not commonly to be seen in Northumbria, but, as noted above, this is because many of them are in private hands or, occasionally, in the care of such bodies as the National Trust. However, a few are on public view, such as the dial in the garden outside the municipal offices at Ferryhill in County Durham. This was commissioned by Ferryhill Council as their civic artefact to mark the Millennium. The original motif was produced by three schoolboys who essentially drew their version of a wild boar on a conventional circular clock face.

The Ferryhill boar is the heraldic beast of the town. It derives from the legend of the Brawn, or boar, of Brancepeth of about 1200. Surtees, the local historian tells us that this was a formidable animal which made its lair on a nearby hill and roamed the ancient forests in undisputed sovereignty. Following the usual course of such tales a valiant knight arrived to deal with it, in this case Roger of Ferie (by some accounts the source of the name Ferryhill). It was a moderately ignoble battle by folk tale standards for Roger simply dug a pitfall and lured the animal in before killing it with his sword.

Surtees concludes his story by saying that *'it is not unusual in England or abroad when a man has slain a boar, wolf or spotted pard, to bear the animal as an armorial ensign in his shield'*. The seal of Roger de Ferie picturing his old antagonist, a boar passant, still remains in the Council treasury.

At this point the Council asked the well-known dialmaker, Tony Moss of Bedlington, to turn the schoolboys' drawings into a working sundial.

The chosen design explored an offset elliptical theme with a wild boar profile derived from one of the sketches and hand-pierced from the web of a substantial gnomon. The dial is of brass with a long axis of about 60cm, and stands on the diagonal of a square stone pedestal. Its robust construction is designed to secure it from mistreatment for many years to come.

Horizontal Dial at Ferryhill <small>AM</small>

HART, Hartlepool

The dial at Hart, while clearly Anglo-Saxon, has been built into the inside of a Norman wall at the west end of the parish church. Clearly it must earlier have adorned an outside wall of an Anglo-Saxon church. It is easily examined, being within the church at eye level. The oblong dial, 45 cm across and somewhat damaged, is inscribed with a semicircular dial plate and a full complement of nine half-tide, (one and a half hour), lines all cut in relief. The time lines are half-round in section, the half-rounds having a diameter of 2cm. The whole is cut from a single stone. Lacking the overlying creatures and barley twist edging of Escomb and Dalton, Hart is nevertheless a fine dial and the raised time lines are remarkable. See also Escomb.

HARTBURN, Northumberland

Scotland in the 17th and 18th centuries was a place of intellectual ferment and there arose a tradition among the gentlemen of that country of

Anglo-Saxon Dial at Hart <small>FE</small>

182

Multi Faceted Dial near Hartburn

church porch in the village of Hartburn but is a little too high up to read. It has three curved faces visible in a block facing east, south and west and beneath is another south facing curved dial surface, the whole being surmounted by a stone sphere. There are traces of gnomons and the sphere may carry hour marks. An opportunity to examine this dial closer would be welcome.

HURWORTH-ON-TEES, Darlington

Precision dialling was becoming fashionable in the 18th century and foremost among Northumbrian dialists of the time was the mathematician William Emerson of Hurworth near Darlington. Emerson (1701-1782), a failed schoolmaster, lived on a small private income and published numerous textbooks for scholars on many aspects of mathematics, all, apparently, beyond the comprehension of the pupils who were his intended customers. He was one of that select group of scientists who have refused election to the Royal Society, in Emerson's case because he was too mean to pay the subscription. He was renowned as choleric, disputatious and miserly. However, he was a great dialist and he and his pupil, Hunter, succeeded in making Hurworth the place with: *'the greatest number of sundials of perhaps any village in the kingdom'*.

Unfortunately Hurworth has now only three surviving dials while there is another close by on a house in Neasham. Three of the four bear a resemblance. The one on a house at the west end

commissioning sundials of many faces for their grounds and parks. These required much skill in the making and in the marking of the hour lines, and were intended to reflect the supposed erudition of the owner. Some of these Scottish-type dials of the period have migrated across the Border into Northumberland. A sad example in Northumberland, now much decayed and no longer visible to the public is an elaborate stone dial with a vertical south face, Polar Dials (where the plate lies east-west and declines backwards towards the pole), east and west Vertical Dials, also an Equatorial Dial with the dial plate in the form of a half round and, most interestingly, a globe dial mounted on the top, with the meridians and hour values carved on it. The whole dial, mounted on a pedestal, bears a fanciful resemblance to a lectern, hence the name, Lectern Dial. There are a very few other Scottish dials in Northumbria, mostly in private hands. One which may be seen by the visitor surmounts the

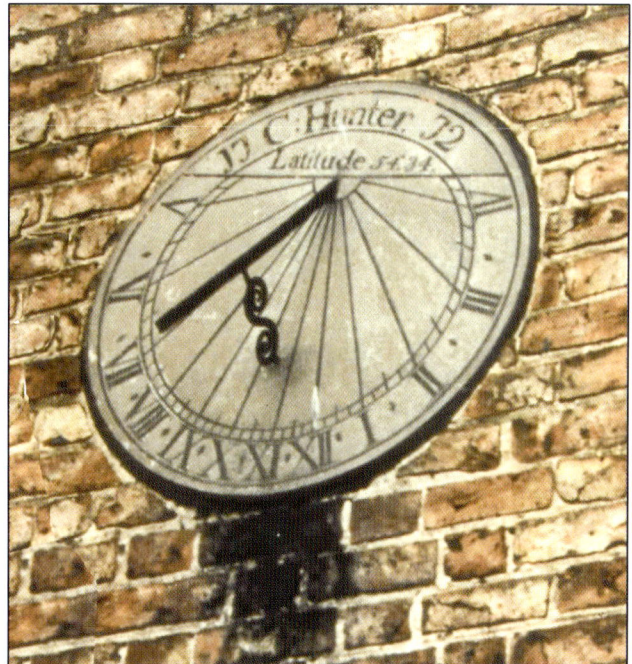
Dial by Hunter at Hurworth

Vertical Dial at the 'Bay Horse', Hurworth ^{FE}

of Hurworth by a known pupil of Emerson is signed **C. Hunter 1772** but the dial on the Bay Horse pub, dated 1739 and with a rather different appearance, is reputed to be by the great man himself. This dial has its hours marked out straightforwardly on a stone dial plate but in addition it bears a number of upward lines which

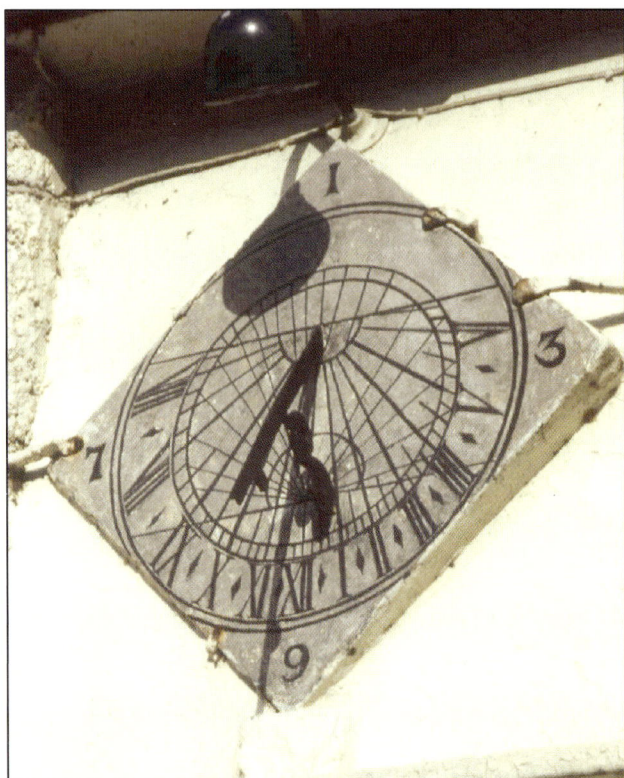

Emerson's Vertical Dial at the 'Bay Horse' ^{FE}

are unintelligible. Expert dialists have examined pictures of the dial but are unable to offer an explanation. However, the following has been plausibly suggested. Not very long ago the brewery owners of the pub contracted a firm of decorators to paint the outside of the pub and they ignorantly painted the dial out. (This is quite possible. It has also happened more recently to the dial at the 'Black Lion' in Sedgefield, but now returned.) On being required to restore this historic dial the painters are thought to have added several meaningless lines of their own for what they imagined was the sake of appearance.

KILLINGWORTH, North Tyneside

The 19th century brought more dials of a secular character. A dial of historic interest was made by the railway engineers, George and Robert Stephenson for their cottage in Killingworth. This dial, to be seen over the door of the cottage, was installed there in 1816. It was said to be one of the achievements that both engineers were most proud of in after-life. Indeed, on one occasion, George Stephenson told members of the British Association that this was one of the finest things that he and Robert had ever done, in view of their sad lack of knowledge at the time.

The most remarkable thing about the dial was that the layout was calculated not by George but by Robert, aged 13. Robert's own account of its preparation read: '*We got a Ferguson's Astronomy and studied the subject together. Many a sore head I had while making the necessary calculations to adapt the dial to the latitude of Killingworth. But at length it was fairly drawn out on paper, and then my father got a stone, and he hewed and carved, and polished it, until we made a very respectable dial of it.*'

The dial is designed for a south wall but the cottage at Killingworth faces about fifteen degrees west of south. There are two ways of overcoming this difficulty. The first is the mathematical way of the skilled dialist, when the gnomon is angled away from the vertical so that it may truly lie in the plane of the Earth's axis, the hour lines being adjusted accordingly. This method requires more mathematical power than that needed for laying out a south-facing dial. The second way may be thought of as the engineer's way. Here a simple solution is found by wedging the dial plate out from the wall until it does face south. Then, a south-facing dial will serve.

George Stephenson's dial at Killingworth

uncommon; there is, for instance, one on the Magistrate's House at Sedgefield (close to the Black Lion) dated 1707.

NEWCASTLE UPON TYNE
Keelmen's Hospital. An early example of a dial associated with a clock is on the tower of the Keelmen's Hospital in Newcastle on Tyne. It dates from 1701. The keelmen of the Tyne were a renowned body of men who worked their keels ('Weel may the keel row') over the shoals of the river from the loading staithes up river, served by wagonways from the numerous coal pits, down to the deep water of the Tyne entrance, where they shovelled their cargoes into the holds of the collier brigs loading for London. It was brutally hard work producing many infirmities among its members. The keelmen famously clubbed together to produce a sheltered home for their aged and ailing brethren; this they did by collecting a levy of fourpence a tide from each keel, which provided the building cost of £2000. The clock tower and dial were an integral part of the construction. The plaque below the dial, besides naming the worthies governing the hospital at the time, proclaims that the keelmen built it *'at their own expense'.*

And that is what the Stephensons did at Killingworth. For a visitor, it is worth examining the Killingworth dial to observe this point. Doubtless the calculations for a south-facing dial were at the extreme of the 13 year old Robert's capacity, which in no way belittles his immense achievement in designing the dial at all. He could indeed be proud of it.

LAMESLEY, Gateshead
Most church dials are on south walls but uniquely for the region the Vertical Dial on the church at Lamesley is west facing. While admittedly the nave of this church is shaded by trees, the church is older than the trees and why the dial should have been placed at the west end of the tower is unknown. It has a slightly damaged gnomon and is very stained but remains nevertheless of interest. Other east or west facing dials are not

Direct West Facing dial at Lamesley

Newcastle, Keelmen's Hospital

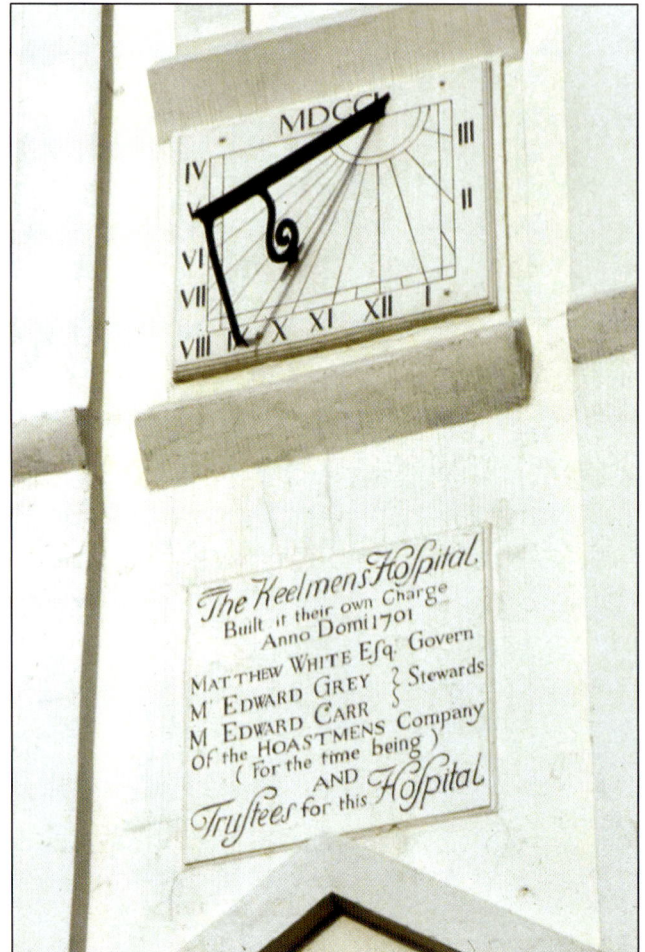

Newcastle, Keelmen's Hospital

Walker. A 19th century dial of historic importance, formerly at Walker, was made by the shipbuilder Wigham Richardson for his shipyard at Walker on the Tyne. The dial is recorded in Mrs Gatty's 'Book of Sun-Dials'. In the fourth edition it is listed as:

Harum dum spectas cursum,
Respice ad novissimam horam
(Watching these fleeting hours soon past,
Remember that which comes at last).
C. C. Walker, 1881, Lat. 54° 58'. W. R.
'On a storehouse of the Neptune Works, Newcastle-upon-Tyne, erected by J. Wigham Richardson, Esq., to whom the motto and its translation are due.'

Wigham Richardson was a prominent shipbuilder on the Tyne. In the present context his importance lies in his making of the above dial and in

his contribution of an appendix to Mrs Gatty's 'Book of Sun-Dials' telling how to calculate the lines for a dial. The initials noted above on his Neptune Shipyard dial but now absent, refer to himself (John) Wigham Richardson and his partner, Charles (John Denham) Christie, latterly known simply as Denham Christie.

Richardson was from a wealthy Quaker family in Newcastle, but he served his apprenticeship as a shipbuilder with a small tug company before becoming an engineering journeyman. In 1869 when he was 23 his father gave him £5000 and he bought the Neptune yard where he began to build ships in partnership with Christie.

In 1881 he himself made the large wooden sundial which he placed on the wall of a storehouse at the entrance to the Neptune yard where, with its prominent motto it remained a feature, being several time repainted.

The yard was amalgamated in 1902 with the adjacent Swan, Hunter yard and Wigham Richardson became Vice-Chairman of Swan, Hunter and Wigham Richardson, taking a 40% interest

Swan Hunter Dial from an old photograph

steeply angled away from south to take the morning sun. It commemorates John Welford, the Guild's Master of the time and is dated 1721.

In 1505 the Guild or Fraternity of the Blessed Trinity of Newcastle upon Tyne became a corporate body of master mariners. It received its royal charter from Henry VIII in 1536 and is now one of the very few survivors of numerous other Trinity Houses, London and Hull being others. It for long controlled the River Tyne, levying fees and licensing pilots but its functions and wealth have dwindled and its prime interest is now in maintaining its historic building. However it continues to licence deep-sea pilots for the North Sea and run a marine safety and training programme. The brethren, still all master mariners, are automatically Freemen of the City of Newcastle and can thus keep their

in the joint venture. Their most famous ship was the 'Mauretania'. Wigham Richardson died in 1908.

When, after 90 years, the storehouse was taken down the dial was set aside. At this point representatives of the Trinity Maritime Centre, a small nautical museum on the Newcastle Quayside, tried to obtain possession of the dial as an item of historic interest, known to multitudes of seamen and shipyard workers who had passed through the Neptune yard. But Swan's refused to release it and it remained, neglected, where it had come to rest in the joiners' shop.

Eventually, when Swan's went into receivership the dial was sent for auction at Christie's in London. A few friends of the Trinity Maritime Centre joined with members of the British Sundial Society to bid for the dial and happily secured it for the museum. By this time its painted mahogany dial plate had become broken into two pieces through rough usage and its gnomon had been wrenched off. The dial has since been repaired and transported back to its rightful location on Tyneside. Since the demise of the Trinity Maritime Centre the dial has passed into the care of the Discovery Museum in Newcastle, where its future is assured. It may be noted that Wigham Richardson also made a Glass Dial which, as he states in his memoirs, was fitted in a window of the directors' luncheon room to be read from the inside, but no trace of it now remains.

Broad Chare, Quayside

In the courtyard of Trinity House is another dial

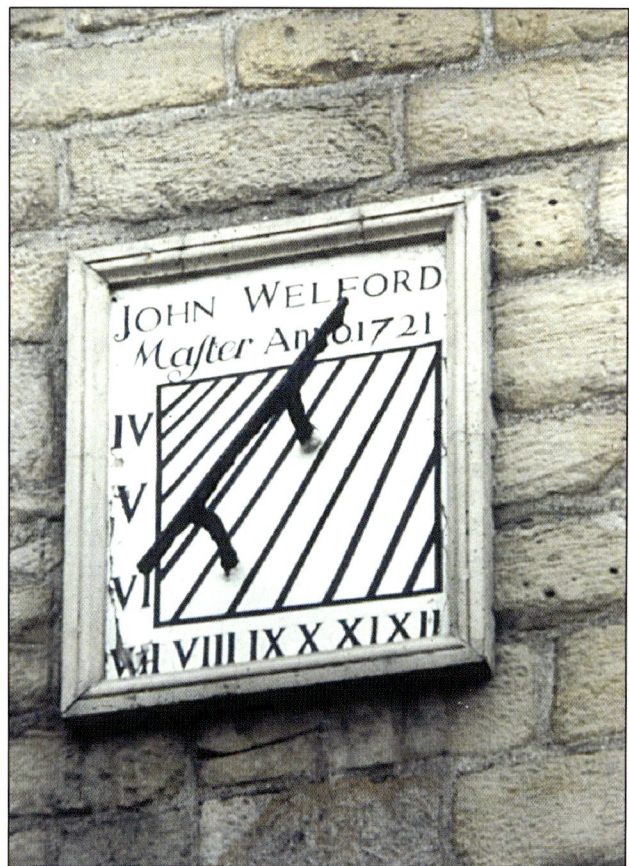

Trinity House, on the Quayside at Newcastle

Dial at North Shields Parish Church

cows on the Town Moor, and they induct apprentices, who are actually serving ships' officers, to enrol in time as new brothers.

NORTH SHIELDS, North Tyneside

Many churches in Northumbria, as elsewhere, have a clock and dial of about the same period. As an example we may cite the dial at Christchurch, North Shields. This Vertical Dial on the Tynemouth parish church had, like many others of its kind, the dual purpose of telling the time when the sun shone and of regulating the church clock. Before the days of railways and, more particularly, of the railway telegraph the only practical way of correcting a clock was by means of a sundial. The fact that this reset the clock to sun time rather than mean time and that no allowance was normally made for longitude east or west of Greenwich was immaterial.

The Christchurch dial was beautifully made by an expert dialist. We are fortunate to know his name for in a manuscript book entitled: 'Gleanings from the Records of Tynemouth Parish' prepared by H A Adamson and now in the Northumberland Records Office we read: *'1793, 2nd April. To Mr. Pringles Bill for drawing the lines and taking the Suns declination for the dial for the church.'* Unfortunately we do not have a price. The dial is finely divided to five minute intervals and although almost south facing has been designed for the small declination found.

It seems that the method used at that the period

for discovering the declination of a wall (not of the Sun as stated) is not entirely clear but it has been suggested that a night-time alignment of the Pole Star with a plumb line positioned sufficiently far back from the south wall of a church to make this possible, together with a simultaneous alignment of a candle beside the church wall would give a north-south line from which the remaining layout would follow.

The Christchurch dial itself with its fine fretted gnomon and handsome sun design has now eroded to the extent that the Latin motto has become indecipherable. Unusually for its time, it seems that allowance has been made in the layout for the longitude of North Shields, this being about six minutes behind Greenwich. North Shields, being a busy seaport town thronged with navigators, makes this more plausible than might otherwise be the case.

OVINGHAM, Northumberland

Besides being painted out as at Hurworth and Sedgefield, many dials suffer other vicissitudes. One which has threatened many Northumbrian dials is of falling down. This fate befell in particular earlier church dials with stone dial plates. The plates were often held in place against the nave or porch wall with iron clamps. In time the iron clamps rusted and the stone tumbled to the ground. At Ovingham a fine Vertical Dial of 1804 lies shattered in the churchyard. This dial is of interest in that in addition to the conventional markings of a south facing dial it has a rare engraved table showing the times of sunrise and sunset for the months of the year. At Stamfordham church the handsome dial has fallen down but has been cramped together and fitted with new clamps so that it is almost as good as new while at Chollerton the potential accident was

The broken dial at Ovingham

188

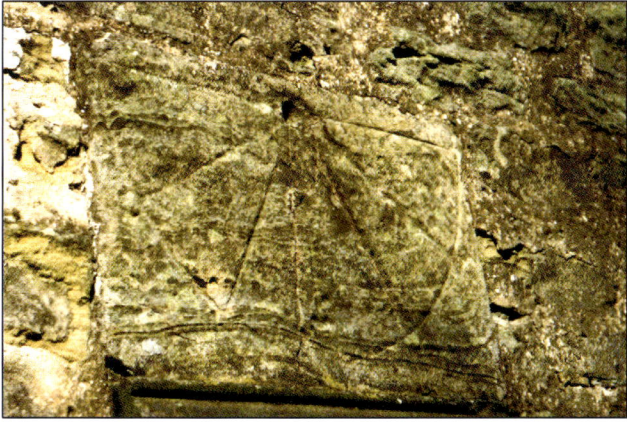
Dial at Pittington

undoubtedly foreseen and the dial was taken down in good time and now rests forlornly on the ground inclined against the nave wall. Many of these eighteenth century stone dials have in any case lost their iron gnomons, which have rusted away, leaving only the lead pellets that once held them in place.

PITTINGTON, Durham

The dial at Pittington parish church is probably in its original situation, although we know that the buttress above which it stands was re-erected during a 19th century restoration. Although approaching 1000 years old it is in far better condition than the dial over the church porch, perhaps a quarter of its age. Nevertheless the stone is cracked and decorations seen by Mrs Gatty in the 19th century, small pock marks on the time lines and engraved squares terminating each line are now almost invisible. Here, remarkably, we have yet another way of dividing the day, in this instance into double hours. The division of the day into simple hours had still not been clearly established in a church whose architecture straddles the Anglo-Saxon and Norman periods. See also the dial at Escomb.

SHIREMOOR, North Tyneside

This large dial was commissioned in 1997 by North Tyneside Council as part of the 'City Challenge' project as the centrepiece for the Silverlink, Shiremoor, biodiversity park. This park is largely artificial. The earth which was removed to expose bedrock for the foundations of a new local microchip factory resulted in a vast hill of spoil rising to ninety four metres above mean sea level. The park's ecological consultant, Dr David Bellamy, suggested a 6 metre sundial on top to take the metropolitan district from 94 to 100 metres at its highest point. The dial was designed and its

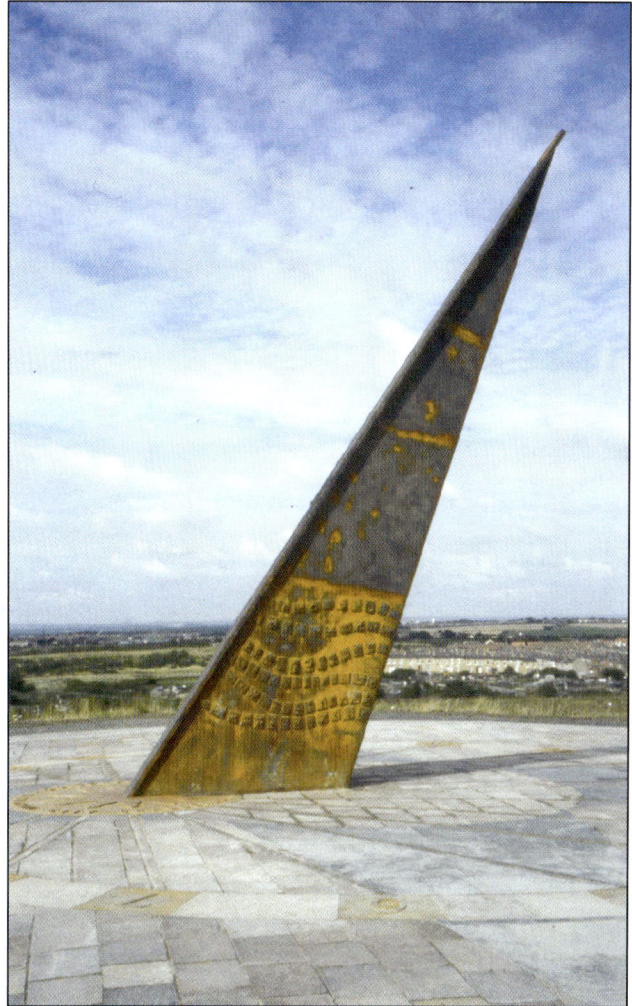
Shiremoor dial's large gnomon

installation supervised by Tony Moss of Bedlington.

Cost constraints resulted in a gnomon in unmachined cast iron weighing four tonnes and with a style width of 12 inches. The casting is of excellent quality and supports a charming feature of decorations in transferred low-relief images of wild creatures, originated by nearby first school

Shiremoor dial dated 1998

189

Shiremoor dial detail showing the designs of schoolchildren on the gnomon ^{FE}

children working with a local artist.

The dial plate is 20 metres in overall diameter with an eight-point compass rose laid out in black Indian limestone and pink Italian porphyry 'granite'. The east marker is in lemon limestone for the rising sun and the west in pink. Hour markers are cast iron with 'Times Roman' numerals in low relief.

Cost constraints again resulted in the replacing of the proposed 1 inch square stainless steel hour lines with old railway lines although the extra work in placing these probably negated any savings.

The dial was inaugurated in pouring rain on 3rd November 1998 by Dr David Bellamy. The view from the dial is very fine and the climb through the ecology park is very well worth it.

SUNDERLAND

In the middle of Sunderland there is a typical Victorian municipal building originally constructed to house the city's library, museum and art gallery. At the rear is a fine park, Mowbray Park, and an ornamental lake. The old winter garden on the back terrace was unfortunately destroyed by bombing during the 1939-45 war but a handsome new winter garden has been built at one end of the terrace. To commemorate this fine restoration the Friends of Sunderland Museums resolved to place an Analemmatic Dial on the terrace itself to mark the transformation of the site.

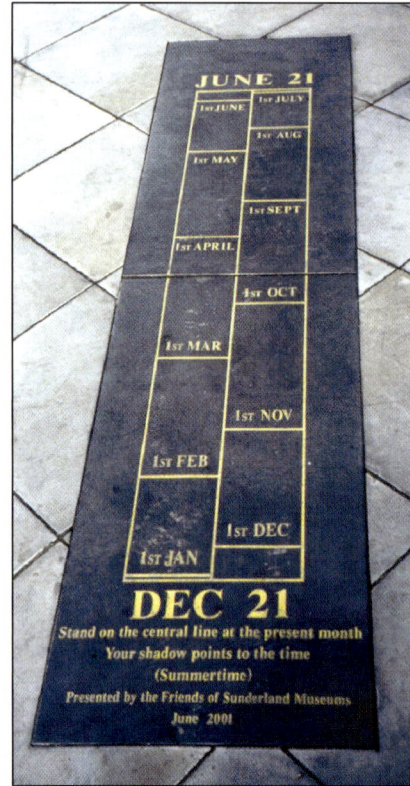

Sunderland, Analemmatic Dial plate ^{FE}

The dial, of Glenabby granite, consists of a footplate 2 metres long and hour disks with coloured engravings all inset into the stone of the terrace. Visitors, acting as human gnomons, are invited to 'Stand on the central line at the present month. Your shadow points to the time (summer time)'. The inscription further declares: 'Presented by the Friends of Sunderland Museums, June 2001'. The makers were the Napper Partnership of Newcastle architects and the delineator, Frank Evans. An interesting point for aspiring dialists was that when the dial came to be laid out on the terrace the north-south line generated by the surveyor's magnetic compass as a base line was found to be in substantial disagreement with the direction given by several later observations of the sun's azimuth. It was clear that deflection to the compass due to steel barriers and other builders' objects was producing substantial errors. The dial was finally correctly laid out by the sun.

SCOTLAND
David Gulland & Ken Mackay

Scotland is a country with a great variety of physical features. Its mainland extends 275 miles north from the border with England, and has a maximum breadth of 150 miles. Its land-area is 30,000 square miles, and its population approaches 6 million. Off its north and west coast-lines, it numbers over 700 islands, of which 150 are inhabited.

The Scottish mainland has three natural divisions:

The Southern Uplands, rolling hill-grazing and

the west and north, but good farmland in the east, leading down to a fringe of thriving cities and towns on the sea-coast; commonly regarded as the 'scenic' part of Scotland, with romantic mountains, lochs and rivers.

The range of latitudes, from 54.6° to 60.85° (only 6° south of the Arctic Circle) has implications for the sundial student, short winter days but long summer days. The weather tends to repeat the British pattern, with moderate temperatures all year round (averages of 5°C in winter and 15°C in

The fine gardens at Drummond Castle with the Obelisk Dial as a centrepiece

KM

forestry up to 800m in height, draining into the Tweed, the Clyde and the Solway.

The Central Lowlands, location of the greatest centres of population, industry and agriculture; drained by the lower reaches of the rivers Tay, Forth and Clyde.

The Highlands, with wild peaks up to 1300 m in

summer) and rainfall around 1.25m though heavier in the west. An optimist might hope to enjoy 3.4 hours of sunshine per day on average.

Up to the mid-20th century, there was a strong tradition of building in sandstone, with many skilled masons. Stone sundials can be widely found associated with mansions, walled gardens, public buildings, market squares and town-houses. Indeed, Scotland appears to have more than its due share of sundials, which has given rise to a search for the reason why. Some have suggested that the Scots are always looking for something for nothing! Others contend that Scots like to know the 'why' of things, and this gives them an excuse to display their knowledge.

Scots seem to go in for multiple dials, whether 'Lectern' types, 'Multi-Facets' or that uniquely Scottish design, the 'Obelisk'. This chapter encourages the visitor to explore and examine some examples of each. May the Sun shine on your endeavours!

ABERDOUR, Fife

Aberdour is a quiet former fishing village on the Forth Estuary, facing the idyllic island of Inch-colm with its ruined Abbey, the Iona of the East. The village is full of history, going back to Robert Bruce and his faithful lieutenant, Douglas, Earl of Morton. Aberdour Castle still stands, partly in ruins, but with its 17th century extension and its Walled Garden (1632) reflecting a period of peace and gracious living. In the Walled Garden stands

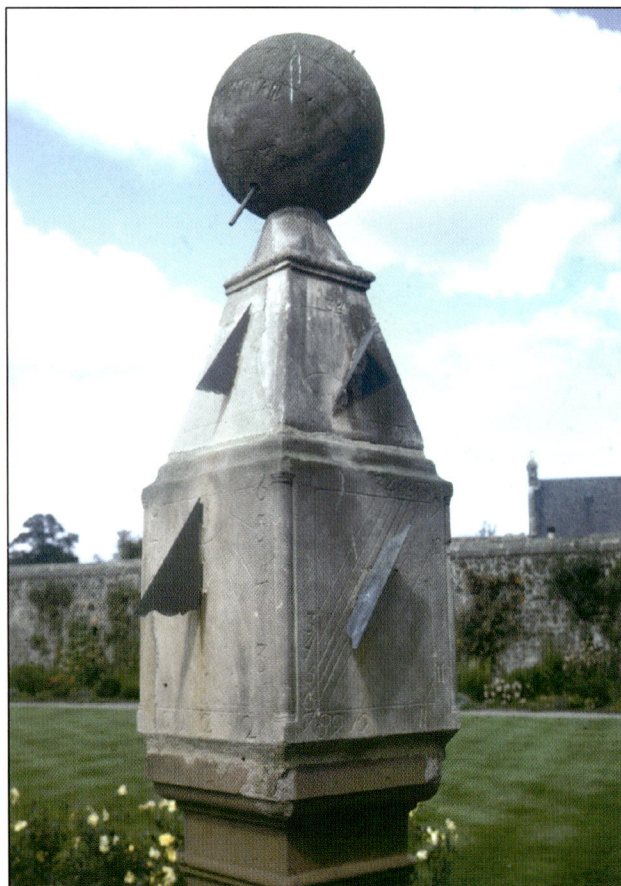

Diptych Dial at the house of Tobias Bauchope

Multi-faceted Dial at Aberdour

an unusual Multi-Faceted Dial. It comes from Wigtown, in fact from Castle Wigg, and it was re-sited here by the 'Ministry' as recently as 1972. Strictly, the latitude for which it was designed, 54.75°, is 1.25° south of that of Aberdour. There are really three parts to the dial. The lowest part is a large cube with gnomons and hour lines on all four sides. The north face shares its panel with an inscribed table giving the Equation of Time, week by week. This means that the dial must be post 1672, the year in which the table was first published in English. Above the cube is attached a pyramid, also equipped as a multiple dial. It is difficult to see what advantage this has over the cube form, but this could be said of any of the dials being described. One suspects that the real purpose was to give a point of support to the uppermost dial, the stone sphere. A sphere is seemingly an ideal model for using as a sundial, since it is just like a miniature Earth. Provision of a polar axis, and alignment of this on the Pole Star, means that inscribing 24 labelled hour-marks, (starting from 12 noon on the meridian), will allow estimation of the Sun's position in the sky. This, it turns out, is not as reliable as you may think. Nor is it so easy to read 'sunrise' and 'sunset' times from the shadows and find the mid-point. Some Spherical Dials are equipped with a captive, rotatable meridian with a narrow blade-like section. To use the dial, one turns the blade until its shadow on the sphere is at its thinnest, giving a precise time on the equatorial scale.

ALLOA, Clackmannanshire

The town of Alloa has been sadly dealt with by the passing years, though there are signs that its heritage is being recognised by those who hold the purse-strings. Its imposing Tower, ancestral home of the Erskines, Earls of Mar, is now a much-praised National Trust for Scotland attraction. One of the oldest dwelling-houses in Alloa, and one with a sense of architectural style, is the house which Tobias Bauchope, built for himself and his wife in the Kirkgait. He was a Master Mason, and had been employed in a number of major projects, including the Tollbooth in Stirling, Dumfries Town Hall, Kinross House on Loch Leven, and St Mungo's Kirk in Alloa.

His design for a sundial on his own house reflects some of the patterns he had to work with at Kinross House, under the eye of Sir William Bruce, the Royal Architect. The idea for the Diptych type of sundial may have come from a visit to Heriot's Hospital in Edinburgh, where William Aytoune had introduced no less than 11 such dials when in charge of Heriot's in the 1630s.

Unfortunately the gnomons are missing from Bauchope's House at present, and the stonework is very discoloured.

CRIEFF, Perth & Kinross

Drummond Castle sits on a low ridge, overlooking the level valley in which the Castle Garden has been established over many centuries. Only 4km south of Crieff, the Garden has been formally laid out with radiating paths, beds and parterres which are a delight to visit.

At the focus of attention in this highly geometric Garden stands a 4metre tall obelisk of warm rose sandstone. It was created for John, Earl of Perth and his Countess Jean, in 1630, by the King's Master-Mason, John Mylne III and his two sons; and its original cost was £32:18s (Scots).

The obelisk is a sundial, or rather at least 60!

Starting at the foot, there is a square shaft carrying four panels on each side except for the north.

Above that comes the 'boss', a carved polyhedron with 26 faces, 24 of which are used as surfaces for sundials. Above that again, comes the tall, tapering 'finial' with seven panels on each side and a ball pierced by a spear to finish.

Each of these sections has employed a different kind of time indication. On the shaft, one is amazed at the variety of the shapes carved into the panels. The most prominent, the upper one on the south side is a hollow hemisphere with 2 gnomons trying to cast the shadow of their tips on the scale behind. On the north side there is a family crest and a Latin verse which relates the colours in which the engravings were picked out with the precise significance of that form of dial.

The upper panels on the finial have pin gnomons perpendicular to their surface. To function precisely, they must have a standard length. However, corrosion of the pins has sharpened and shortened them and any attempt to correct these

Boss of Obelisk Dial at Drummond Castle [KM]

Heart-shaped Scaphe Dial at Drummond ^{MC}

individual dials is going to be fraught with difficulty. Halfway up the finial there are small panels bearing family crests.

Two more fine Horizontal Dials are to be found at Drummond, placed either side of the steps leading down to the gardens. These are both very similar and were made by *Johannes Marke* in 1675.

CUMBERNAULD, North Lanarkshire

Cumbernauld House is one of the finest architectural structures in North Lanarkshire. It is an Adam design, and is 'A'-listed. For the past 50 years or so it has been the Headquarters of the Cumbernauld New Town Development Corpora-

tion, but that body has been wound up, and the future of the House is in limbo.

The Obelisk Sundial used to stand facing the main entrance to the house. With the need for car-parking space around the house, the sundial was re-sited to a quieter area, but a permanent location has not been finalised.

Re-siting has its hazards. The obelisk is shown in photographs taken in the 1930s to have had a two-step platform on which stood the shaft with its five full panels on each side. Currently it stands on a kind of pyramid of whinstone setts, with the platform on top cemented to halfway up the first panel. In the absence of clear instructions, the boss has been replaced 180° out, and the finial out by 90°. Sadly, such displacements have been noted on number of re-sited obelisk dials.

The boss is basically an octagon of square panels, supported from below by four proclining panels, and supporting four reclining panels above. The corner recesses are well-defined. There are no

One of a pair of fine Horizontal Dials made for Drummond Castle by John Marke in 1679 ^{MC}

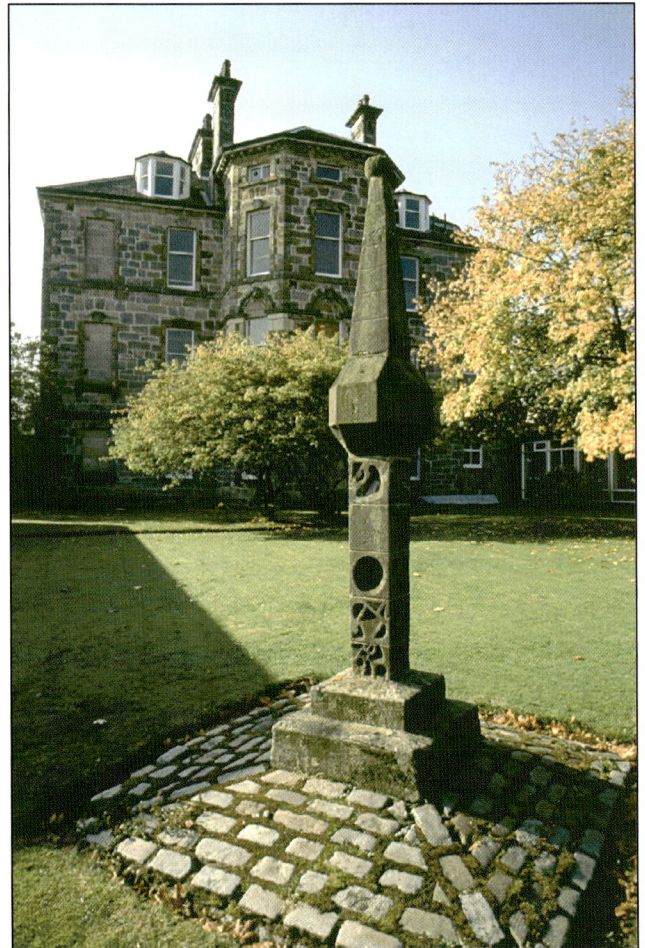

Obelisk Sundial at Cumbernauld ^{KM}

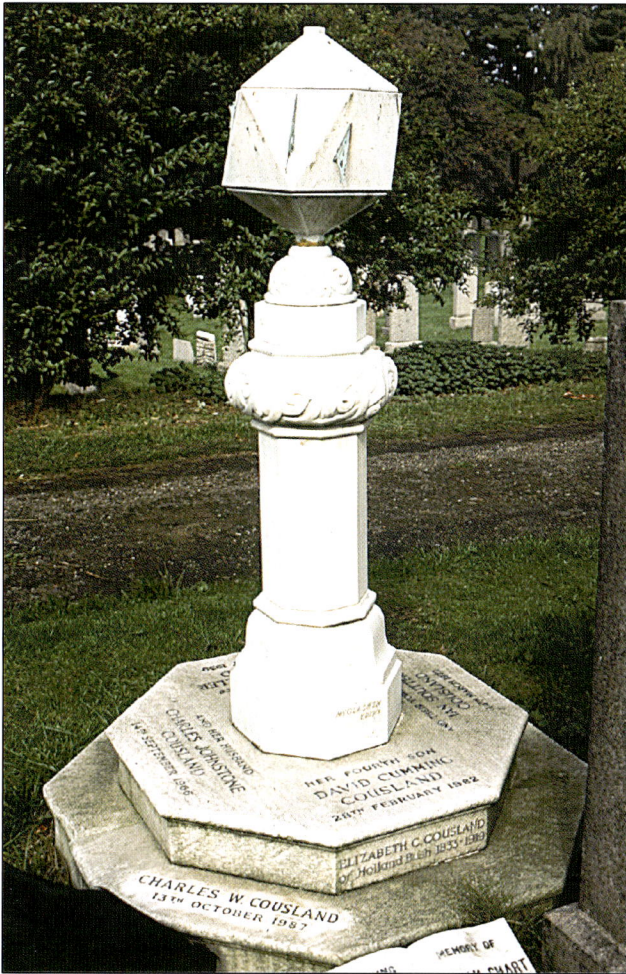

Multiple Dial at Warriston Cemetery ^{KM}

the memory of Elizabeth C Cousland (1833-1919) of Holland Bush and the names of her husband and four sons have been added, the most recent being in 1987.

An octagonal pedestal of two steps supports an octagonal pillar of elegant design, on which is fixed a Polyhedral Dial with 20 equilateral triangle facets, an icosahedron. The ten vertical faces have been brought into display as dial surfaces, and four of the ten attractively designed bronze gnomons have so far survived, although one has been damaged.

The name **McGLASHAN, EDINH** appears near the base of the shaft. If a trifle tardy, one would like to congratulate the firm for introducing such an interesting and refined note into a garden of remembrance.

Lauriston Castle. The castle was first built by Sir Archibald Napier, father of the inventor of logarithms, John Napier (born 1550). Enlarged in 1845, it stands in a lovely park, on the way to Cramond village. It is now one of Edinburgh's period museums and contains a rich collection of Victorian furnishings. It is also famous as the early home of John Law (born 1671) who was an early proponent of the use of paper currency to bolster national finances.

The Lectern Dial, which stands in front of Lauriston Castle, does not share their history. About 1870 it was brought here from Cartsburn House, Greenock, which had belonged to the Crawford family. The dial bears the date 1684.

gnomons, but their fixing locations are visible. There are few signs of hour-lines.

The finial appears to have had seven or eight panels, and a carved stone ball on top. The overall height is 3 metres.

There is one panel which has not yet been mentioned. This is the upper panel on the shaft which carries the date **1731**, the initials **EJW CMW** and a coat-of-arms. Certainly John, sixth Earl of Wigton, built Cumbernauld House in 1731, but by then he had outlived two wives, Margaret Lindsay who died in 1708, and Mary Keith who died in 1721.

EDINBURGH
Warriston Cemetery. A cemetery is perhaps an unusual location for a sundial, but here are two, plus a 2 metre obelisk in memory of one of Edinburgh City's Astronomers and sundial-makers, Sir William Peck (1862-1925).

This Multi-Faceted Dial is particularly striking, being executed in white marble. It was erected to

The Icosahedron Dial at Warriston Cemetery ^{KM}

195

Lauriston Castle Lectern Dial

The form of the Lectern is best understood if one starts at the star desk, characterised by an equatorial plane on which lies an eight pointed star, the points of which cast shadows on the sides of the stars' indentations. The upper gnomon records the hours of summertime on the Equatorial Dial.

In front, a large hemispherical scaphe with a gnomon pointer denoting its centre permits one to read the time throughout the year, as well as to measure the sun's declination in the various seasons. A long hemi-cylindrical hollow links the scaphe with the upper surface. Other surfaces enable Polar Dials to be constructed, and there are two large heart-shaped hollows which add variety to the design.

The sundial is in good condition for its age, and its gnomons have been renewed.

Holyrood Palace. Here is one of the earliest dated Facet-Headed Dials, and at the same time it is one of the most accomplished. It stands on a high base of three moulded steps. The support is

hexagonal, and is delicately carved. The dial-head is in the form of a regular icosahedron (a solid shape with 20 triangular faces). Each triangle can be studied on its own.

CR with a crown represents King Charles I, while **CP** means Prince Charles (the future King Charles II). The sundial was commissioned in 1633 by the King on the occasion of his visit to Edinburgh to be crowned 'King of Scotland'. His Queen, Henrietta-Maria, is commemorated by the **MR** symbol on the underside of the dial head. The dial was created by the King's Master Mason, John Mylne, and his sons John and Alexander, the cost being recorded as £408:15s:6d (Scots) !

The symbols used to decorate the dial include the national symbols (Rose and Thistle), several Crowns, the Royal Coat of Arms, and the Collar and Badge of the Order of the Thistle, which had been founded by King Charles I.

It may be difficult to distinguish individual dials, but one at least should stand out. A flaming sun-like face casts a shadow of his nose on a pattern of hour-lines.

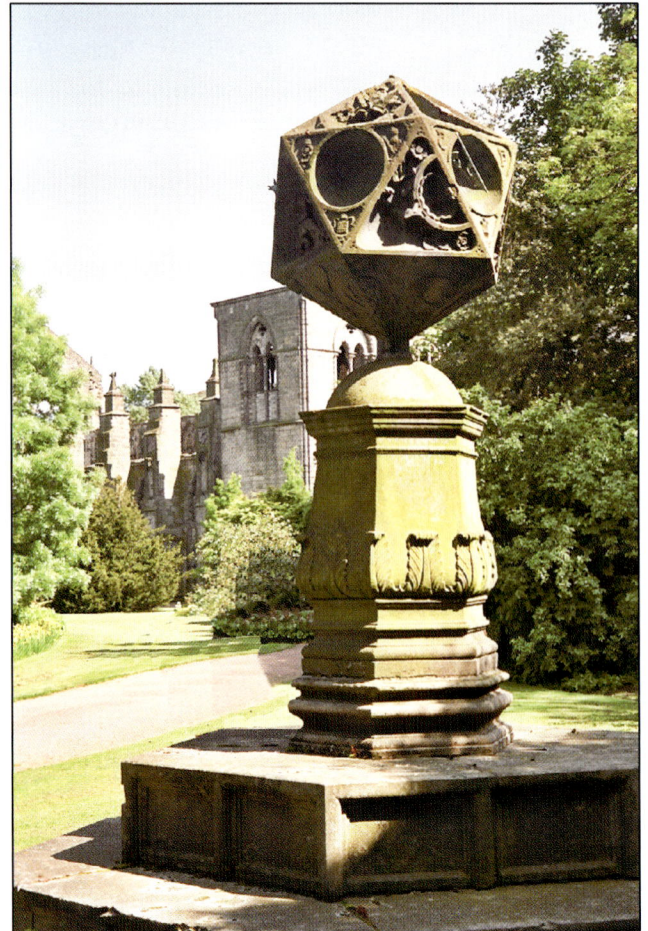

Facet-Headed Dial at Holyrood House

John Knox's House.
Walking along the Royal Mile one passes the old house popularly known as John Knox's House. (He is supposed to have died here in November 1572.) Angled on the corner projecting into the street, there is a 'corner-dial' of a rather striking form. Above, a golden Sun with the name of GOD in three languages emerges from some curly clouds. Immediately below, the figure of Moses turns appealingly, while grasping with his left hand his remaining Tablet of the Law. He is standing on a ledge under which appear the two sundials. Their planes seem to make an angle of 120°. Both dials have apparently been recut recently in slate. The left one is a 'direct west' dial, while the right one declines perhaps 30° west.

John Knox's House in The Royal Mile, Edinburgh ^{KM}

Royal Museums of Scotland. The Museum in Chambers Street has several interesting dials on display. In the collection may be found a Mass Dial, a Lectern Dial, a Multi-Facet dial, and brass Horizontal Dial. It is exciting to be able to draw attention to a modern dial of unusual design on the outside wall of the Museum. Moreover this dial was designed by the present Patron of the British Sundial Society, Sir Mark Lennox-Boyd. The late Earl of Perth, the immediate Past Patron, was a trustee of the Museum of Scotland, and obtained permission to commission a new dial. It was duly unveiled on his 94th Birthday in 2001. The dial is on the south-west of the Museum of Scotland, where Forrest Road and Bristow Place meet. It is engraved in the honey-coloured sandstone of the new Museum, at a height of about 2.5 metres above street level. Its horizontal stainless-steel 'pin gnomon', 72.5mm long, first catches the Sun when the latter is at an angle of 42.5° east of south, i.e. about 11am BST. The time is read from the *tip* of the shadow of the nodus-pin. The dial

John Knox's House in The Royal Mile ^{KM}

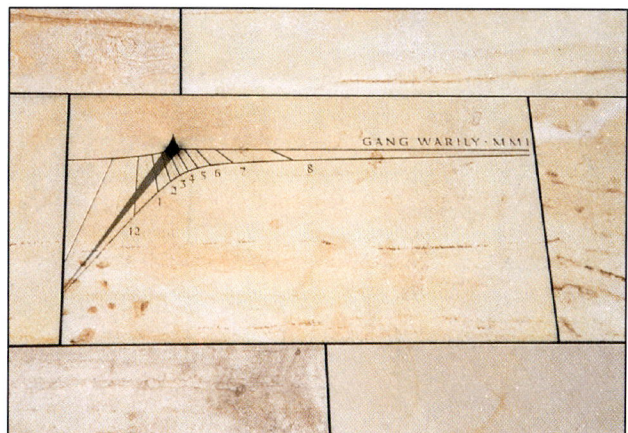

Vertical Dial at the Museum of Scotland ^{KM}

197

One of the pairs of Dials at Heriot's Hospital ^{KM}

Modern dial at Heriot's Hospital ^{MC}

will read the time between 11am and sunset. The inscription:

GANG WARILY

is the motto of the Drummonds of whom the Earl of Perth was clan-chief.

Heriot's Hospital. A remarkable number of dials can also be found at the George Heriot Hospital. George Heriot was goldsmith and banker to King James VI (James I of England). When James moved to London in 1603, 'Jingling Geordie' went with him. When Heriot died in 1624, he was very rich. His legacy to his native Edinburgh was the School which bears his name. The building was designed and built between 1628 and 1660, the architects being Wm Wallace followed by Wm Aytoune. The latter was instructed to *'mak all sort of dyallis as shall be fund fitting'*; which explains why there are 11 pairs of vertical declining sundials (eight on the outside walls, and three on the courtyard walls). These Diptych-type dials and their gnomons appear to be in excellent condition. McKean rates Heriot's as Scotland's finest earliest Renaissance building. A further treasure nearby is a 25-dial great

Multiple Dial at Heriot's Hospital ^{KM}

Cammo House Multi-faceted Dial ^{KM}

rhombi-cub-octahedron which stands on a low slab in a quiet corner to the north-east of the main building. It has four scaphe dials without gnomons but with clear hour-marks; all the flats have hour-marks and traces of their gnomon sockets. This appears to be the dial referred to by Ross as having been gifted to the Hospital in 1679 by Mr Alexander Burton *laitly ain of the Doctors of the High School'*.

Nearby is a junior school with an interesting modern Direct South Dial on it. The dial features

a hawk, sitting on a ball. He is looking at a mouse who is sitting on the end of an architect's Tee square. It is inscribed:

REMEMBERING BOB CLUNAS, ARCHITECT

**PLEASURES ARE LIKE POPPIES SPREAD
YOU SEIZE THE FLOW'R ITS BLOOM IS SHED:
OR LIKE THE SNOW FALLS IN THE RIVER,
A MOMENT WHITE, THEN MELTS FOREVER**

Cramond. Near to Edinburgh, Cramond has an unusual and decorative Multi-Facetted Dial which has recently been returned to its 19th century home, though not its precise location. Cammo House is no more than a ruin, and the terrace where this sundial sat is surrounded by trees, and would possibly be a target for vandals. So the dial temporarily occupies a bright bay window in the visitor centre looking up the drive towards the old House.

The dial is in the form of a rhombic dodecahedron, or 12-faced polyhedron, supported on an apex, on an ornate pedestal, which bears the inscription CW1795. This is consistent with the owner of Cammo Estate at that time who was Charles Watson. (Ross maintains that the dial is considerably older than its pedestal, and that it originally stood in Minto House Gardens, on the other side of Edinburgh.)

The dial presents the viewer with four vertical facets, each containing a Scaphe Dial. These face north-east, south-east, south-west and north-west respectively, and though the gnomons are lost, their locations are obvious and their hour-lines and numerals are clear. The other facets, four above and four below are marked as Reclin-

Obelisk Dial at Gartmore (described on page 200)

ing or Proclining Dials, but lack gnomons except for the south-facing recliner, whose 12° gnomon is still in position, but sadly bent.

The dial-block is held vertical by an iron dowel in a socket at its lowest corner. It is free to swivel about this dowel, a point of weakness which has led to damage (and skilled repair) some time in the past.

GARTMORE, Stirling

Gartmore Estate belonged to the Grahams', Earls of Menteith, and their descendants from the mid-16th to the mid-19th centuries, when increasing debt forced the latest owner, the renowned R B Cunningham-Graham, to sell. The sundial then stood on the terrace-walk of Gartmore House, itself a 19th century structure. The Estate was bought by Sir Charles Cayser, founder of Clan Line Steamers Ltd, who did much to improve the village. Sir Nicholas Cayser relinquished the Estate in 1950, and took the sundial to his Suffolk home. Then in 1963, the Cayser family decided to return the sundial to their family burial-ground at

Multi-Faceted Dial at Glamis Castle ^{KM}

the rear of Gartmore Parish Church.

The obelisk itself is undated, but may have been constructed in the mid- or late- 17th century. The years had been unkind to the gnomons and inscribed lines, so the Caysers decided to have them renovated. The sundial stands on a four-step pedestal. The shaft has four panels to each side, with recesses which vary from circles to squares, to quadrants, even to heart shapes - very unusual, but not helped by vertical gnomons!

Above the shaft, there is a collar of eight circular hollows, and then the boss. This has four vertical planes each with a large Scaphe Dial facing one of the intra-cardinal points. Between the scaphes four proclining surfaces point downwards, and four reclining surfaces point upwards, giving eight more dials. (Gnomons seem to sprout in abundance and in all directions, except the right ones!)

Above the boss, the finial carries five sections with gnomons to the summit pineapple - a traditional good-luck wish, two hundred years ago.

The local tradition is that this was a 'Moon-Dial' and was used for necromancy, something to do with the unusual collar of hollows!

GLAMIS, Angus

The Multi-Faceted Dial at Glamis Castle has been classed with those of the facet-head type, as it has their distinguishing feature in a very pronounced form. Its overall height is almost 6.4 metres. It may be regarded as certainly one of the finest monumental dials in Scotland, befitting the majestic castle beside which it stands.

It consists of an octagonal base, on which there are four rampant lions, each holding a Vertical Dial in his fore-paws. Between the lions there are twisted pillars with carving in the hollows, which support a canopy from which a carved neck rises up bearing the sphere-facetted globe, the facets of which are arranged in three tiers.

There are one or two other considerations to be added. The date of the dial may be about 1690, since around the platform on which the lions stand, are carved monthly tables of the Equation of Time, the correction to be applied to 'sun-time' to get 'clock-time'. These were first published in England in 1672 by John Flamsteed. Patrick, third Earl of Kinghorne, made extensive improvements to the Estate, between 1677 and 1695,

Glamis Castle Dial

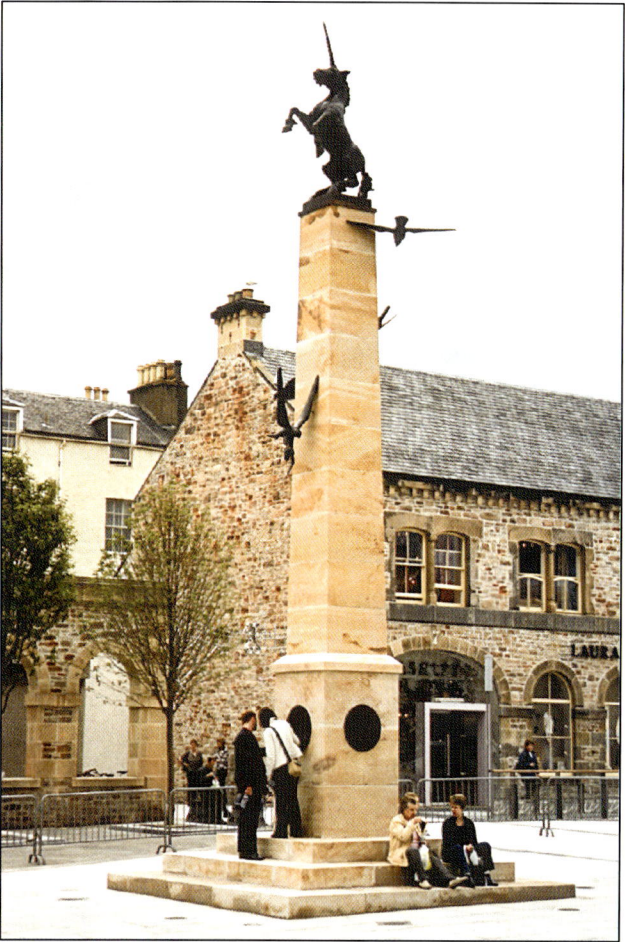
Inverness Market Cross

possibly with the advice of Inigo Jones.

There is a name for the extraordinary geometrical shape which tops the sundial. It is a stellated rhombi-cub-octagon. The 24 facets mentioned by Ross each have a three or four faced pyramid attached to them, giving 80 facets on which to fix gnomons and inscribe hour-lines, and thanks to the late George Higgs and the late Earl of Strathmore, they are all there!

INVERNESS, Highland

Designed as a centrepiece for Falcon Square, as part of the Eastgate Development in Inverness, this 'market cross' was commissioned from the renowned sculptor Gerald Laing. The base of the cross consists of a 2.2 metre high granite plinth set in a stepped surround of Caithness slabs, and surmounted by a tapering stone column 6 metres high. A bronze unicorn 2 metres tall stands on top of the column, a heraldic supporter of the Royal Arms of Scotland. Around the sides of the column appear four bronze peregrine falcons, or rather one falcon in four stages of a hunting attack on its prey, a woodpigeon.

Near the base of the column, about eye-level, are inset four bronze hemispherical Scaphe Dials, each with a rigid vandal-proof gnomon marking

Scaphe Dial at Inverness

201

its precise centre. The latitude of Inverness is 57.5°, which had to be allowed for in setting the graduations. Since the cross is rotated 40° with respect to due north, the task of marking the hour-lines and declination curves had to be undertaken with care. In the event, the internal graduations resemble those of an Armillary Sphere Sundial, sectioned through the centre, and impressed into the bronze of the hemispherical hollows.

On a sunny day, the shadows of two neighbouring gnomons should be visible in their respective scaphes (reading the same time!) the other two hollows being in shadow. As the hours pass, the shadows should travel along the declination line for that date, for which purpose the zodiacal signs for each month are inscribed. The overall effect is impressively simple.

After the Sun has set, another piece of magic comes into play. Spotlights aimed at each falcon light up in sequence, giving the visual impression of a falcon stooping and seizing its prey. Nothing to do with sundials, but a touch of activity when most sundials are asleep!

KIRKCUDBRIGHT, Dumfries & Galloway

In the historic Tolbooth in Galloway's Artist Town of Kirkcudbright there is a fine memorial to George Higgs which was installed in 1995 by members of the British Sundial Society in recognition of his distinguished work in the design and restoration of sundials in Scotland.

This window dial is one of many designed and engraved by David Gulland who collaborated with George Higgs in the production of about twenty such dials from 1988 - 1994. The use of a double glazed panel, where the gnomon is an engraved spot on the outer pane and which casts its shadow on the abraded surface of the inner pane was an original concept designed by Higgs. He discovered subsequently that this idea had first been suggested in 17th century France.

The window dial in Kirkcudbright is divided into four panes, three of which depict Galloway sundials associated with Higgs. The engraving in the fourth pane is of a Vertical Declining Dial and graph of the Equation of Time to enable the correction for Greenwich Mean Time to be applied. The logo of the British Sundial Society and an

**Engraved Glass Sundial at Kirkcudbright,
a memorial to diallist and engineer, George Higgs**

appropriate monogram complete the design.

David Gulland continues to make window dials and presently co-operates with John Higgs who has inherited his father's skill in dialling. His use of modern computer programmes enable calculations to be produced with a speed and accuracy which would have surprised and delighted his father. Most of these window dials are in private homes and whilst public access to them is not available they are a source of pleasure to their owners and a constant subject of conversation among their guests.

Another Window Dial by David Gulland is in the Public Library in Leominster Herefordshire.

MAYBOLE, South Ayrshire

A Lectern Sundial stands at the centre of the Walled Garden to the south of Culzean Castle. It is sculpted from a massive block of grey sandstone, in a superb exercise of design and workmanship. The composite dial stands about 0.75 metres high, and is mounted on a square plinth which brings its top to more than 2 metres above the base slab, well above eye-level.

The structure can be considered in three zones; a cube, above that an octagon, and above that a lectern. The cube has four vertical faces, set to north-east, north-west, south-east and south-west, and four reclining faces above the north, east, south and west junctions of the vertical faces. The octagon has eight Vertical Dials set to the eight principal directions, with four Proclining Dials below the north, east, south and west dials. All 20 dials mentioned have hour lines and bronze gnomons with their indicating edges correctly aligned on the Pole Star.

The lectern part features a more complex structure, supported on the south side by a south-facing Polar Dial, with a similar north-facing dial on the north side. To east and west sides the tilt of the lectern allows two hemispherical Scaphe Dials and two miniature heart-shaped recesses to be brought into play. The inclined star-desk has four semi-cylindrical dials facing the cardinal points, and four right-angled corner notches, all engraved to allow neighbouring edges to act as shadow-casters. The equatorial plane on top carries a polar gnomon, but the engraved hour-lines resemble those on a Horizontal Dial. Being well above eye-level, it is difficult to read, and may in

Lectern Dial at Culzean Castle

fact be accurately inscribed.

In all, there are 51 reading surfaces served by 49 gnomons, of which 25 are in copper and 24 in stone.

The sundial was described by Ross in 1892, when it stood in the grounds of Mid-Calder House, the property of Lord Torphichen, and it was reckoned to date from the mid-17th century. In 1971 Midlothian County Council offered the sundial as a gift to the National Trust for Scotland. Culzean Castle Gardens were seen as suitable, being similar in latitude (55.5° north) to Mid-Calder. Only three or four of the original gnomons had survived, and these were bent or broken.

In 1984, on the advice of Mr William Hean, Principal of Threave School of Gardening, the services of retired engineer and sundial expert George Higgs of Kirkcudbright were sought to advise on restoration of the dial. He completed his renovation in 1985, and the device represents one of the best and most complete examples of a Lectern Dial in Scotland.

SCOTLAND

Peebles Market Cross

PEEBLES, Borders

In the centre of the prosperous market-town of Peebles, a slender 4 metre shaft rises from a rather massive two-step stone cylinder. Looking upward, from across two traffic lanes, may be seen a small, dark, cuboid sundial, supporting a metal weathervane, which is perforated with the date 1662. The dial itself is said to carry the date 1699, and it is on record to have been renovated in 1895 and 1965. It carries four gnomons, but these are not in the cardinal directions as might be expected. Because the cube and its gnomons are so dark, it is difficult to gain much information from this public dial. According to Ross, the shaft at one time rose from the top of an octagonal building about 10 feet high and 12 feet across. An internal stair led up to the platform, from which official announcements could be proclaimed. At one stage (last century) the cross and its little building were taken down *so as not to obstruct the traffic on the street of Peebles*.

There are a number of similar market crosses throughout Scotland, including Airth, Cumnock, Duns, Fettercairn, Galashiels, Houston, In-verkeithing, Lochmaben, Melrose, Nairn, Pencaitland, Swinton and Wigtown.

Neidpath Castle. About 3km west of Peebles, the Castle stands on the north bank of the River Tweed. It is substantially complete, but the Castle itself is not inhabited. The sundial is on display without a pedestal, and without explanation, in a ground-floor chamber.

It has the customary star-desk, with four hemi-cylindrical hollows providing edges for recording the time throughout the day. On the corners, square notches provide similar surfaces, and on the top surface a perpendicular gnomon would have offered equatorial time-readings during the summer months.

On its south aspect, the dial possesses two Scaphe or Hemispherical Dials, one small, the lower large. Both are likely to have had gnomons indicating the exact centre of each sphere. The lowest level of all is inscribed with a south facing Polar Dial.

The east and west aspects have provision for vertical direct east and west dials, along with decorative heart-shaped depressions.

Neidpath Castle Dial

204

There is no indication of the dial's date, but 1650 is regarded as possible. There is an interesting reference in Ross: *'This dial belonged to Neidpath Castle, and about the time (1795) when 'Old Q.'* [the 4th Duke of Queensberry] *begun his work of desolation there, his gardener, Mr Spalding, fortunately got possession of the dial; his son, a nurseryman in Peebles, erected it in his grounds, where it remained for many years, till it was presented to the Chambers Institute there a few years ago, and where it now remains.'*

In 1961, the dial was discovered in Hornel's Museum in Kirkcudbright, and restored to its original home at Neidpath.

PITMEDDAN, Aberdeenshire

In 1675, Sir Alexander Seton undertook the creation of a Great Garden on his estate of Pitmeddan in Aberdeenshire. This took the form of four formal or 'knot' gardens within a vast walled garden. At the centre, as a focus, he placed this impressive Multi-Faceted Dial. When presented to the National Trust for Scotland in 1952, it had been neglected for decades. The National Trust for Scotland redesigned the flowerbeds, using authentic 17th century plans.

The dial stands on a square pedestal, about 1.5 metres high, curving inwards to offer support to the dial itself, in the shape of a deeply incised sphere. Around the 'equator' are eight hollow hemispheres, with gnomon-tips denoting the centres, and meridians as hour lines. Above and below these, are a total of 16 hollows, either square or triangular, cut square into the stone so as to allow further dial surfaces to be marked out. The sundial is surmounted by a ball-finial.

In the northeast corner of the Walled Garden stands a stone building whose walls must have been laid out precisely north-south and east-west. This has allowed the sundial maker to construct a direct south and a direct west dial on the corner.

ROTHESAY, Argyll & Bute

The Isle of Bute is a lush and peaceful place, away from the bustle of the mainland. Mountstuart House, the seat of the Marquis of Bute, dates only from the turn of the 20th century, and rivals Cardiff Castle in its magnificence; they were, after all, the brainchildren of the same individual.

There was an earlier visionary, who planned a garden, and avenues, with plantations and vistas -

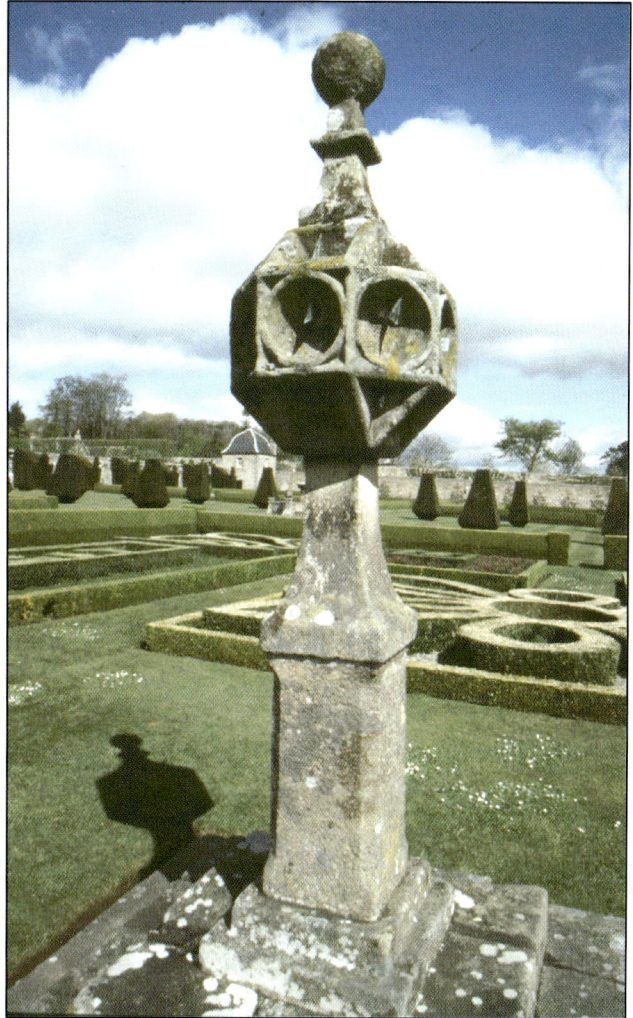

Pitmeddan Multi-Faceted Dial <small>KM</small>

the second Earl of Bute, who constructed Old Bute House around 1712, and who may have been responsible for installing the Obelisk Dial described here.

The Mountstuart archives indicate a temporary move to Rothesay, Bute's only town, since in 1807 *'the sundial was moved from Mr Blain's Garden in Rothesay'*. Now situated in a sloping semi-formal garden some hundreds of metres from Mountstuart House, the sundial - only 3.4 metres tall-invites closer inspection.

The shaft carries five panels on each side, with a variety of patterns such as quadruple heart depressions to which metal gnomons have been added, and circular hollows like that second from the base on the south side which bears a strange black button.

The boss has a zone of eight Vertical Dials, supported by four Proclining Dials below, and carrying four Reclining Dials above. The main dials

205

Dial at Mountstuart House KM

square. The wall to which it is fixed has a bearing of about 30° west of south and this would have to be very precisely measured by William Peck, before embarking on the involved calculations necessary to prepare the dial 'furniture'.

The dial displays a fantastic amount of information, all derived from theory. Much of the information to be read from the dial requires the viewer to consult, not just the gnomon's shadow, but also the shadow of the nodus (the notch on the gnomon). It shows apparent time, apparent Greenwich time (allowing for Stirling being 4° west of Greenwich) declination curves for each month of the year, longest day, equinoxes, shortest day, times of sunrise and sunset and carries the names of the donor, the designer and the Chairman of the School Board! All this was calculated the hard way, a century before the advent of the computer!

have pairs of spherical hollows or other dials (marking the cardinal directions) and simple designs on the linking vertical faces.

The finial has seven panels on each side, with gnomons showing common direction towards the Pole Star, but unfortunately no sign of hour-lines or other inscription. This, it must be admitted, could be due to the presence of extensive lichen growth ... an indication of local atmospheric purity!

STIRLING

Stirling Highland Hotel. Originally the High School of Stirling, the building has a very fine Victorian astronomical observatory and an impressive bronze wall-dial. Both instruments were designed and constructed by William Peck (later Sir William) City Astronomer of Edinburgh, and donated to the School by Laurence Pullar, a local businessman.

The engraved bronze plate is about 1.5 metres

Detail of Dial at Mountstuart House KM

Vertical Dial at Stirling Highland Hotel

Stirling-Smith Museum. An interesting dial by Richard Melville (also known as Melvin) is in the museum. He was an itinerant Irish sundial maker whose speciality was engraving slate Horizontal Dials. He flourished in the period from 1840 to 1871, though most of his dated Scottish dials appear to have been made between 1845 and 1850. He executed orders for individual customers, apparently at reasonable prices, since his dials may appear in cottage gardens.

Melville dials have a characteristic design. They are about 40cm square, smoothly finished, and around 1cm thick. The central dial may be 12cm in radius, but there may be four, six, or even eight subsidiary dials neatly spaced, so as to occupy the available surface. The corner dials might be 6.5cm radius, and the smallest (in between) 3.5cm.

The gnomons, of copper or bronze, are not fixed into the slate itself, but are cut with tangs, which fit through neat slots cut through the slate. The tangs have holes through which bronze pins are forced to lock the gnomons vertically, and these joints are then soldered under the slab, giving a neat finish. The dial is supplied with an under-slab, which has recesses to receive the solder joints, and to ensure the dial maintains its orientation.

There is lot of detail to absorb on the engraved side of the dial. The central dial with Roman numerals can be read from 3.30am to 8.30pm to the nearest quarter-hour. It bears the name and the location of the customer, plus a motto:

Memento Aeternitatis
(Remember Eternity)

and the latitude used in the design. Around the dial face is a scale of 360°, and a list of some 60 ports - eastern hemisphere to the right, western hemisphere to the left.

The secondary dials are designed to show the time around the world, at locations with significance to the customer who perhaps had a say in their choice; *'If it's noon here, it's 7am in New York, and 11pm in Sydney'*. With eight outer dials, you could devise time shifts for as many as 12 overseas locations, using double rings round the medium size dials. There is even space for a version of the table of the Equation of Time.

Examples of Melville dials are on display in public museums in Stirling, Dunblane and Dollar. The Stirling one, dated 1848, is the most complete.

Melville Slate Dial

Wynne Dial at top of staircase, Drumlanrig [KM]

THORNHILL, Dumfries and Galloway

Drumlanrig Castle near Thornhill was built between 1679 and 1691 by the first Duke of Queensferry, and is still occupied by the same family today. During construction, its cost escalated so much that the Duke felt he could not afford to live in it, and he is reputed to have slept only one night there.

On the south balcony, outside the dining room window, stands a fine Double Horizontal Dial dated 1692, made by Henry Wynne, Master of the Clockmakers Company of London. The dial plate is 85cm in diameter, and the double gnomon is 67.5cm long. The whole instrument is solidly mounted on a robust plinth, and gives an impression of great complexity.

The outer two rings are engraved as on an ordinary Horizontal Dial, reading the edge of the shadow of the sloping part of the gnomon. Within these are two rings with marks for noon at about 50 places around the world. Then follow 15 rings for use as a Moon Dial, each one marked with the age of the Moon in days, and with hour numbers staggered by 48 minutes per day, giving a spiral appearance. This is followed by a ring of compass points.

The centre of the dial plate is occupied by a grid of curved lines, which form a second independent sundial, served by the, lower vertical edge of the gnomon. These lines represent a projection of the upper celestial hemisphere on the plane of the horizon. The outer circle of the grid represents the horizon, and is graduated in degrees. The curved lines radiating from the gnomon are hour lines, marked with Roman numerals at each end. These are crossed by a series of declination lines, the outermost being the tropics of Cancer and Capricorn (23.5° north and south) and the central one the equinoxes. Note that for any particular day of the year, the Sun follows the declination line for that day.

Since the gnomon is vertical, the time is read by the intersection of its shadow with the appropriate declination line for the date. (A calendar is inscribed inside the ring of degrees.) The Drumlanrig dial has one line every four days. The direction of the shadow projected onto the ring of degrees gives the compass bearing (or azimuth) of the Sun. The intersections of the current declination line with the horizon give the times of sunrise and sunset, and hence the length of the day.

Double-Horizontal Dial by Henry Wynne at Drumlanrig Castle [MC]

IRELAND
Owen Deignan & Michael J Harley

The first people to arrive in Ireland came during the Mesolithic Stone Age around ten thousand years ago. Five thousand years later, during the Neolithic Stone Age, a massive 200,000 ton sun-aligned stone cairn was built in the Boyne Valley at Newgrange. This was a thousand years before Stonehenge in Britain and 500 years before the first pyramids were built in Egypt.

for secular purposes, each monastery provided itself with a suitably delineated sundial by which to divide the day into appropriate periods.

Of the many hundreds of these ancient time-pieces, there are just nine remaining in the country, and these are located as follows, working in a clockwise direction from the northern counties of Ireland: Bangor and Nendrum, Co. Down,

Newgrange, Co. Meath

MH

Early Irish Monastic Sundials

In the years following St Patrick's conversion to Christianity of the Irish people, literally hundreds of monastic settlements, both large and small, were established throughout Ireland. Patrick's mission commenced in the year 432 and so great was his influence that the monastic movement became one of the most powerful sources of enlightenment and culture in medieval Europe. The very early monks were dedicated to an isolated existence and established their communities well away from the towns and villages where they could follow their apostolic rules and live frugal lives in accordance with the tenets of their new-found faith.

In order to properly regulate their times of prayer and, to a lesser extent, mark the passage of time

Clogher, Co. Tyrone, Monasterboice, Co. Louth, Clone, Co. Wexford, Peakaun, Co. Tipperary, Kil malkedar, Co. Kerry, Inishcaltra, Co. Clare and Kilcummin, Co. Mayo. There was a 10th monastic sundial located at Saul, in Co. Down, which was recorded and drawn by the historian George V Du Noyer in 1868 but which is now missing, presumed stolen from the graveyard of the old church at Saul.

These early Christian sundials date from the 5th to the 9th century and are unique to Ireland, and to a few locations in England and Wales which were visited by Irish monks, in that they are free-standing vertical stones, not unlike grave-stones but of narrower and sometimes taller construction. Consequently, they differed greatly from the so-called Mass Dials, which were traditionally engraved directly on the walls of churches throughout the British Mainland and Northern Europe. The nine surviving sundials of monastic Ireland also differ from each other in shape, construction and decoration but they do have in common, clearly carved on their various faces, the principal canonical hours or times of prayer, including Terce, Sext and None.

Where decoration or embellishment is in evidence, it usually takes the form of Christian symbols such as crosses, images of fish or the occasional example of intricate, early Celtic interlacing design. Linear or angular patterns were also used, as in the case of the lower part of the Nendrum dial.

It is significant that no two surviving monastic dials bear much close resemblance in shape or decoration, but that they do agree remarkably in their time-telling or prayer-indicating properties. Thus, there would appear to have been a fairly strict observance of ecclesiastical rule, combined with a rather relaxed acceptance of local artistic expression.

These stone sundials were always set into the ground in open areas, within the boundaries of the monasteries, well clear of all shadows, with their faces orientated due south. The iron or timber gnomon was let into a horizontal hole in the stone, from which the hour lines radiated, casting a shadow along the lines in accordance with the time of day, always indicating the time of noon by the vertical line running down from the base of the gnomon.

The Kilmalkedar monastic sundial, is a fine example of these ancient Irish canonical dials.

Irish Instrument Making
- The Lynch Family

In the 18th and 19th centuries, Dublin was noted for the number of family firms engaged in the development and manufacture of scientific and mathematical instruments of every description - optical, horological, surveying, meteorological and what were referred to at the time as 'philosophical' instruments.

Some of these firms specialised in the making of sundials, many of which survive to this day. The most prominent of the families were the Masons, Lynches, Yeates and Spears, all of which produced high standard solar timepieces for mainly Irish clients, but were also engaged in the export trade. Perhaps the best known of these sundial makers were members of the Lynch family who were active as quality instrument makers at the time. We are indebted here to the authors Alison

Substantial Gnomon on Lynch dial from ^{MH} Dublin Botanical Gardens

Morrison-Low and John Burnett in their excellent book 'Vulgar and Mechanick - The Scientific Instrument Trade in Ireland 1650-1921' for much of the following information on this family.

Members of the Lynch family are known to have been active in the trade between 1767 and 1846. Three generations appear to have been involved; unfortunately there are three James Lynches, one in each generation, which makes chronological identification rather difficult. However, the eldest James Lynch is identified from correspondence with Trinity College, dated 1767. James Lynch II was assistant to the Professor of Natural Philosophy at Trinity. Records imply that he was employed there from about 1785 until his death, and that James III worked in the College from 1833 to 1840. Also, James III's brother George was assistant in Professor Humphrey Lloyd's magnetic observatory from 1838 to 1856. As far as can be judged James Lynch II was the leading supplier of experimental apparatus to Trinity College during his lifetime.

One Lynch family bill is of particular interest - the earliest extant for work by James Lynch II, other than that for assistance with lectures. It covers work done between 1788 and 1793, and is endorsed by the Professor of Natural Philosophy:

'The above work has been done in the Philosophy School and the charges appear to be reasonable'. Although the bill includes a variety of repairs and alterations, it also lists a number of instruments which have been made, such as *'a large horizontal dial for ye Garden'.* The bill does not specify which garden but as the Dublin Society (later the Royal Dublin Society) was planning the establishment of the Botanic Gardens during this period, it could well refer to the fine Lynch sundial that still stands by the Palm House in those Gardens.

From Lynch's work at Trinity arose another post: that of lecturer in Natural Philosophy to the Royal Dublin Society, in which capacity he served until his death in 1833.

James Lynch II was a maker, vendor and inventor of instruments, laboratory assistant and public lecturer. His business, on his own and later with his son and other partners, lasted until 1846. The third generation of the family was represented by James III and George; though the final name of the firm - 'Lynch & Co.' - suggests that at least one other principal was involved in the end.

Horizontal Dials, an Armillary Sphere and a large Ring Dial carrying the Lynch name, still survive.

AHAKISTA, Co. Cork
On the seashore at Ahakista village, near Durras, on the Sheep's Head peninsula in west Cork there is a 1500mm diameter modern stone sundial by Cork sculptor Ken Thompson. It was delineated by Owen Deignan. It is a memorial to the 329 people who died in the Air India crash in 1985. The sundial marks the time of the explosion on board the plane - 08.13 hours on June 23rd, 1985.

The site for the monument was chosen both for its proximity to the off-shore location of the dis-

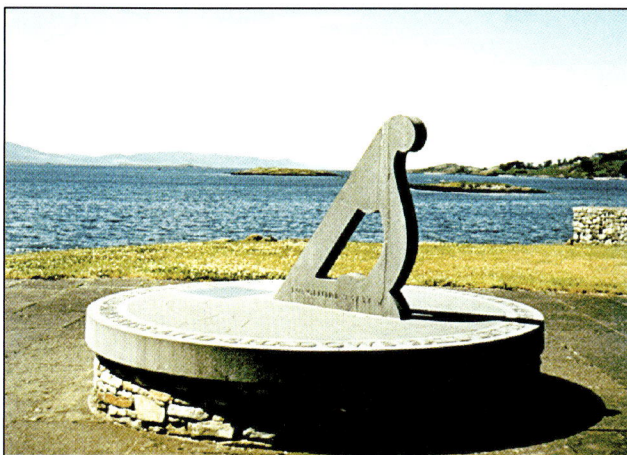

Memorial Dial at Ahakista KT

aster and the beauty of the surrounding area. In laying out the overall design, Ken Thompson divided the available space into two parts, one consisting of an attractive garden surrounded by traditional low stone walls and containing plants and shrubs such as fuchsia, heather, gorse and wild rose which flourish in these exposed conditions but grow so low as not to obscure the monument.

On the seaward side of the garden, with the sundial as centrepiece, is a special paved area, bordered on the landward side by a 12 meter long crescent shaped stone wall carrying a central inscribed stone panel commemorating, in English, French and Hindi, all those who died. This wall also carries 14 separate bronze plaques, recording in beautiful raised lettering the 329 names of the deceased.

It is now a place of annual pilgrimage on the date of the disaster, and justifies in full the designer's intentions that it should be *'a place that in some way brings solace to the bereaved and is also of permanent interest to visitors; a place where people like to be'.*

ARMAGH
Armagh is the ecclesiastical capital of Ireland. Patrick established his bishopric and built his Cathedral here in 445 on the spot where St Patrick's Church of Ireland Cathedral now stands.

The history of the Cathedral is one long record of burnings and plunderings. In the year 832 the Vikings sacked it. Partially burned by a fire caused by lightning in 995, the Cathedral lay for

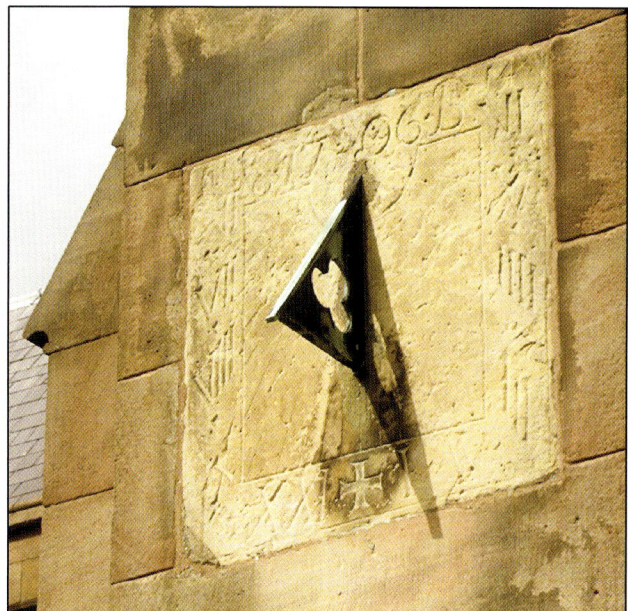

Vertical Dial at St Patrick's Cathedral MH

the most part unroofed for some 130 years. After further burnings it was rebuilt in 1261 and again in 1365. Once again, fire destroyed the Cathedral and it was restored again in 1428. During the Reformation, St Patrick's Crozier and other relics from the Cathedral were publicly burnt in the High Street, Dublin. Of the surviving relics, St Patrick's Bell is in the Royal Irish Academy, and the Book of Armagh is in the library of Trinity College, Dublin.

In the late 16th century the Cathedral suffered several more burnings in the war between Elizabeth and the O'Neills of Ulster. Restored again in 1613 it was to be destroyed once more in 1641. Further restorations were made in 1729 and again in 1765. In 1802 further works were carried out which tended more than any of the previous ones to alter the fabric of the building.

The Cathedral as it stands today is the result of its restoration by the English architect, L N Cottingham between 1834 and 1837.

In the graveyard outside the Cathedral under a massive granite boulder are the remains of the three premier Irish saints, Patrick, Brigid and Columba. Buried for 500 years in different locations they were re-interred together by the Norman, John de Courcy about 1180 in thanksgiving to the saints for keeping Chrstianity alive during the Dark Ages. The body of Brian Boru, the High King of Ireland, who was killed fighting the Vikings at the Battle of Clontarf, in 1014, also lies in the vicinity.

Five metres up on a buttress of the south transept of the Cathedral there is a 600mm square, vertical sandstone sundial. Badly eroded it shows the time from 6am to 6pm in Roman numerals with a maltese cross for twelve noon. It has a bronze gnomon with a 'shamrock' perforation. The date **1706** at the top has what appears to be an **E** and a **B** on either side of it. The dial was reputedly restored c 1931.

BALLINTOY, Co. Antrim
Six kilometres west of Ballycastle on the Antrim coast is Ballintoy, one of the most northerly points in Ireland. The present Ballintoy Church of Ireland Parish Church was built in 1813 replacing a 1663 building which had in turn replaced an earlier church on this site. On the south-west corner at the eaves of the church there is a 350mm sandstone corner stone with Vertical Dials on the two exposed faces. The south facing dial has **Lat 55° 14' North   Long 6° 16' West** inscribed, a pierced bronze gnomon, full hour and short half hour lines with Roman numerals.

Cube Sundial at Ballintoy
MH

There was a mistake made with the Roman numerals - both sides appear to have been inscribed as a mirror image of the **VI, VII, VIII** side and the resultant effort to correct the mistake is evident. The pad stone that supports the overhanging dial stone bears the date 1817. The west face has hour lines with Arabic numerals **1** to **8** and is inscribed:

HIGH W--------
The stone is badly eroded but looks like:
HIGH WAY TO GOD
It has a solid bronze gnomon inscribed with **Lat 55° 14' North** on one side and **Long 6° 16' West** on the other.

BLACKROCK, Co. Louth
The 7.3 metre diameter Blackrock sundial, one of the largest in Ireland, was erected during 2000 to mark the Millennium year. The 3 metre high gnomon, created by local artist Tanya Nyegaard, consists of a bronze sculpture of a female diving figure on a hexagonal stone pedestal. Gold numerals engraved on black granite blocks form the face of the sundial. Inscriptions in English and Irish on the pedestal read:

Anno Domini 2000 - To commemorate the jubilee of the birth of Christ, and to mark the beginning of a new millennium

The year is also expressed by the Roman numerals **MM** and in binary notation **11111010000** to reflect today's computer technology. An Irish seanfhocal (proverb) on the face of the sundial emphasises the linkage between the sea and time:

Ag tuile 'is ag tra a chaitheann an fharraige an la
(Ebbing and flowing is how the sea spends the day)

Sundial on the Promenade at Blackrock

The name 'Aisling' was assigned to the sculpture by a poll in the village in 2001. 'Aisling' was a form of 18th century Irish poetry in which a female figure in various guises appears to the poet. Larry Magnier did the calculations and delineated the dial.

BLESSINGTON, Co. Wicklow

Russborough, said to be the finest house in Ireland open to the public, was built for Joseph Leeson, later Baron Russborough and then Earl of Milltown, by the German architect Richard Cassels between 1741 and 1751.

Cassels was the leading country house architect of his time in Ireland and designed other houses, the most important of which are Powerscourt and Carton. He also designed the Rotunda Hospital in Dublin as well as Leinster House, the present seat of the two Houses of the Oireachtas

(National Parliament) comprising Dáil Éireann (the House of Representatives) and Seanad Éireann (the Senate). Russborough consists of a central seven-bay, two-storey over a basement block which is extended on each side by curved colonnaded quadrant wings leading to seven-bay, two-storey pavilions on either side. It is then further extended on each side by plain walls, each pierced by a high centrally placed arch to give access to the kitchens and stableyards. The walls terminate in single storey pavilions giving the entire facade an overall length of 215 metres, making it the longest in Ireland. Sir Alfred Beit bought Russborough in 1952 as a home for the Beit Collection of Dutch, Flemish and Spanish old masters as well as English, Scottish, Italian and French paintings. In 1976 he established the Alfred Beit Foundation and in 1978 the house was opened to the public.

The house is beautifully maintained and furnished, with fine displays of silver, bronze, porce-

Sundial over gate at Russborough House

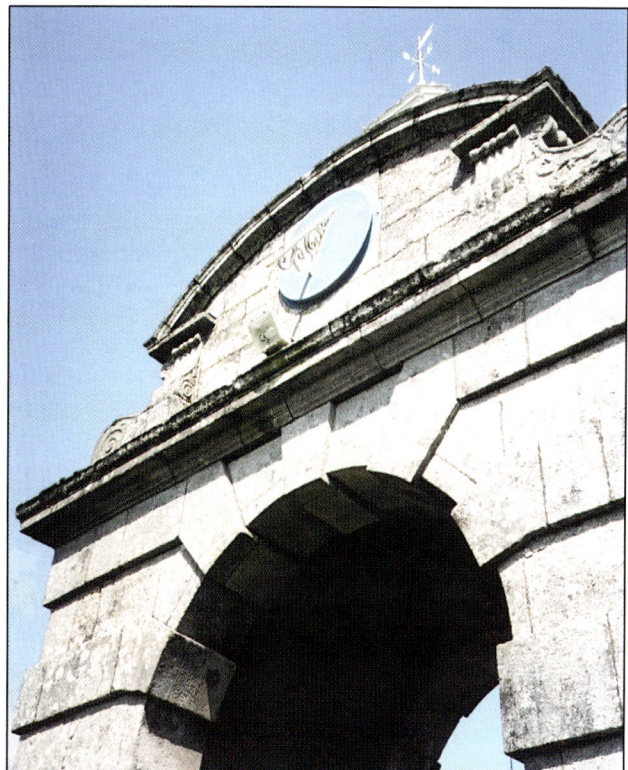
Sundial at Russborough House

IRELAND

Walled 'Time Garden' at Carnfunnock

Jester's Face Sundial at Carnfunnock

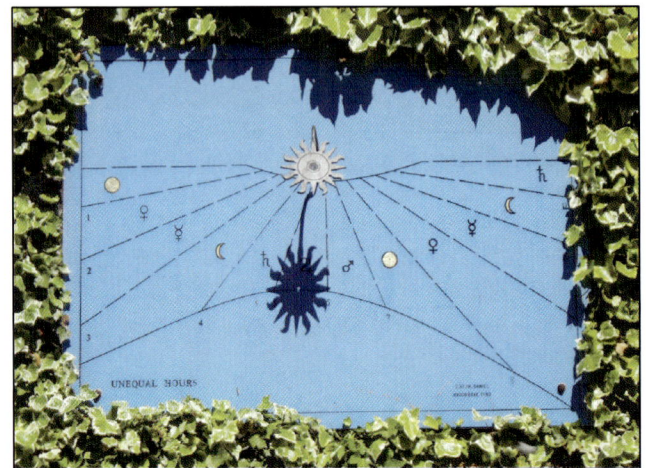

Unequal Hour Dial at Carnfunnock

Multiple Dial at Carnfunnock

lain, carpets, tapestries, furniture and *objets d'art* from around the world. Russborough House has not escaped the attention of the art thieves. The first major robbery happened in 1974, when a gang stole 19 paintings. The haul included a Vermeer, a Goya, two Gainsboroughs and three Rubens. The Gardaí recovered all the paintings and captured the thieves within eleven days. Twelve years later in May 1986, paintings to the value of £30million were stolen, most were recovered later in Britain and Belgium.

Located over the west arch entrance to the stableyard there is an undated, painted, vertical stone, south east declining sundial. It is approximately 900mm in diameter and has a scroll iron gnomon and a fractional time ring reading from 4am to 5pm with the hours marked in Roman numerals. The dial, hour lines and numerals are painted the same colour making it difficult to read from the ground.

CARNFUNNOCK, Co. Antrim

The 191 hectares Carnfunnock Country Park has 13 modern sundials within its walled 'Time Garden'. The various dials trace the history of time-telling starting with the Ancient Babylonians who divided the time from sunrise to sunset into 12 equal parts taking no account of the difference in day length thus giving rise to hours of different length, shorter hours in winter – longer hours in summer. Our modern day is divided into 24 hours of equal length. There are similar dials for Equal Hours (longitude corrected to show Greenwich Mean Time) and another that is showing the number of daylight hours in the day with the Sign of the Zodiac in which the Sun is situated.

There is a vertical bronze dial in the form of a disc with a Jester's face breaking through from

214

Armillary Sundial at Carnfunnock ^{MH}

the human becomes the shadow casting device and two multiple dials, one using gnomons and the other using the cut surface of the stone to cast the shadows.

In the centre of the garden is a magnificent Armillary Sphere.

CASTLECAULFIELD Co. Tyrone

There are two sundials on the tower at St Michael's Church of Ireland Church, Castle-caulfield, Co. Tyrone. The church was built in 1685 by William Caulfield, Lord Charlemont, to replace the old Parish Church at nearby Donagh-more. The south windows and some of the other carved stonework of St Michael's came from the old parish church. The chancel, vestry and tran-septs were added in the 1860s. The direct south Vertical Dial is made from several dressed stones set flush with the random rubble stone of the tower. It is dated 1685 the year the Church was built. The **16** and **85** are separated by a mono-gram or symbol which is now unrecognisable because of the erosion of the stone. The dial seems to have been restored at some stage as the style of the gnomon is a piece of copper pipe tied with wire to the simple horizontal iron support. It shows the time from 6am to 6pm in Roman numerals with no divisions of the hours.

behind. His hands can be seen grasping the disc at both sides. He has a flute in his mouth which casts a shadow to indicate the time on the Arabic numerals around the bottom edge of the disc.

The garden also has a Horizontal Dial to show summer time, when our watches are set one hour ahead of Greenwich, an Analemmatic Dial where

The direct west dial is also made from several dressed stones set flush with the random rubble stone of the tower. It is undated and has a 'T' shaped gnomon with a serrated edge on the bot-tom of the cross bar. It shows the time in hours from 1pm to 8pm in a mixture of Roman and Arabic numerals, presumably because there was insufficient space to use all Roman. It shows the hours **I, II, III, 4, V, 6, 7,** and **8.**

**Vertical Dial on South Wall of the tower
at St Michael's, Castlecaulfield** ^{MH}

**Dial on West Wall of the tower
at St Michael's, Castlecaulfield** ^{MH}

215

Kilmalkedar Sundial ^{PR}

DINGLE, Co. Kerry

This remote Early Christian site of Kilmalkedar on the Dingle Peninsula, associated with St Brendan, has at its centre a 12th century Romanesque Church, an Alphabet stone, a holed Ogham stone, two Bullaun stones, a large stone cross, St Brendan's Oratory and numerous cross slabs. Located in the old cemetery in the ruins of the Church is a vertical stone goblet-shaped sundial. Dated to the 8th century, it is 1200mm high by 300mm wide by 150mm thick. The elegantly carved bowl-shaped dial face is divided into four equal sections by four double and one triple-line dividers that radiate from a gnomon hole with each divider terminating in a semi-circle. The gnomon is missing. The stem is engraved with vertical lines terminating in a rectangular geometric pattern.

DONABATE, Co. Dublin

St Patrick's Church. This charming little church, surrounded by its ancient churchyard, stands on the southern side of The Square in the village of Donabate, which community is also

St Patrick's Church Sundial, Donabate ^{OD}

referred to in the account of the sundial at nearby Newbridge House. It was built on the site of earlier churches dating back, as far as records can be traced, to a period shortly after the Anglo-Norman invasion of Ireland in 1169, evidenced by the fact that it is joined on one side to a Norman tower house. Also, there is written proof that it was in the patronage of the Archbishop of Dublin in the year 1235.

Over a low gothic doorway into a small porch, which was added in 1706, on the southern side of the church, is a rather large timber-faced Vertical Dial with an iron gnomon. The oral history of the parish indicates that, going back many generations, there has been a succession of sundials of timber construction in this precise position, and that as each one of them eventually succumbed to decay and wood-rot it was replaced by a new successor, while retaining its iron gnomon for more prolonged periods of service.

216

The present custodian of the sundial is Mrs Deirdre Unger whose father before her had accepted the responsibility for it as a young man and eventually passed it on to her in his latter years. Along with her responsibilities, she inherited a very old delineation of the dial's original hour lines, on paper, with which to ensure exact reproduction at renovation times. The timber element of the sundial has been renewed once during her stewardship, to date, and she continues to paint it herself on an approximate ten-year basis. However, the iron gnomon is now becoming very wasted and brittle, and will soon have to be replaced with a new one of exactly the same design, thus continuing this cycle of perpetual renewal under the watchful eye of its designated personal minder.

The sundial is a south-west decliner, delineated in 15 minute sub-divisions and measuring 79cm high by 65cm wide. It continues to be a remarkably accurate time-keeper for a dial with so many individual restorers over the years, and stands as a visual reminder of their on-going care and attention.

Newbridge Demesne. The Demesne is situated close to Donabate, just north of Dublin and adjacent to the sea coast.

Its centrepiece is Newbridge House, a very fine Georgian mansion built by Archbishop Charles Cobbe after he purchased the lands at Donabate in 1736. Newbridge was family home to several generations of the Cobbe family before it was acquired by Dublin County Council in 1985. It is now managed by Fingal County Council and is open to the public, although the family also continue to reside there from time to time as part of the agreement of sale.

A five acre walled garden, which was formerly used as a vegetable and fruit garden, was converted to an orchard in the mid 20th century. Part of the later restoration work in this garden was the establishment of a Horizontal Dial in 2001. A suitable pedestal in the style of the gardens period was acquired at that time and Silas Higgon of 'Connoisseur Sun Dials' was commissioned to produce a compatible brass dial plate and gnomon. The 375mm diameter plate carries the Signs of the Zodiac and the Equation of Time, and Local Apparent Time can be read from a one-minute interval, diagonal scale. The heavy-gauge brass gnomon is pierced in a pattern and style of the period, giving the entire structure a very pleasing appearance, which enhances the general ambience of the walled garden.

DUBLIN

Iveagh Gardens. The gardens are among the finest and least known of Dublin's parks and gardens. They lie on a site just south of St Stephen's Green, screened on all four sides by houses and secluded from the noise of the city. They were designed by Ninian Niven in 1865 as an intermediate design between the 'French Formal' and the 'English Landscape' styles. They demonstrated the artistic skills of the landscape architect of the mid-19th century and display a unique collection of landscape features which include rustic grottos and a cascade, sunken formal panels of lawn with fountain centrepieces, wilderness woodlands, maze, rosarium, American garden, archery grounds, rockeries and rookeries. The conservation and restoration of the gardens commenced in 1995 and, to date, most of the features have been restored; for example the maze in box hedging with a Horizontal Dial as a centre piece. The recently restored cascade and exotic tree ferns all help to create a sense of wonder in this 'Secret Garden'. The pedestal for the sundial, which was established in 2001 and based on a typical mid-19th century design, consists of three different types of stone. The base is of Wicklow granite, the pedestal in Portland

Fine Horizontal Dial at Newbridge House [OD]

Horizontal Dial in Iveagh Gardens, Dublin ^{OD}

Dial at the Royal Hospital, Kilmainham, Dublin ^{MC}

stone and the dial plate in Kilkenny limestone. A carving of an eight-pointed compass, with a fleur-de-lys indicating north, forms a centre piece for the dial plate, which is 370mm in diameter, with the hour lines sub-divided at five minute intervals.

The limestone plate was carved by Tania Mosse, the design of the pedestal and the delineation of

Iveagh Gardens, Dublin ^{OD}

the dial was provided by Owen Deignan, and Ted Sweeney constructed the brass gnomon.

The Royal Hospital. The hospital in Kilmainham, was founded by the first Duke of Ormond as a rest home for infirm and maimed officers and soldiers of the army of Ireland in 1680. The architect was Sir William Robinson and while basing his design on work by Sir Christopher Wren he was also influenced by the layout of Les Invalides in Paris. The building consists of four unbroken ranges enclosing a courtyard. At the time of its construction the Hospital Grounds were part of the Phoenix Park. The Royal Hospital continued in use as a retirement home for soldiers until 1927, it then lay empty for many years. Following restoration the building was reopened in 1991 as the Irish Museum of Modern Art. There is a wooden Vertical Dial 1000mm wide by 1200mm high on the south wall of the courtyard over the entrance to the dining hall. It has a blue painted background with gold Roman numerals and hour lines and an ornate scroll gnomon.

The National Botanic Gardens. Founded in 1795 by the Royal Dublin Society the gardens cover a total area of 19.5 hectares, part of which is the natural flood plain of the River Tolka. The gardens contain a large plant collection which includes approximately 20,000 species and cultivars. There are four ranges of architecturally notable glasshouses including the recently restored curvilinear palm house. Notable features include herbaceous displays, rose garden, rockery, vegetable garden, arboretum, extensive shrub borders and wall plants. Gardens are accessible for people with disabilities but there are some steep gradients. There are two sundials in

IRELAND

218

National Botanic Gardens, Dublin

the gardens. Outside the Alpine Yard there is an undated (c 1800) brass Horizontal Dial. It is 350mm in diameter with a 250mm high pierced gnomon and it bears the makers name **Lynch 26 Capel St Dublin**. The centre of the dial around the base of the gnomon is badly corroded and the Equation of Time values and the names of the geographical locations are barely legible. The time rings near the edge of the dial are showing signs of wear probably from being polished by numerous visitor's hands. The chapter ring shows 4am to 8pm in Roman numerals. There is a minute ring outside the chapter ring around the outside edge of the plate. It has a 16 point compass at the base of the gnomon and there is an anti-clockwise Equation of Time. There are 32 lines on the dial plate to indicate noon in distant cities and countries. The dial, which is

still quite capable of displaying local time, is inset into the top of an 1150mm high modern granite pillar. It was originally located in the Rose Garden but was moved to its new location for security reasons.

A modern black iron Armillary Sphere now sits on the original Lynch fluted sandstone column.

Sandymount Strand. A Polar Sundial is located here, so called because the gnomon or style inclines towards the celestial pole. It is situated in a park at the northern end of Sandymount Strand on the southern shore of Dublin Bay. It is one of a pair of granite sculptures erected in 1983 to celebrate the centenary of the birth of the Irish writer, James Joyce, who refers extensively to the Sandymount area in his seminal work of litera-

Sundial by Lynch at the Botanic Gardens

Modern Dial at Sandymount Strand

219

**'Standing Stone of the Sun'
at Sandymount Strand**

^{OD}

ture, 'Ulyssses'. The co-alignment of a second stone - the Standing Stone of the Sun - which is 4.3 metres high and lies about 150 metres southeast of the sundial, indicates the direction of the rising sun on the morning of the annual Winter Solstice.

The Polar Dial was cut from a smaller granite block, just 1.5 metres high, in the shape of an inverted, diamond-based pyramid, with corners precisely hewn, and parallel hour lines finely engraved to mark the Solar Time. Both stones were carved by the sculptress, Cliodna Cussen, for which composition and craftsmanship she won an award in the International Sculpture Exhibition in Gorey, Co. Wexford. Dr Ian Elliott of Dunsink Observatory was instrumental in providing all the solar calculations required for the declination of the dial and the solstitial orientation of the two stones. The four corners at the base of the dial correspond to the cardinal points of the compass, and the faces of the pedestal are engraved with motifs inspired by the Four Forces of Nature. The carving illustrating nuclear power is intriguing, as its figurative expression is the three equal circles enclosed by a large circle, a motif, which is frequently found in early medieval books, the Book of Kells in particular. There is here a further astronomical relevance to the year 1983, in which the two sculptures were erected. The Cycle of Saros occurs just once in every 18.6 years, when the Sun sets in the west at precisely the same time as the Moon rises in the

east. Consequently, it was fitting that the unveiling ceremony performed by the Lord Mayor of Dublin, should take place in that particular cyclical year. Finally, there was a fervent hope expressed at the time by Cliodna Cussen and local conservationists, that the erection of these commemorative sculptures would help put a stop to any unsightly interference with the natural beauty of Sandymount Strand by the introduction of commercial development or ring-road construction. Happily, these hopes have been realised and Sandymount Strand still retains its ethereal vista, approaching which, James Joyce once felt that he was *'walking into eternity'*.

Trinity College. Trinity College was founded by Royal Charter of Queen Elizabeth I in 1592, its present Georgian architecture dates from the 1700s. It now has over 15,000 students from more than 90 different countries. In the Memorial Garden there is a modern bronze Vertical Dial, 400mm wide by 500mm high on the gable wall of Residential Block 40. Longitude corrected, it has full hour lines and ticks for the half hours. It shows the time from 7am to 7pm in Roman numerals and is marked Long 6° 25'.

St James's Gate. On the last day of December 1759, 34 year old Arthur Guinness entered his newly acquired brewery at St James's Gate, Dublin City. In addition to ales, Arthur Guinness brewed a beer relatively new to Ireland that contained roasted barley which gave it a characteristically dark colour. This brew became known as 'porter', so named because of its popularity with the porters and stevedores of Covent Garden and Billingsgate in London. Arthur tried his hand at porter and the rest is history. Today ten million glasses of Guinness are enjoyed daily in countries

Modern dial at Trinity College ^{MH}

Obelisk at St James's Gate ^{TL}

Harbour. It was built in 1586 in expectation of an attack on the area by the Spanish Armada. There had previously been a Celtic fort and a Norman castle on the site. Because of its prominent location the fort has been of strategic importance throughout the centuries. Under British rule it was described as 'The Second Fort of the Realm' and bore the title of 'The Royal Fort of Dun Cannon'. The threats of the Spanish Armada, Napoleon and Hitler were responsible for the frequent refurbishment and strengthening of the fort over the years. The fort remained in the control of the British Government until the War of Independence in 1919. It was set alight by Republican forces in 1922 and lay in ruins until the outbreak of World War II. It was then rebuilt and occupied by the Army. When the garrison was withdrawn at the end of the war it was used

all over the world. A 10 metre high obelisk, with a drinking fountain at the foot and four Vertical Dials near the top, one on each face of the obelisk, was erected in James's Street, outside St James's Gate, in 1790. It was designed by the architect Francis Sandys as one of a series erected under the patronage of the Duke of Rutland, Lord Lieutenant of Ireland 1784-7. None of the others have survived. The obelisk was restored by Dublin Corporation in 1995 and the north facing dial bears the date 1790.

DUNCANNON, Co. Wexford

Duncannon Fort is a star-shaped fortress on a strategically important promontory in Waterford

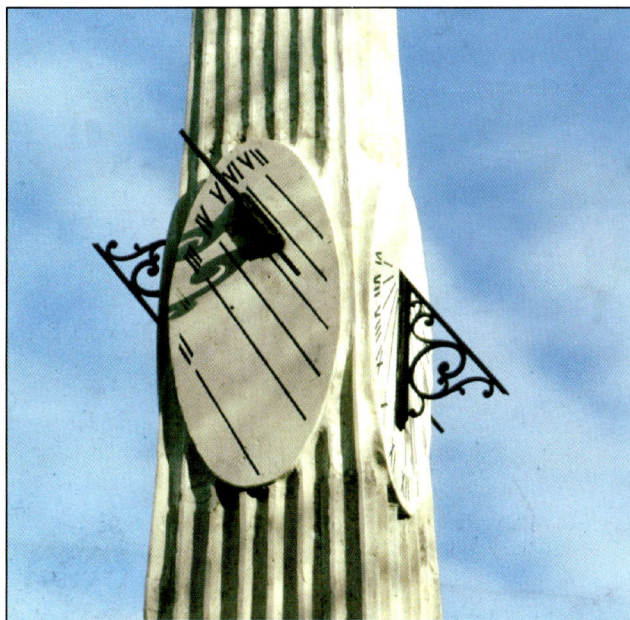

Detail of Obelisk Dial at St James's Gate ^{TL}

The Comoradh Sundial at Duncannon Fort ^{MC-}

221

as a summer training camp on a diminishing scale until 1986. In 1993 it was handed over to Wexford County Council and is currently open to the public. In the centre of the parade ground stands the Comoradh Sundial. This was especially commissioned and executed, in Kilkenny limestone, in 1998, as a permanent memorial to the people and the events of 'The Rising' of 1798, during which Wexford played a leading role in that tragic conflict. The sundial pedestal, which is of square construction, has an Irish harp carved on one side along with the words:

Commemorating the People of 1798

On the opposite side is carved a twelve-month graph of the Equation of Time, with instructions on the conversion of Local Apparent Time to Greenwich Mean Time. The time graduations on the dial plate are at five minute intervals and there is a line of declination carved on it which indicates, by the shadow of the nodus on the gnomon, the time at which 'The Rising' commenced: 7.30pm on 26th May 1798. The dial plate and all the graphics and lettering on the pedestal were carved by the sculptress Tania Mosse, Owen Deignan provided the solar calculations, and Ted Sweeney fabricated the brasswork.

DUNGANNON, Co. Armagh
The 'Argory' was built in 1819/24 by Walter MacGeough Bond. His great grandson gave it to the National Trust in 1979.

The house is just as it was at the turn of the 20th century without electricity. The courtyards have fascinating displays; the oxy-acetylene plant which made the gas with which the house was lit; the old laundry; the coach house and the stables. There is an octagonal brass Horizontal Dial in excellent condition in the rose garden. It measures 355mm across the flats and bears the Bond Armorial Shield with their motto:

Nemo Me Impune Lacessit
(No one provokes me with impunity)

Dial by Lynch at the Argory

The dial itself was made by Lynch & Son Dublin, and is dated 1820. There is an Equation of Time table in the form of four concentric rings and an 8 point compass rose. It shows the time in Roman numerals from 4am to 8pm. There are 38 lines on the dial plate to indicate noon in distant places. Each line is labeled with the name of the appropriate place, from **Batavia** in the east to **California** in the west. The dial is recessed into the circular capital top of a sandstone pillar. There is a circular inscription in the stone between the dial and the edge of the capital, in a curious mixture of capital and small letters:

**~HERE Reader.Mark_THE silent steps
of never-standing TIME E.B.**

The Argory near Dungannon

The Argory Dial by Lynch of Dublin

222

E.B. may have been Edward Bond (1842-1891) one of Walter's sons.

DUNMORE, Co Galway

The town of Dunmore, in north Co. Galway, prides itself in having the largest brass Horizontal Dial in Ireland. It measures 30 inches in diameter and stands on a stone pedestal, surrounded by a protective iron railing, in the centre of the Market Square.

Its existence had its origins in 1948 when a local man, Tom Keaveney, came into possession of a 12 inch diameter bronze sundial which had been made for Black Rock Lighthouse, Co. Sligo, in 1833. This was the only survivor of 24 similar dials, made by David Scott of Dublin, which were supplied by the Commissioners of Irish Lights to the major Irish lighthouses for the purpose of correcting and rating the Lightkeepers' clocks - this being the only method of doing so at the time as it was long before the introduction of telegraph or wireless communication.

Subsequently, in the late 1990s, Tom Keaveney decided to have a memorial, in the form of a large sundial, made and dedicated to his wife, Margaret, who had died some years earlier, and to use the beautifully designed 1833 lighthouse dial as a pattern for its construction. Consequently, he commissioned Galsworthy, the London graphic engravers, to carry out this work.

By happy coincidence, around the same time, the Dunmore Tidy Towns Committee were considering the erection of a suitable monument to mark the arrival of the new Millennium, and accepted the offer of the Keaveney family of the donation of their dial, with the result that the two objec-

tives were simultaneously achieved - the Committee and the townspeople had their splendid monument and Tom had the perfect site for his memorial to his beloved wife, Magaret.

The unveiling took place at the equinox, the 22nd September 2000, and revealed a brass sundial of unusually generous proportions, with a pierced gnomon and deeply-cut hour lines and furniture, which included sub-divisions of very readable two-minute intervals, a comprehensive table for the application of the Equation of Time and a centrepiece depicting an eight pointed Mariner's compass.

ENNISKERRY, Co. Wicklow

The German architect Richard Cassels built Powerscourt House, in 1740 for Richard Wingfield, Viscount Powerscourt, on the site of an earlier 14th century castle. The house was completely gutted by fire in 1974 just after it had been completely renovated. Only the facade survived and it now houses shops, a cafe and an exhibition centre dealing with the history of the estate. The 19th century gardens were based on formal renaissance designs. It took 100 men 12 years to finish the five sweeping terraces of the Italian Garden which drops down to Triton lake, a circular pond with a 30 metre high fountain jet

Large brass Horizontal Dial at Dunmore [DI]

The 'Spitting Men' fountain at Powerscourt [SS]

Vertical Stone Dial at Powerscourt ^{SS}

Equatorial Dial at Limerick University ^{KT}

flanked by two gigantic zinc winged horses based on the supporters of the Wingfield coat of arms. Other features include statues of Apollo, Diana, Fame and Victory and the 1770 Gate from Bamberg Cathedral in Bavaria. There are over 200 varieties of trees and shrubs in addition to Italian, Japanese and Walled gardens. There is a Vertical Dial carved from stone 1000mm high by 750mm wide located in the 'Spitting Men' fountain. Both ends of the hour lines are marked with Roman numerals **V** to **XII**. The 'Spitting Men' were bought in Paris in 1872 and the dial would date from that period. The bronze plaque above the dial reads:

HORAS NON NUMERO NISI SERENAS
(I count not hours that are not serene)

LIMERICK

In 1985, when the National Institute for Higher Education in Limerick (later to become Limerick University) had undergone considerable structural renovation and extension, the principal architect involved, Patrick Whelan and his fellow consultants, offered to fund the provision of a suitable piece of sculpture to mark the successful completion of the contract. To this end, he commissioned the Cork sculptor, Ken Thompson, to submit a suggestion for an appropriate work, in stone, for the enhancement of the Campus.

As the Institute at the time was specialising in the sciences and in the promotion of high-grade technology, the sculptor came up with the suggestion that, *'if you know where you coming from, you have a better idea of where you are going'*, and proposed the erection of mankind's oldest scientific instrument - the sundial. His proposal was accepted and, having designed the

project in full, he set to work on a ten ton block of limestone, from which he fashioned an Equatorial Sundial of truly massive proportions, in one single piece. It measures 2.1 metres high and almost 1.2 metres square.

An Equatorial Dial is unusual in that it has two face plates, the upper one for use in the summer and the lower one for the winter period when the Sun is below the equator. Consequently, both faceplates lie parallel to the equator with the gnomon or shadow pin lying parallel to the Earth's axis. The Limerick University sundial has four inscribed panels, each carved into the homogenous stone. On the south side are the words from Shakespeare:

WHILST PHOEBUS ON ME SHINES,
THEN VIEW MY SHADE AND LINES
The north panel carries the Psalm:
I HAVE CONSIDERED THE DAYS OF OLD
AND THE YEARS THAT HAVE PASSED
On the western side is carved the Equation of Time, which is required for converting Solar Time to Standard Time, and the eastern side of the stone bears the name of the donors of the sundial.

LISSUMMON, Co. Armagh

Three metres up on the wall of the re-built Roman Catholic Church of the Immaculate Conception at Lissummon, 5km south-west of Poyntzpass, is a 750mm square slate Vertical Dial. It is inscribed:

Vertical Slate Dial at Lissummon

MADE BY THOMAS M'CREASH
THE REVEREND BERNARD LOUGHRIM P.P.
LOWER KILLEAVY
ANNO DOMINI 1826
LATITUDE 54^D 12^M N

There is a Latin motto:
MEMENTO MORI
(Remember you will die)
and two mottoes in English:

   Years following years steal something everyday
   At last they steal us from ourselves away
lines from Alexander Pope (1688-1744) and:

   This plainly shows to foolish man
   That his short life is but a span
from an unknown author.

The dial shows the time from 4am to 8pm in Roman numerals. Inside the chapter ring there are 5 minute, 15 minute, 30 minute and hour time rings. There is a split noon to compensate for the thickness of the bronze scroll gnomon. The dial is decorated with a small sun with a human face at the top, two rising suns with floral background, one in each bottom corner, and two half moons with a human face in profile at each side. The dial has obviously been restored probably when the chapel was rebuilt, c 1933.

Thomas McCreash was the first schoolmaster in the Poyntzpass Chapel School from around 1813 until 1826. Another three of his dials adorn other local churches; St Joseph's Roman Catholic Chapel, Poyntzpass, the Church of Ireland, Poyntzpass, and St Patrick's Ballyargan Roman Catholic Chapel at Mullaghglass. Another is in a private collection.

MALAHIDE, Co. Dublin

Malahide Castle, set in 250 acres of parkland beside the seaside town of Malahide in North County Dublin, was both a fortress and a private house for nearly 800 years. The Talbot family lived here from 1185 to 1973, when the last of the male line, Lord Milo Talbot de Malahide, died. He had taken a keen interest in the Castle's large walled garden and literally transformed it into a botanic garden. A great traveller, he collected many of the species from the wild, placing a particular emphasis on southern hemisphere plants, particularly Australian and Chilean species. It is in this fine garden, which is open to the public, that a conventional Horizontal Dial now stands. Its rather low, sandstone base was originally a pedestal for a garden statue which had been missing for some time. Subsequently, in 2001, it was decided by the current owners of the Estate, Fingal County Council, to give it a new lease of life in its present function.

The 375mm bronze sundial is by Silas Higgon and has a pierced gnomon, based on a 1718 design by William Dean. The dial plate is deline-

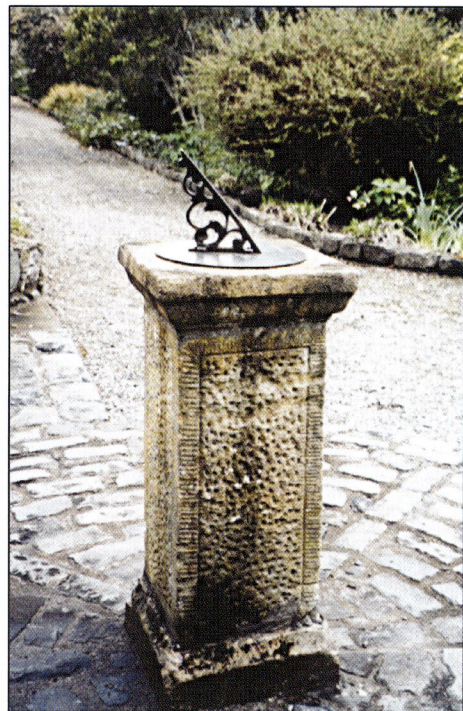

Horizontal Dial at Malahide

225

ated in five minute intervals, carrying an image of the traditional 16 pointed mariner's compass card, and has the Equation of Time engraved in graph form on its most southern sector.

MITCHELSTOWN, Co. Cork

This thriving town nestles under the Galty Mountains, just 33 miles north of Cork on the main road to Dublin. It has a fine, spacious market square in its centre, which is known as New Square, and it was on the upper, or western side, of this area that it was decided to establish a Timepiece Sculpture to mark the arrival of the new Millennium.

Mitchelstown - Market Scene and St Michael the Archangel ^{WP}

The Mitchelstown Heritage Society, with the support of Cork County Council, had at first considered the erection of a conventional, Horizontal Sundial but on further thought opted for an Analemmatic or Human Dial, a type that is rare in Ireland and would ideally suit the space available. It is laid out at ground level as a large ellipse, measuring 6 metres along its major axis, using hour points rather than hour lines. These points are in the form of 280mm diameter bronze discs which carry Arabic numerals, in relief, and are flush with the surrounding ground, which is paved in brick, laid in a circular pattern. The scale of dates in the centre of the dial, on which a person can stand in order to read Local Apparent Time by the direction of his or her shadow, is cut in limestone. A bronze panel showing the Equation of Time and conversion

instructions is also let into the adjoining paving. A further practical feature included in the Timepiece Sculpture plan was the provision of a semicircular, limestone bench overlooking the analemmatic sundial. This bench has a carving at one end of St Michael the Archangel, the patron saint of Mitchelstown, complete with sword and scales. At the other end is carved a buxom young woman, symbolising the original function of the Market Square by selling carrots, onions and a sheep - all the ingredients of a good Irish Stew. Across the length of the bench is carved the legend:

The learned line showeth the City's hour

being a timely nod to the function of the sundial. The design of the sundial and bench and all the carving involved in the Timepiece Sculpture was the work of the Cork Sculptor, Ken Thompson. The solar calculations and delineations were by Owen Deignan.

The dial was described by RTE's 'Rattlebag' programme as the best and most original public sculpture erected in Ireland during 2001.

MONASTERBOICE, Co. Louth

Monasterboice was founded by Saint Buite who died in 521. It was in use until 1122 and only its ruins remain today. Two of the finest High Crosses in Ireland, dating to the 9th century are to be found here. One of these, the 5.5m high 'Cross of Muiredach' is highly decorated with detailed biblical carvings and not far away is the equally decorated 'Tall Cross''. Nearby is the Round Tower which when burned in 1097 contained the monastic library and other treasures.

Human Analemmatic Dial at Mitchelstown ^{WP}

In the graveyard there are two medieval churches.

Early Dial at Monasterboice

Other items of interest include a fragment of another High Cross and a sundial, whose function was to mark the hours of prayer. The semicircular sundial face has a horizontal diameter line across the top of the slab; a hole for the gnomon at the centre of the circle, and three radii, one vertical, the other two flanking it and sloped at about 45°. The times shown by these lines are those for the canonical prayer hours of; Prime (6am), Terce (9am), Sext (12 noon), None (3pm) and Vespers (6pm).

The drawing shows that below the dial there is a circle containing a cross, and underneath that, another circle containing a cross in relief. The cup shaped hollow further down the stem is thought to be a natural fault in the stone.

NENDRUM, Co. Down
Located on Mahee Island in Strangford Lough, the ruins of Nendrum Monastery consist of three concentric dry-stone walled enclosures. The central enclosure has the remains of a round tower, the ruin of a church and a graveyard. The middle enclosure contains the remains of huts and workshops. Little is known about the outer enclosure but there is evidence for industrial work outside, including a tidal mill and landing places. A vertical stone sundial, 1900mm high 400mm wide by

Drawing of Monasterboice Sundial

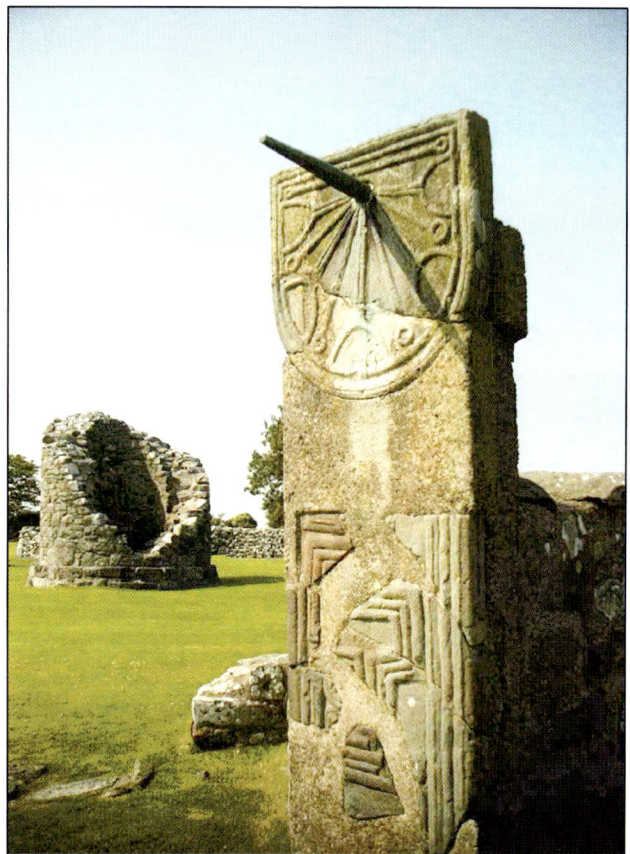
Early Dial at Nendrum

227

The Roofbox at Newgrange

The sundial enthusiast visiting Knowth has an opportunity to see kerbstone SE4 - the so called 'sundial stone' and kerbstone NE4 which has an easily recognised sundial carved on its upper face. This sundial is now permanently in shadow due to an overhanging concrete ledge which was installed during reconstruction work to protect the kerbstones from damage. The sundial is divided into 8 segments, the dividing lines terminating in four dots. The space between each line is further divided by four dots. It has two other holes for locating a gnomon in addition to the centre hole on the east-west line.

100mm thick was reconstructed from pieces found during excavations in 1924. Tradition has it that the Monastery was founded in the 5th century by St Machaoi, but the dial is thought to date from the 9th century. There is a visitor centre on site with interactive displays and models. The island is accessible by car across a very narrow causeway.

NEWGRANGE, Co. Meath

The 11 metres high, 80 metres in diameter cairn, covers a passage, which leads to a dark cruciform shaped chamber. A continuous circle of 97 large kerb stones, many of them decorated, form the base of the cairn. Over the entrance to the passage is a 1 metre square opening known as the 'roof box'. The Stone Age builders of Newgrange positioned the roof box so that at the Winter Solstice the dawn sunlight would shine through it and enter the mound. On the morning of the solstice as the Sun rises, its rays shine through the roof box and creep slowly along the passage, across the chamber floor illuminating the back wall, and then gradually filling the entire chamber with sunlight. After about ten minutes it retreats back down the passage and darkness returns to the chamber for another year. This precisely aligned solar monument has been signaling the beginning of each New Year for over 3000 years.

In addition to Newgrange there are two other large cairns in the Boyne valley, Knowth and Dowth and aproximately 40 other smaller satellite mounds.

Knowth Sundial Stone, SE4

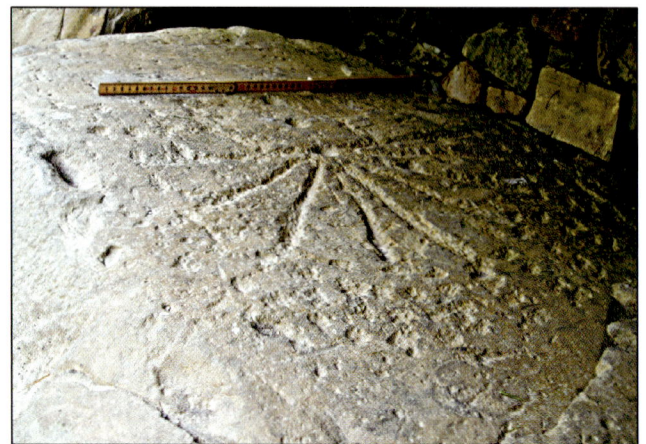
Knowth Sundial on Kerbstone, NE4

Modern Vertical Dial on house at Portaferry

PORTAFERRY, Co. Down

There is a 1990 brass Vertical Dial on the exterior wall of a dwelling house in Ferry Street, Portaferry.

The engraving of a seascape shows a sailing ship on the waves, a lighthouse, a castle and the sun with the motto:

**TIME'S BUT A SHIP THAT BEARS
THEE NOT THY HOME...**

LAT LIV XXIII N   LONG V XXXIII W

DUNLOP FECIT MXM

It has an engraving of a crest with a cat rampant and the motto:

TOUCH NOT THE CAT OUT OF GLOVE

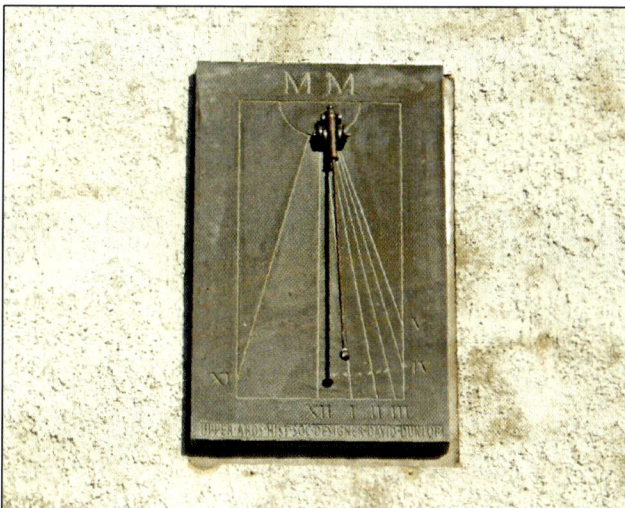
Millennium Dial at Portaferry

There is an Equation of Time graph with **BRITISH SUMMER TIME** above it. Also in Portaferry there is a millennium **MM** vertical slate dial 7 metres up on south wall of The Market House. It has a brass gnomon in the form of a cannon firing a ball, in remembrance of the cannons fired in defence of the Market House during the 1798 Rising. Inscribed **Upper Ards Hist. Soc** and **Designer David Dunlop**, the hour lines are marked with Roman Numerals from **IX** to **V**.

POYNTZPASS Co. Armagh

In 1792 the lease for the land on which St Joseph's was to be built was signed making it one of the first Roman Catholic churches to be built in Ireland since the Reformation. Tradition has it that a bottle of whisky was built into a wall of the church. It would seem that the builders were fond of a drop of alcohol while working but not wishing to be caught drinking on the job they concealed the bottle in the partly built wall. When the priest arrived unexpectedly they had to continue building and so entombed their spirituous refreshment. The 600mm square vertical south slate dial is 4 metres up on the external south wall:

**MADE BY THOMAS M'CREASH
for THE REVEREND Mr HENRY CAMPBELL
1817**

There is a Latin motto:

MEMENTO MORI
(Remember you will die)

The dial is decorated with a small sun with a human face at the top, two rising suns with floral background, one in each bottom corner, and two half moons with a human face in profile at each side. The dial shows the time from 4am to 8pm in unorthodox Roman numerals. The angular layout of the hour lines on a direct south Vertical

Vertical Slate dial at St Joseph's

Vertical Slate dial at Acton Parish Church ^{MH}

Modern Dial at Ardgillan Demesne ^{OD}

Dial is the same as that for a Horizontal Dial of the same latitude. A mirror image of a Horizontal Dial will show the layout for a Vertical Dial and vice versa. It would appear that M'Creash used the mirror-image of a Horizontal Dial as his template and meticulously copied exactly what he saw and so produced his unconventional numerals. He also included the hour lines for times before 6am and after 6pm which is correct for a Horizontal Dial but not for a direct south Vertical Dial.

He did the same on the almost identical dial he made for the nearby Acton Parish Church two years later. We have no way of knowing how he explained this to his Reverend clients who must surely have recognized the error. Thomas M'-Creash was the first schoolmaster in the Poyntz-pass School from around 1813 till 1826. Another two of his dials adorn local churches at St Patrick's Ballyargan Roman Catholic Church at Mullaghglass and the Church of the Immaculate Conception at Lissummon.

SKERRIES, Co. Dublin

Although referred to as a castle, Ardgillan Demesne is a large country-style house with castellated embellishments. Originally named 'Prospect', the central section was built in 1738 by Robert Taylor, with the west and east wings added in the late 1700s.

The 200 acre Demesne, which is an area of great beauty overlooking the adjoining coast, is situated mid-way between the seaside towns of Skerries and Balbriggan, in north Co. Dublin.

The gardens at Ardgillan are in two main parts, the formal Rose Garden to the west of the house and the 2 acre Walled Garden, in which a Horizontal Dial is established to the north-west. In 1982 work began on the replanting of these formal gardens, using the layout from the Ordnance Survey map of 1865, on which it was noted that a sundial, long since missing, had been depicted. As sundials were common garden features up to and including Victorian times, it was decided to include a replacement dial in the evolving scheme of things. Consequently, in 1992, Fingal County Council, which now owns and manages the estate, commissioned the making of a new sundial to be erected in the Ornamental Garden, which forms an integral part of the general garden complex. As the surrounding walls are of old mellow red-brick it was decided to incorporate the same material in the construction of a 1100mm high base and pedestal, while having the 400mm square face plate of the dial cut on Kilkenny limestone. Two brass plates, one engraved with a graph of the Equation of Time and the other indicating the procedure for converting readings to Standard Time were affixed to the pedestal.

Tania Mosse carried out all the stone carving involved, Owen Deignan delineated the hour lines and Ted Sweeney provided the brasswork requirements.

STAINED & PAINTED GLASS SUNDIALS
Christopher Daniel

Perhaps the most outstanding class of sundial and the most beautiful is that of the uncommon vertical stained-glass sundial, normally to be found in the windows of historic buildings. This form of dial is constructed in such a manner that the gnomon is fitted to the dial on the outside of

private individuals. Heraldry was an attractive alternative to religious imagery, since those who had wealth and power often commissioned heraldic windows to portray their family tree and their connections to other important families. On receiving a substantial and lucrative commission,

Two small Stained Glass Sundials by John Oliver originally installed in Northill Rectory, [MC] **Bedfordshire now kept in Northill Church Museum**

the window, although its shadow is viewed from the inside, indicating the time in the usual way. Most stained and painted sundials date from the 17th century, when the Art of Dialling, as it was called in England, reached its zenith. However, the construction of stained-glass dials received impetus from the puritanical attitude of the time to religious glass-painting, which obliged 17th century glass-painters to turn away from the church for their main source of employment. They sought work from universities, city livery halls, civic authorities and the stately homes of

a glass-painter might sometimes include a window sundial *gratis*.

The earliest recorded stained-glass sundial seems to have been one dated 1518 in the castle of the Kurfürst von Sachsen at Altenberg, in the Principality of Sachsen, in Germany. Altenberg, once noted for the production of fine leadwork, is a small city, close to what is now the border of the Czech Republic. At that time, this was the country of Bohemia, once a wealthy and powerful kingdom, with the beautiful city of Prague as its capital. Enclosed by mountains or high hills, covered by forests, Bohemia had an abundance of natural resources and, in the Middle Ages, the region's silver mines were the richest in Europe. Furthermore, the country was known for the quality of its glass-work, for its fine texture and lovely colours. Germany also had a reputation for its glass-work, as well as for instrument-making in the 16th century. Indeed, the earliest use of coloured glass in windows was in Augsburg at the close of the 10th century, whilst Augsburg later became a centre of instrument-making. Thus, no doubt with the exchange of ideas and skills with Bohemian craftsmen, there is no real surprise in the fact that stained-glass sundials first made their appearance in this part of Germany.

It is not known when the first painted or stained-glass window sundial was made in Britain; but the idea may possibly have been introduced into England from the thriving German or Flemish schools of glass-painting in the mid 16th century. Nicolas Kratzer, the astronomer and mathematician, who became 'deviser of the King's horologes' at the Court of King Henry VIII, may have known of such dials in his native Germany and might well have extolled their virtues. Nevertheless, the earliest known stained-glass sundial, still extant and in situ, dated 1585, is set into the magnificent south-facing heraldic window in the Great Chamber of Gilling Castle, in Yorkshire. The splendid Elizabethan painted-glass panels in this window portray the genealogy and heraldry of the Fairfax family and are the work of Bernard Dininckhoff, who is thought to have been a refugee from Bohemia, which had been annexed in 1526, becoming part of the great Austro-Hungarian Empire. Dininckhoff soon gained an excellent reputation, was admitted to the York school of glass-painting and became a much respected person. Indeed, in 1586, a year after he had completed his great heraldic window for Sir William Fairfax, he was made a freeman of the city of York. His sundial was the final embellishment to this monumental work and takes the form of a small glass roundel, just 70mm in diameter, below which is the tiniest self-portrait imaginable (about 7mm in height!), underneath which is the artist's signature and the date.

Signature of John Oliver from the 'Housefly Dial' at Northill

Nevertheless, it was during the 17th century that the stained-glass sundial became fashionable in England, if not in other parts of the British Isles. To the best of my knowledge, only one stained-glass dial has ever been recorded in Scotland, whilst there is only one of which I know that is still extant in Wales. The Scottish dial, dated about the year 1672 or a little later, is attributed to John Oliver (1616-1701), a surveyor, who served under Sir Christopher Wren in London, after the Great Fire, and a well-known member of the London Company of Glaziers and Glass-Painters. Remarkably, the Welsh sundial, at Tredegar House in Newport, is actually dated 1672 and is attributed to Henry Gyles (1645-1709), the celebrated glass-painter of York. Gyles was one of the most notable glass-painters of his day and many dials are credited to him, with or without foundation! Nevertheless, he found it difficult to make a living, since the earlier effects of the Reformation and the later climate of puritanical prejudice, which emerged with the gathering storm of the Civil War, had put an end to the production of religious glass-painting for the windows of churches and ecclesiastical establishments. During these periods, much beautiful glass-work in cathedrals and churches throughout the country was destroyed in acts of what can

Two flies as painted on the Northill Dials by John Oliver; a housefly and a fruitfly

only be described as wanton vandalism. Thus, with the sudden drop in demand for religious work, the glass-painters of the day sought to make a living from other sources, producing illuminated family trees, coats of arms and other heraldic work in stained-glass, not to mention decorative windows reflecting nature or local industry.

In being obliged to turn away from painting traditional religious subjects, despite the difficulties encountered in making a living, the glass-painters of the 17th century, in some senses, gained a new artistic freedom, which is born out in the sundials that have survived to this day. They included fine scroll-work, images of the Sun (sometimes called 'sun-bursts'), various illustrations of the changing seasons, cherubs, winged hour-glasses and other features that are associated with sundials, reminding one of life and death. In many stained-glass dials, their mottoes reinforce the fact that life is short and that 'time flies'! However, the 17th century glass-painter also introduced the image of a fly into the artistic aspects of the sundial, which is regarded as the glass-painter's pun on 'time flies'. These little life-like images were sometimes painted in such a way that the body and wings of the fly were on one side of the glass, whilst the legs were on the other side, giving a distinct three-dimensional effect! Such was the delight in the use of this particular image that it was employed by various glass-painters, to the extent that there are some 30 stained-glass sundials extant that feature the image of a fly. Nevertheless, there are also six dials containing images of spiders, three with images of birds, two with butterflies, one with a dragonfly, one with a fruit fly and one with a wasp!

In the following centuries, interest in stained-glass sundials seems to have waned and few were commissioned by comparison to the number that must have been made in the 17th century. Fur-

The 'Goldfinch' Dial from the Museum of the History of Science in Oxford

thermore, neglect and damage to these beautiful but fragile works of scientific art, more often than not caused by the gnomon being broken off and the glass shattering, has drastically reduced what must have been a popular form of dial in its day, to the rarity that it is now. There are only some fifty or so historic stained-glass sundials still extant in the British Isles. These few remaining sundials are mostly to be seen in private houses, in castles, farmhouses and civic buildings, although there are some churches where such dials may still be found. The Convocation Hall in Oxford and the Oxford Museum for the History of Science also have a number of interesting stained-glass dials that may be seen, the latter mostly having been donated to the Museum when their original site was likely to be destroyed.

To-day, with the 20th century sundial 'renaissance', that has brought about a remarkable resurgence of interest in sundials and in their construction, it is pleasing to note that there are a number of modern stained-glass sundials that have been commissioned, which are in keeping with the art of making such dials. There are examples at Buckland Abbey in Devon, in the church at Toller Porcorum in Dorset, in the church at Old Basing in Hampshire and in the chapel of the Merchant Adventurers' Hall in York. There is also a remarkable modern glass window sundial in the entrance hall of the QinetiQ building at Farnborough in Hampshire, constructed as a mean-time noon-mark, which is accurate to a matter of a few seconds, made in 1996. It is illustrated on page 9. This last dial is not a stained-glass sundial as such; but it makes use of

Spider and fly from the Merton Dial in Norfolk

The dial at Arbury Hall

the same principles that enable it to be read from inside the building and it is a dial that deserves recognition as an accurate work of scientific art.

ARBURY HALL, Warwickshire

Stained-glass sundials dating from the 18th century are exceptionally rare. Indeed, only two such dials, of which I know, date from the mid-18th century, both of which are by the same maker, namely John Rowell of Wycombe (or High Wycombe) in Buckinghamshire. John Rowell (1689-1756) evidently moved with his family from London to High Wycombe about the turn of the century and started business as a plumber and glazier. However, he was a talented man and developed his skills to the extent that he soon became recognised both as a water engineer and as a fenestral artist. He seems to have taken up painting on canvas about the year 1718, although it was not until the 1730s that he started painting on glass. His enthusiasm for his new found occupation led him to advertise in 1733 that he was reviving 'The Ancient Art of Staining Glass'. He became well-respected and much in demand in every aspect of his business. One of his commissions in 1733 was for a stained-glass sundial and although it is not clear as to who commissioned him, or for which property the dial was made, it appears that the glass-work, including the sundial, was later purchased for Arbury Hall near Nuneaton in Warwickshire. Arbury Hall is the ancestral home of James Newdegate, the 4th

Viscount Daventry and has been the family seat for over 400 years, since Tudor times. The dial, taking the form of a square, is set in the bottom central light of a large south window, facing onto a courtyard, and is delineated as declining from south by about 8° towards the west. It is a simple construction, comprising a central rectangular translucent white panel, featuring a realistic ornamental fly and also a butterfly. This is contained by a yellow stained border on three sides, with Roman numerals denoting the hours of the day. Outside this, there is a matt white border delineated with hour-lines, half-hours and quarter hours. The whole sundial is represented as being supported by a painted pedestal, which bears the maker's signature. The window, in which the sundial is situated, in addition contains much other beautiful glass, which seems to be of the same period and which might also be Rowell's work.

BERKELEY CASTLE, Gloucestershire

Set in the lovely countryside of Gloucestershire, overlooking the River Severn and the Welsh border, Berkeley Castle is not only one of the most beautiful and unspoilt of all castles, but is one of the most historic. It has been in the possession of the Berkeley family, from which it takes its name, for some 850 years and is still their family home. It witnessed the brutal murder of the unfortunate King Edward II, at the hands of Sir John Maltravers and Sir Thomas Gurney, in 1327;

Berkeley Castle Dial with its fly at the centre

234

it was celebrated by Shakespeare in his play *Richard II*; and it was besieged by Oliver Cromwell during the Civil War. In 1645, after a siege of three days, the massive wall of the Keep was breached and Berkeley was captured by the Parliamentary forces. This damage to the Keep remains un-repaired as evidence of this historical event. One is told that the damage inflicted on the Castle was not as much as it might have been; but it is likely that the one time Chapel of St Mary, now the Morning Room, would have suffered from the same puritanical vandalism as other churches and chapels. The religious glass-work would almost certainly have been destroyed and evidence to support this is to be found in the beautiful stained-glass sundial, that is now set into one of the windows. A rare example of the 17th century's glass-painting art, now lacking its gnomon, the dial is a south-east declining sundial, in that it declines from the south towards the east. It is embellished with fine ornamental scroll-work and in the centre of the rectangular glass 'dial-plate' it features a fly, seemingly the glass-painter's pun for 'time flies'.

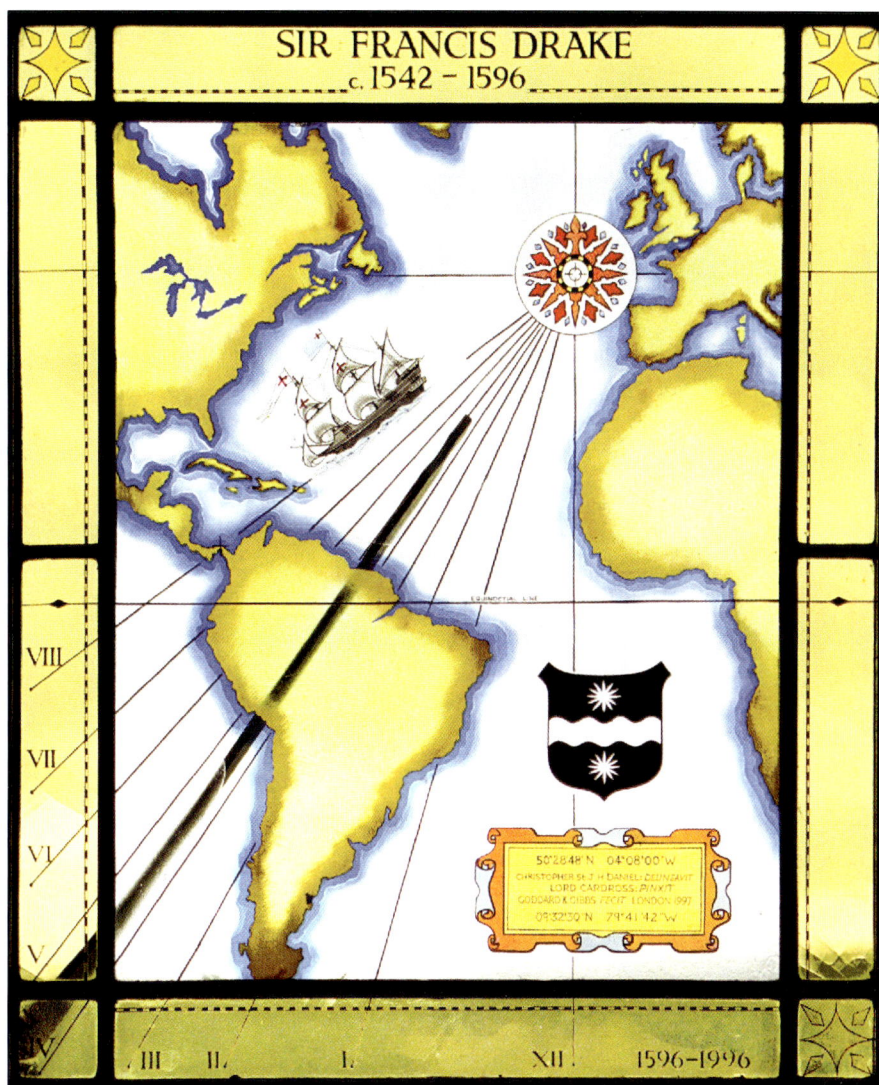

Buckland Abbey, 'Sir Francis Drake' Dial

BUCKLAND ABBEY, Devon

Buckland Abbey lies about 8 miles due north of the City of Plymouth, in wooded countryside, on the eastern slopes of the River Tavy valley. Following his successful circumnavigation of the world, in his famous ship the *Golden Hinde*, in the years 1577-1580, which brought him fame and fortune, Francis Drake was knighted and acquired lands, including a fine family home in Buckland Abbey. In 1595 he ventured on an expedition to the West Indies, which was to be his last. His fleet came to anchor in Nombre de Dios bay, when, despite torrential rain, he was obliged to put his small army ashore in an attempt to seize Panama. It was ambushed and forced to retreat through the jungle, back to the waiting ships. The expedition was stricken with fever, with men dying like flies, as the fleet set sail in a westerly direction. Drake himself succumbed and, on the morning of Wednesday 28th January 1596 (NS), off the island of Buenaventura, at the entrance to Porto Bello bay, he died, where he was buried at sea with full military honours. In 1996 a stained-glass sundial was commissioned for Buckland Abbey by the National Trust, who cares for the property, to mark the 400th anniversary of Drake's death. The dial is situated in a window on the west side of the building, in the upper gallery, which declines from south by 66° towards the west. This large declination was fundamental to the design of the sundial, as it allowed the dial to be constructed in the style of a 16th century chart of the Atlantic Ocean, where the hour-lines are delineated as *loxodromic* lines, radiating from a compass-rose. An image of the *Golden Hinde* is featured, as if sailing on course from Plymouth towards Porto

Merton Church, Norfolk

the bottom, at right angles to the extremities of the hour-lines. The black and white segments of this border mark the quarter hours. Outside this border, Arabic numerals along the top and Roman numerals along the bottom denote the hours of the day, with small black intermediate dots marking the half-hours. This feature is contained within a black rectangular frame which identifies the parameter of the dial-plate. Within the top and bottom of this frame, there may be seen the small holes that once held the supporting structure of the gnomon, which would have been fixed to the outside of the dial-plate. Curiously, the dial-plate seems to have been overlaid on a yellow-bordered square, which, in one corner, contains a life-like fly, that appears to have just settled onto the web of a predatory black spider, whilst, in another corner, there is the motto:

Dum Loquimur Fugit Hora
(While we speak, the Hour flies)

These aspects of the sundial have been 'sandwiched' into a square leaded frame, which, remarkably, cuts off the corners of the dial-plate! Surrounding this square frame there is some strange ornamental glass-work, with four small yellow and brown cartouches, that seemingly depict figures from classical antiquity. The whole of this arrangement appears to rest on an ornate dark orange circular 'wreath', of the kind worn on a knight's helm, with just the glimmer of a blue background. This stained-glass panel, containing the sundial, was damaged some years ago; but now appears to have been restored to its former glory. It is evident from its design that it was intended to balance a similar panel in the adjacent light, which contained a square that featured two large heraldic crossed keys. Thus, the sundial appears to have been cut down to size so that the two squares matched each other!

Bello, on the evening 8 o'clock hour-line, a mythical link between Buckland Abbey and the position of Drake's burial. Drake's heraldic shield is displayed to provide formal identity with the chart, whilst a cartouche below it records that it was designed by Christopher St J H Daniel, developed as a cartoon by Lord Cardross in the studios of Messrs Goddard & Gibbs Ltd of London, where the dial was executed in stained and painted glass by Norman Attwood, and manufactured by their craftsmen.

MERTON, Norfolk

The village of Merton lies in Norfolk, about 20 miles to the west-south-west of the City of Norwich. In St Peter's Parish Church, which is situated in the grounds of Merton Park Hall, there is to be found an intriguing direct east-facing 17th century stained-glass sundial. The rectangular dial-plate is delineated with black painted parallel hour-lines (which also lie parallel to the Earth's polar axis) on a white matt background, with a black and white border running along the top and

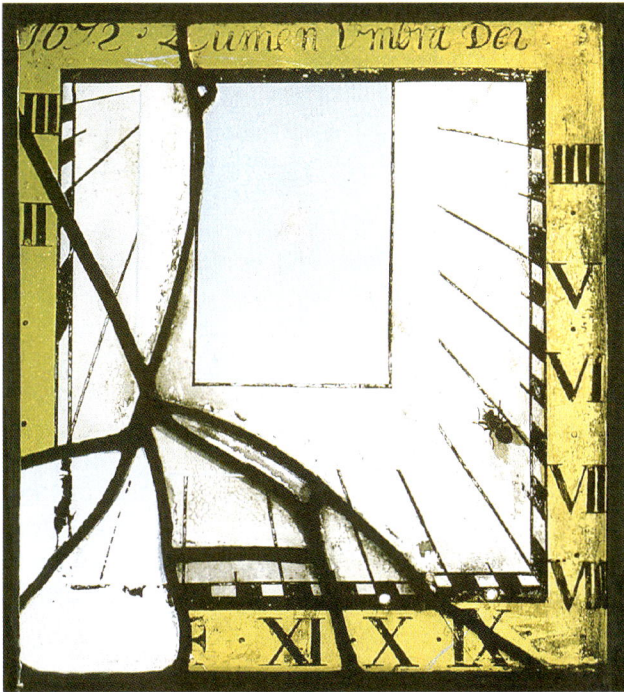

Dial at Tredegar House, South Wales _{CD}

NEWPORT

In a window in Tredegar House at Newport, South Wales, there is the only known stained-glass sundial still extant in Wales. It is a simple south-east declining dial, in the form of a rectangular yellow-stained bordered panel, featuring a house fly, a motto and the date 1672. Tredegar House was the ancestral home of the Morgan family, later the Lords of Tredegar. The house was substantially rebuilt by Sir William Morgan between about 1664 and 1672. Since the Morgan family was wealthy and powerful, it is likely that they would have employed the best artists and craftsmen of the day to embellish their property. Windows, in what was once the dining room, contain armorial panels, which include the heraldic achievements of the Morgan family. These are thought to have been the work of Henry Gyles (1645-1709) of York, which suggests that the sundial might have been his 'finishing touch'.

TOLLER PORCORUM, Dorset

Toller Porcorum is one of two villages that lie some nine miles west-north-west of Dorchester. The other village is that of Toller Fratrum. Both villages take their name *Toller* from the River Toller, now called the River Hooke. The Latin affixes evidently date from the early 14th century, *Fratrum* meaning 'of the bretheren', referring to the possession of the manor by the Knights Hospitallers, whilst *Porcorum* literally means 'of the pigs', referring to the herds of swine or wild boar that once frequented the surrounding woods.

Both villages have historic and interesting Norman churches; but, to mark the Millennium, the residents of Toller Porcorum decided to commission a stained-glass sundial for their church of St Peter and St Andrew. Accordingly, they commissioned John Hayward, the stained-glass artist who has designed and made many beautiful windows for churches in the City of London, as well as creating new glass for the great west window in Sherborne Abbey. With the aid of some technical expertise from a member of the British Sundial Society, Hayward produced a plain direct south-facing stained-glass sundial, decorated in an elegant and colourful baroque style, which manages to be in keeping with both the past and the present.

YORK

The Company of Merchant Adventurers' of the City of York is a similar body to the great livery companies of the City of London. Originally a religious fraternity of influential men and women, founded as a guild in 1357, the Company was an organisation that endeavoured to make their members prosperous through a system of mutual government and fair trading. They sought to regulate commerce and to maintain standards, concerning themselves with the quality of merchandise and the honesty of those involved. The Company embraced most of the richer traders of

Stained Glass Dial at Toller Porcorum _{CD}

237

Merchant Adventurers' Hall Dial in York

Their knowledge of oceanic navigation enabled their ships to venture out of sight of land for many days and the ability to safely reach their destination. The resulting prosperity is manifest in the splendid Merchant Adventurers' Hall that still graces this magnificent city. The *Great Hall* is still in its original state and is still used for banquets on civic occasions. There is also the *Undercroft*, once used as a hospital, and the *Chapel*, at the southern end of the undercroft, which, before the Reformation, was richly decorated with stained-glass windows, fine stone carvings and beautiful tapestries. All these appear to have been swept away with the subsequent purges on the finery of such ecclesiastical establishments. Most of the present furnishings of the Chapel date from 1661; but recently, through the generosity of certain individuals, the plain glass in the lights of the great south window has been adorned with some appropriate stained-glass in a 17th century style. The 'centre-piece' of this new glass-work is the circular panel of the sundial, which declines from south by 35° to the east. The hour-lines are delineated to resemble the deck-planking of a 17th century merchant vessel, on which a young navigator may be seen making a noon-day observation of the Sun with a cross-staff. The sundial is set in a larger rectangular panel depicting a ship of the period under sail.

the city, whose principal source of wealth was derived from the export of wool, and they built their Hall with its own quay on the River Foss, a tributary of the River Ouse. Thus, the Company's merchants could sail their ships out from the heart of the city on trade routes around the world.

Commissioned by the Company of Merchant Adventurers', the dial was designed by Christopher St J H Daniel, painted by Dav Bonham and constructed by the York Glaziers Trust in 1998.

PORTABLE SUNDIALS

Mike Cowham

A Brief Look at some of the Wide Range of Portable Sundials made in Britain

The tradition of making sundials goes back many years. The earliest dials found in Britain date from Saxon times but even earlier dials are known from Roman and Greek civilisations. At the same time that these early monumental dials were being created there was a need for dials small enough to be carried around to tell the time whilst travelling, giving the user an idea of day

shadow, with an estimate being made of the day's progress from the length or the position of the shadow formed. The first 'technical' dials were Altitude Dials, measuring the height of the Sun above the horizon. In order for these to work correctly, it was necessary to know the time of the year, because in summer the Sun is much higher in the sky than in the winter. Therefore allowance

A Universal Equinoctial Mechanical Dial by Thomas Wright, 'Maker to Ye KING'

length and in particular how much time remained until dark. Timekeeping as we know it was unimportant at that time and any appointment would be very vague such as 'in the morning'. As civilisation progressed, man would have become more aware of time and would probably have planned his day by reckoning to the nearest hour or two. Later, as mechanical clocks developed, fractions of hours and even minutes eventually became important to him.

The earliest Portable Dial may have been a stick placed in the ground, or even a man's own

had to be made for the change of season. Dials of the Roman period include two simple types of altitude dial, the Pillar Dial and the 'Ham' Dial, the latter so called from its shape. The Pillar Dial, often made from wood, was a simple device and easily portable. It consisted of a cylinder with a gnomon stored inside. In use, it would be placed on a flat surface or suspended with its gnomon folded out at right angles, pointing horizontally towards the Sun, to create a shadow on a scale inscribed on the outside of the cylinder. In order to compensate for varying seasons the gnomon

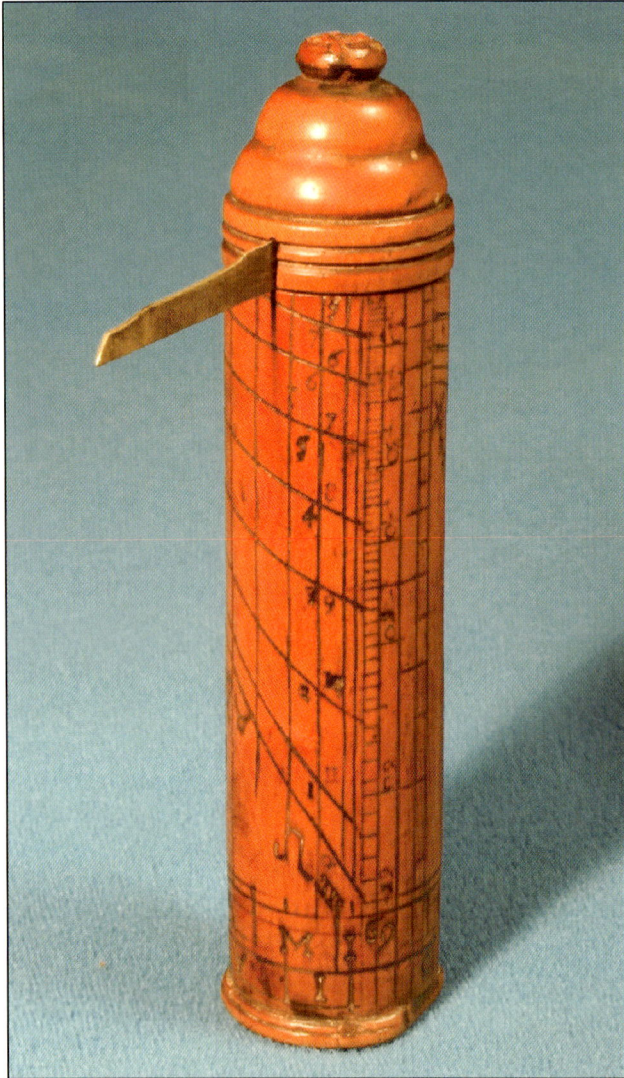

A typical boxwood Pillar Dial with its gnomon extended

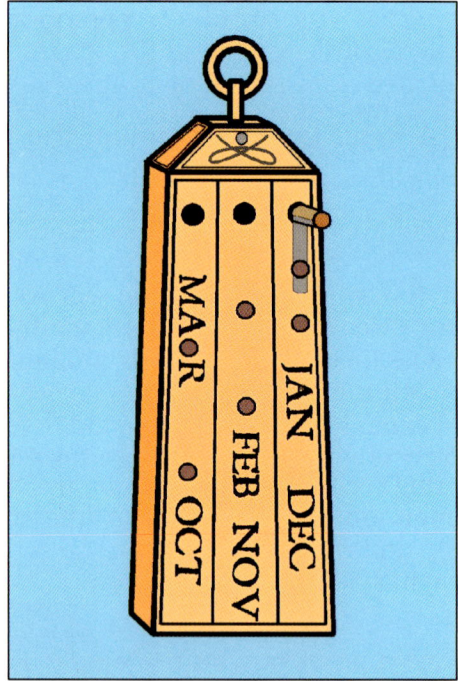

The Saxon Dial found at Canterbury

The art of dialmaking in Britain was revived in Elizabethan times when several good makers were working in London. They laid the foundations of the scientific instrument making tradition that was to flourish in the 17th to 19th centuries.

The altitude dial remained popular as it was simple to use and its principle was used in early Disc Dials. Here the gnomon was attached to an

would be rotated against a date scale moving it above the correct part of the scale.

A simple form of altitude dial of Saxon origin has been found in Canterbury. It uses a small pin for the gnomon that may be inserted into the hole above the appropriate month and the shadow, when vertical, falls onto a simple scale indicating either mid-morning/afternoon or midday. There were no smaller time divisions indicated. The reverse of the dial is similar but is calibrated for the remaining six months.

Very few Portable Dials existed until about 1500 when quantities of simple wooden compass dials, made in Nuremberg, flooded across Europe. The nine small compass dials found on the Mary Rose, which sank in 1545, were probably from Nuremberg. Other dials of this type have been found in Britain and careful examination shows that they also originate in the same place.

A silver Altitude Disc Dial c1600

arm which was set at the appropriate month, engraved on the disc, and the shadow would fall on an hour scale on the opposite edge of the disc. The one illustrated, in silver, was probably made in Britain around 1600. Like many dials this one also includes a Perpetual Calendar on its reverse with the dates of various saints' days, month lengths etc.

Several other types of altitude dial became popular over the centuries, one of the simplest being the simple ring or 'Poke' Dial. Poke is the old word for a pocket or pouch, the place where the dial would be carried. In this dial the gnomon consists of a small hole in a sliding band around the ring's circumference. The position of the gnomon hole is adjusted to the appropriate month and the small spot of sunlight transmitted through the hole will fall onto the internal hour scale.

The main problem with the simple altitude dial, measuring the height of the Sun, is that the Sun's altitude also varies as its owner travels to different latitudes. Therefore the dials described become less accurate as the latitude changes, particularly towards the middle of the day. Its other problem is that the dial can have the same reading in the morning as in the afternoon, so its user needs to know which half of the day it is. If in doubt, two readings can be taken separated by perhaps ½ hour to determine if the Sun is still rising or if it is falling, but around noon there is very little change.

Brass Universal Equinoctial Ring Dial by John Naish, dated 1707 MC

To overcome the problems due to the variation of latitude the Rev William Oughtred published details of a new type of ring dial in 1652. His description of the Universal Equinoctial Ring Dial encouraged instrument makers to concentrate on this new design. This dial also uses a small pin hole gnomon, but this has now been set across the centre of the ring supported by a bridge. The outer ring holds the whole assembly together and its suspension point can be slid around the outside to the appropriate latitude. The pin hole gnomon still needs to be adjusted for season, but now the time can be determined exactly on a linear hour scale. When not in use the whole assembly of rings fold flat and may originally have fitted into a small round case. These dials were quite soon being made in Britain by the finest instrument makers and the technology rapidly spread to Europe. The new dials were now really made to travel and so their makers also engraved on them the latitudes of some of the larger cities of Europe to make the latitude setting much easier. These dials were normally made from brass but silver or silver-gilt versions were produced as up-market models.

Most of the pocket dials to be found will be similar to the regular Horizontal Dial as found in our gardens, but they need to be correctly set so that the gnomon is in line with the Earth's axis. To aid this a magnetic compass is built into the dial. In the earlier days this compass was very

A typical 'Poke' Dial MC

Compass Dial by J Simons, London ^{MC}

Underside of the Whitehead Dial showing ^{MC}
its list of towns and their latitudes

small and inaccurate but as greater accuracy was required, the compass gradually got larger. By the mid 18th century the compass on some dials had become as large as the dial itself, the plate of which was now virtually a skeleton. The silver dial by Simons is typical of these.

Perhaps the most well-known portable dial of all is the 'Butterfield' Dial. This is the dial with the small bird used to support the gnomon showing the latitude setting with his beak. On the dial plate are three or four different chapter rings, each for a different latitude and on the underside is a list of towns and their latitudes. The 'Butterfield' Dial was produced in Paris in quantities from about 1680 up to the French Revolution of 1789 by many makers. Michael Butterfield, after whom the dial is commonly named, was an Eng-

lishman who set up his trade in Paris and his workshop produced thousands of these dials, mainly working in silver for his rich French clients. Although always thought of as his design it is probable that the design originated in England. Certainly, several English makers were producing the design at the same period. The octagonal silver dial by Richard Whitehead was made about 1685. It is of the highest quality and equal to the finest produced by Butterfield and other French workshops. The British market and the French market were not identical and there were differences in the designs. Most of the English dials were made from brass, a much more suitable material for showing the contrast of the shadow, and generally these dials were more accurately made, often with allowance made for the gradual change in magnetic declination. The well-known maker Thomas Heath made several 'Butterfield Dials'. One rather different version made by him was fashioned in the shape of a figure eight. This design

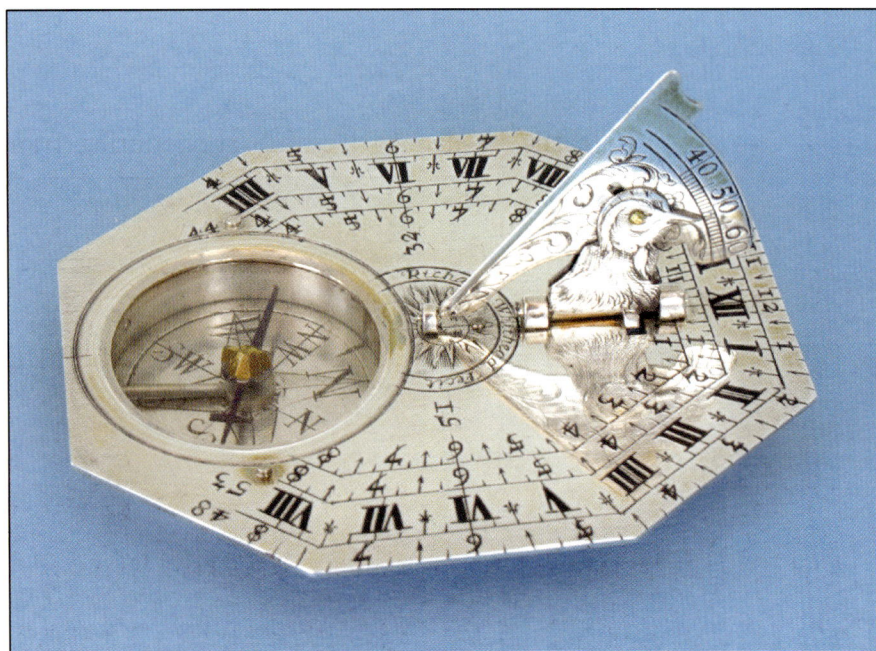

English made 'Butterfield' Dial by Richard Whitehead, c1690 ^{MC}

Unusual figure eight shaped 'Butterfield' Dial by Thomas Heath

There was a small market for precision timepieces and occasionally so-called 'minute dials' were produced. Theoretically these dials could tell the time to the nearest minute. Very few of these were really pocket dials but the one illustrated on Page 239, a Mechanical Equinoctial Minute Dial by Thomas Wright, seems to serve the purpose. It is an Equinoctial Dial, fully adjustable for a wide range of northern latitudes. Its gnomon is a small vane or finger that casts its shadow onto a small fiducial line. The user has to rotate the central arm until this shadow is exactly on the line and the time can be read in hours from the pointer on the main scale and the minutes from the smaller scale. It seems that Thomas Wright is the only London maker of these dials with only about 3 extant but several more are known to have been made by Irish makers in Dublin. The design was much more commonly produced in Europe, particularly in France, Germany and Austria.

allowed the compass to be large and the dial plate was kept intact showing much clearer markings, the engraving of which was filled with different colour waxes.

The 'Butterfield' Dial was very attractive but was somewhat limited in its geographical range due to its small range of latitude settings, typically between 40° and 60°. A similar looking dial known as an Inclining Dial was introduced, again probably simultaneously in Paris and London. The dial plate was generally delineated for a fixed latitude of 60° north, this being about the most northerly latitude generally required. However, this plate was hinged such that it could be tilted by up to 60° allowing the dial to be used right to the equator. A latitude arc was attached to one side to make the setting simple. The fine dial illustrated is by Edmund Culpeper working in London around 1720.

In about 1800 a simple to use dial was introduced that had a Horizontal Dial mounted onto a light compass card. Underneath was a small bar magnet that was supported by the pivot point. Therefore the dial was self-aligning to north and would always tell the correct time. It had several small problems. It was designed for a fixed latitude and therefore unsuitable for the traveller. It was relatively fragile and the needle could so easily be damaged by vibration during transit.

Inclining Dial by Edmund Culpeper

Shadow of Gnomon on fiducial line

Magnetic Compass Dial by Fraser

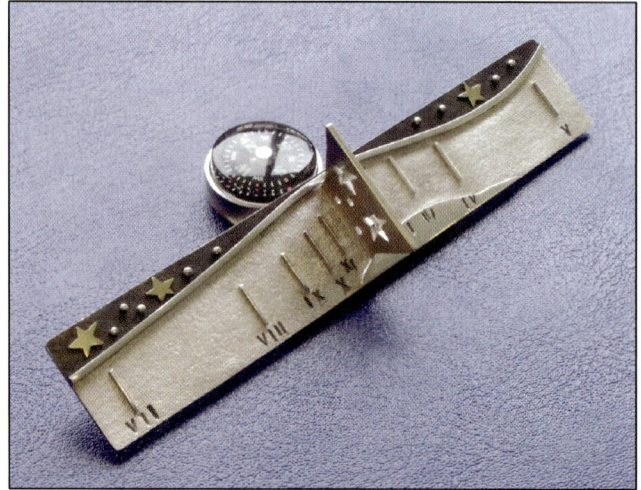

Modern Equatorial Dial in silver made by Jackie Jones

With the gradually changing magnetic declination, approaching a maximum of 24° west in about 1824 it needed to have the compass magnet so that it could be turned to the current figure. The dial by Fraser had an adjustable magnet but the cheaper dials had theirs firmly fixed to the dial plate.

The Inclining Dial is one that was frequently made by British instrument makers. It was quite simple and could be used over a wide range of latitudes. The dial illustrated was made in Bristol by Joshua Springer in about 1780. It is similar in concept to the earlier mentioned Inclining Dial by Culpeper, but of simpler construction.

With the advent of the affordable watch the pocket dial's popularity started to decline and by the late 19th century it was almost extinct.

Dials like those illustrated in this chapter may be seen in several good museums around the Britain and Ireland.

Modern Portable Dials
Today we have seen a resurgence in making replicas of many Portable Dials. These are normally fully functioning devices and allow us to

Inclining Dial by Joshua Springer of Bristol

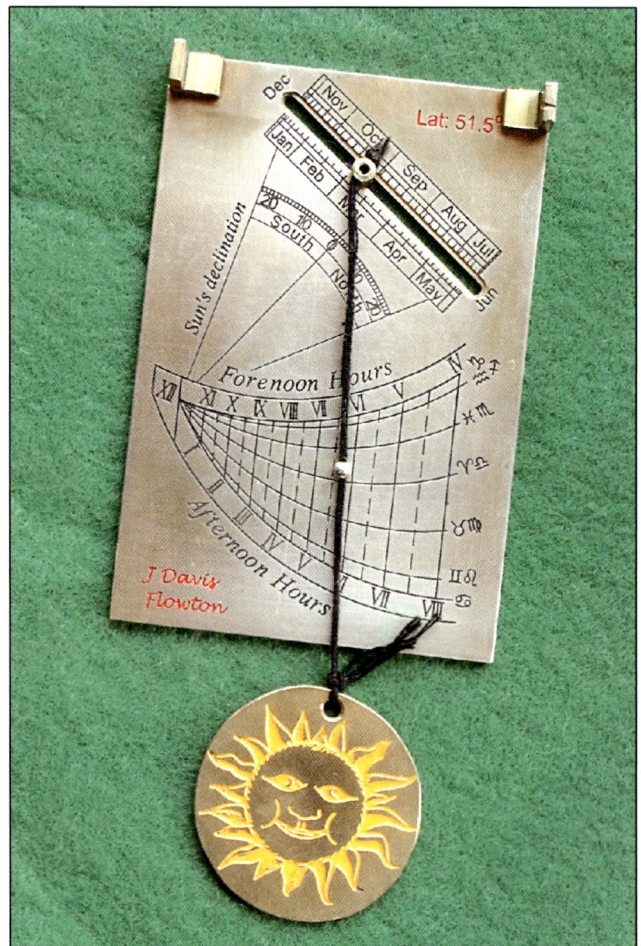

Replica of a Capucine Dial by John Davis

appreciate how these little objects were used and to understand some of the difficulties and inaccuracies involved. They are often to be found in museum shops and make interesting and instructive souvenirs.

244

APPENDIX 1 - GLOSSARY OF TERMS
John Davis

Altitude (of the Sun)
The angular height of the Sun above the local horizon. It varies between 0° at sunrise and sunset, and about 65° at noon on the summer solstice (in the UK).

Analemma
The shape which the Sun's position, or its shadow, traces out over a year at the same clock time each day. It has the appearance of a long, thin figure-eight.

Analemmatic Sundial
A sundial in which a movable vertical gnomon indicates the time against a set of hour points laid out around an ellipse. The gnomon is often the observer standing on a date scale in the middle of the sundial.

Armillary Sphere
A type of equatorial sundial in the form of a skeletonised globe.

Azimuth (of the Sun)
The horizontal direction to the Sun. In sundial literature, this is usually measured from south (with west being 90°), although in astronomical terms it is measured from the north.

British Summer Time
Clock time in the summer, one hour ahead of *Greenwich Mean Time*. Not shown by most sundials which can thus seem to be slow by about an hour in summer.

Declination (of the Sun)
The angle between the equatorial plane and the direction of the Sun as seen from Earth. The declination varies between +23.5° (mid-summer in the northern hemisphere) and -23.5° (mid-winter) over the course of a year.

Declination (of a Wall or Sundial)
The angle that the wall makes to due south, due west being 90°.

Declining Sundial
A sundial on a vertical wall which does not face directly south, east, west or north. A 'great decliner' is one where the dial nearly, but not quite, faces east or west.

Direct Sundial
A vertical dial which faces one of the cardinal points of the compass, north, south, east or west.

Double Horizontal Sundial
A specialised and much-prized form of horizontal sundial, mainly from the 17th and early 18th centuries. Its markings are similar to those of an astrolabe and it can provide much astronomical information.

Ecliptic
The plane traced out by the Earth over one year as it circles the Sun.

Equation of Time (EoT)
The correction, usually in minutes, which must be applied to solar time in order to obtain mean time. It has extreme values of about 14 minutes (sundial slow) in February and 16 minutes (sundial fast) in November, although in the summer months it is usually less than 5 minutes. The EoT arises because of variations of up to 30 seconds per day in solar day length (compared to the exact 24 hour mean solar day) at different times of the year. The difference is cumulative, and is due to the Earth's orbit is elliptical rather than circular and the plane of the ecliptic being inclined to the equatorial plane by 23.5°.

Equatorial (or Equinoctial) Sundial
A sundial in which the dial surface is a circle (or part of a circle) arranged to be parallel to the Earth's equator. The hour points are laid out around the circumference of the circle with equal spacing.

Equinox
The date and time when the Sun's declination is zero. The equinoxes are the dates when there is equal daylight and night (to within a few minutes). The exact dates of the Vernal (Spring) and Autumn equinoxes can vary by one day due to the leap year cycle.

Furniture
Any designs on a dial plate not directly used for time telling are referred to as dial furniture.

Gnomon
The physical structure which casts the shadow on a sundial. Its origin is Greek meaning literally 'one who knows'.

Greenwich Mean Time

The basis of British civil timekeeping (now superseded). Calculated from the time of noon at Greenwich by a hypothetical mean Sun.

Hour Angle (HA)

The time of day expressed as an angle of the Sun. The Sun appears to rotate around the equatorial plane once a day, or at 15° per hour. It is usually expressed with the origin at noon.

Horizontal Sundial

The common or garden sundial with a horizontal plate and a sloping (polar pointing) gnomon.

Latitude

The position of a location north or south of the equator. The angle which the gnomon makes to the horizontal is equal to the latitude on most types of sundial. The British Isles has latitudes in the approximate range 50° to 56°.

Longitude

The position of a location east or west of the Greenwich, or Prime Meridian. Solar time at any location west of Greenwich is later than Greenwich time by 4 minutes per degree of longitude.

Mass (or Scratch) Sundial

A medieval form of sundial usually scratched on the south wall of a church using a horizontal stick gnomon (invariably missing) in the centre of a circle of lines or dots. It did not indicate modern hours but showed the times of masses.

Mean Time

Normal 'clock time'; it is an artificially-created time on the basis of each day having a length of 1/365.24... of a year, the length of a mean solar day.

Nodus

A point on the gnomon (usually a small sphere or a notch) which casts an identifiable shadow on the dial plate. The position of the shadow depends on both the Sun's altitude and its azimuth and can indicate the date as well as the time.

Noon

The local time when the Sun is due south (azimuth = 0°). It is also when the Sun is at its maximum altitude for that day, and is (approximately) half-way between sunrise and sunset.

Polar Sundial

A sundial in which the flat dialplate slopes so that it is parallel to the polar-pointing gnomon.

Proclining Sundial

A sundial similar to a vertical one but with the top leaning towards the observer.

Reclining Sundial

A sundial similar to a vertical one but with the top leaning away from the observer.

Saxon Dials

Sundials from the Anglo-Saxon period, usually finely carved in stones which are set into church walls or form free-standing monuments.

Scaphe

A type of dial in the form of a shallow dish or a hemisphere.

Scientific Sundial

A dial in which the gnomon lies perpendicular to the equatorial plane. It points to the north celestial pole in the northern hemisphere.

Seasonal (or Temporary) Hours

An ancient system of timekeeping in which the period from sunrise to sunset was divided into 12 'hours', the length which varied with the time of year.

Solar (or Sun) Time

Time measured at any given location, relative to local noon. It is the time indicated by most sundials. It varies from clock mean time *(GMT)* due to the *Equation of Time* and the effects of *longitude*. Known technically as Local Apparent Time.

Solstice

The date and time when the Sun's declination is a maximum (mid-summer in the northern hemisphere) or a minimum (mid-winter). It corresponds to the maximum day or night length respectively. The actual dates of the solstices can vary by one day due to the leap year cycle.

Style

The line in space which generates the time-telling edge of a shadow. It is typically an edge of the three-dimensional gnomon.

Sub-Style

The line on a dial plate which is perpendicularly below or behind the style.

Zodiac

The imaginary band around the *ecliptic* in which the constellations lie. The signs of the zodiac are an old astronomical method of indicating the time of the year.

APPENDIX 2 - BIBLIOGRAPHY
Mike Cowham

The subject of Sundials has been covered by various authors for many hundreds of years. Books in the English language started to appear in the 16th century. These were not books *about* sundials as much as *how to make sundials*.

It was not until 1872 that Margaret Gatty published her well-known book of sundial mottoes. She had been collecting these since childhood and had amassed a large collection from her own observations and from her many friends. This book was mostly of mottoes but it did have a few pages at the back with sketches of interesting dials. Margaret was to die shortly afterwards and a second edition followed in 1889 edited by her daughter, Horatia Gatty, (later Horatia Eden). This book was considerably extended and included sketches done by their family friend Eleanor Lloyd. The fourth and last edition of 1900 was considerably expanded with numerous line drawings and substantially more mottoes. It is this book, usually known as 'Eden & Lloyd' that is frequently referred to in the chapters of this present volume. It has become almost the 'bible' of sundials and is still in frequent use today by diallists old and new.

Other books followed in the early 20th century and still appear from time to time. Many of these are listed below. This listing should not be considered as comprehensive.

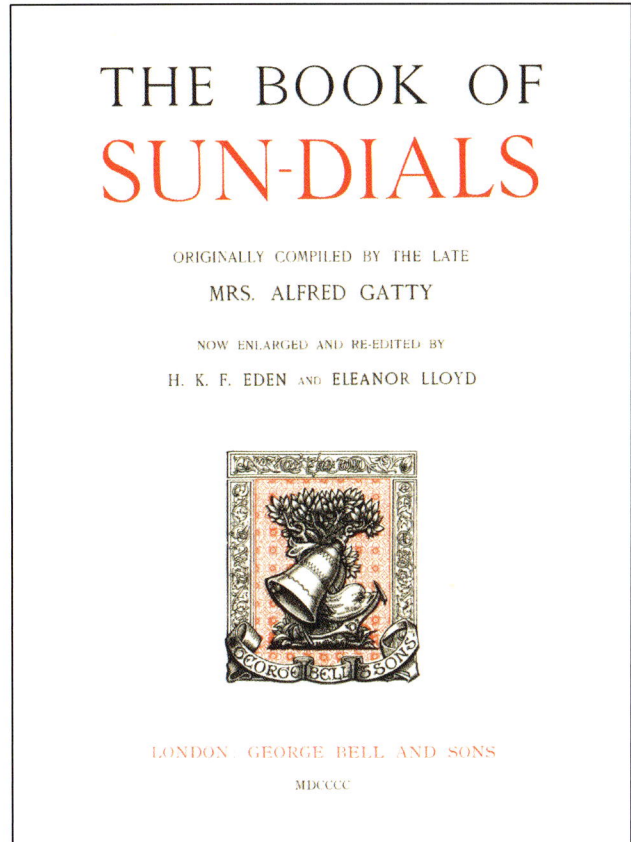

Title Page of the 4th edition of Mrs Gatty's Book of Sun-Dials

ARNALDI, Mario
The Ancient Sundials of Ireland
BSS, Crowthorne 2000

BION, Nicolas (Ed Stone)
The Construction and Principal Uses of Mathematical Instruments
London 1758

BOTZUM, Richard & Catherine
Scratch Dials, Sundials and Unusual Marks on Herefordshire Churches
Botzum, Lucton 1998

BROOKES & STANIER
Cambridge Sundials
Brookes & Stanier, Cambridge

COWHAM, Mike
A Dial in Your Poke
Cowham, Cambridge 2004

CROSS, Launcelot
A Book of Old Sundials and their Mottoes
Foulis, Edinburgh & London, 1922

DANIEL, Christopher St J H
Sundials
Shire Publications, Princes Risborough 2004

DAVIS, John
BSS Sundial Glossary
BSS, Crowthorne 2000

DRINKWATER, Peter
The Art of Sundial Construction
Drinkwater, Shipston-on-Stour 1985

EARL, Alice
Sun-dials and Roses of Yesterday
Macmillan, London 1902

EDEN & LLOYD, Horatia & Eleanor
The Book of Sun-Dials
Bell, London 1900

FALE, Thomas
The Art of Dialling
London 1593

GATTY, Mrs Alfred
The Book of Sun-Dials
Bell, London 1872

GREEN, Arthur
Sundials
SPCK, London 1926

HAIGH, Daniel
Yorkshire Dials
Yorkshire Archæological & Topographical
Journal, Bradbury, Agnew & Co.,
London 1878

HENSLOW, Geoffrey
Ye Sundial Book
Henslow, London 1914

HERBERT, A P
Sundials Old and New
London 1967

HIGTON, Hester
Sundials at Greenwich
OUP, Oxford 2002

HORNE, Ethlbert
Scratch Dials
Simpkin Marshall, London 1929

HYATT, Alfred
A Book of Sundial Mottoes
Wellby, London 1903

LANDON, Perceval
Helio-tropes or Posies for Sundials
Methuen, London 1904

LEADBETTER, Charles
Mechanick Dialling
London 1756

LEYBOURN
The Art of Dialling
London 1669

MARTIN, Carolyn
A Celebration of Cornish Sundials
Dyllansow Truran, Redruth 1994

MAYALL & MAYALL, R Newton & Margaret W
Sundials
Sky Publishing, Cambridge Mass. 1938

PATTENDEN, Phillip
Sundials at an Oxford College
Roman, Deddington 1979

MOXON, Joseph
A Tutor to Astronomy and Geography
4th ed. London 1686

SOMERVILLE, Andrew
The Ancient Sundials of Scotland
Rogers Turner, London 1990

STANIER, Margaret
Oxford Sundials
Oxford

STIRRUP, Thomas
Horometria
London 1652

WAUGH, Albert
Sundials, their Theory and Construction
Dover, New York, 1973

WELLS, Edward
*A Young Gentleman's Astronomy,
Chronology and Dialling*
London, 1725

WILSON, Jill
*Biographical Index of British Sundial
Makers*
BSS, Crowthorne 2003

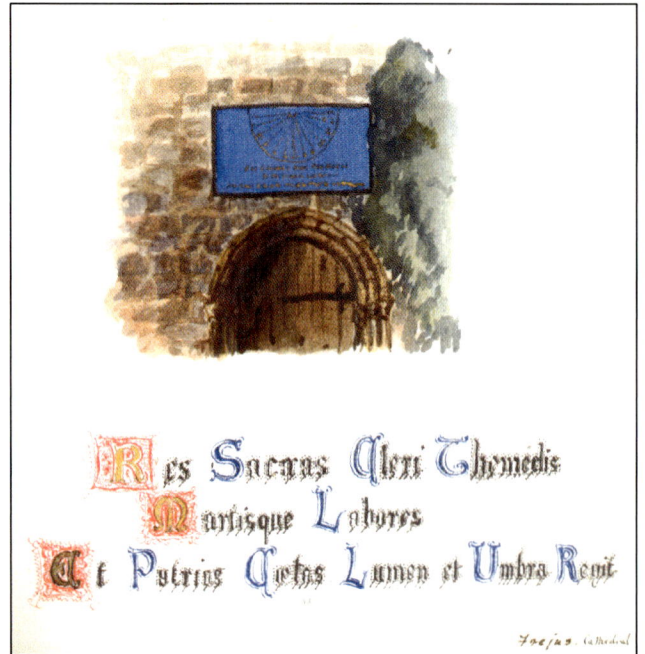

**A page from Eleanor Lloyd's sketchbook with
her watercolour of the Vertical Dial at Frejus
Cathedral in France with its motto:**

Res Sacras Cleri Themidis

Martisque Labores

Et Patrios Cœtus, Lumen, et Umbra Regit
(The sacred work of the church,
the toils of Themis and of Mars,
the councils of the nation too,
light and shadow rule)

APPENDIX 3 - THE BRITISH SUNDIAL SOCIETY
Douglas Bateman

The British Sundial Society was formed in 1989 by a small group of enthusiasts. It grew rapidly as many people joined who had been fascinated by the subject but had never before met anyone else who was interested! The Society is now larger than any other similar society and has a membership of around 500. It is a remarkable testimony to both the international flavour of sundials, and to the quality of the regular Bulletin, that a quarter of the members live outside the United Kingdom.

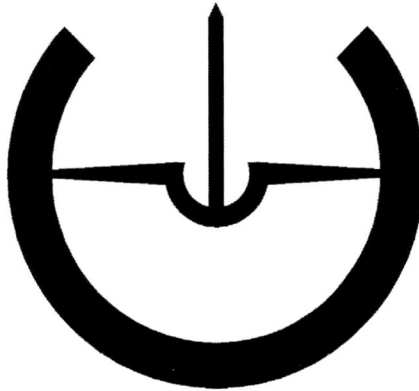

Objects of the Society
• To advance the education of the public in the science and art of gnomonics and the knowledge of all types of sundial.

• To catalogue the dials which still exist in the British Isles and research their history.

• To advise on the preservation and the restoration of old sundials and the construction of new ones.

• To publish a Bulletin containing original articles, reports from other societies, news and other items of interest to members.

The Society gained charitable status in 1992, and in 2004 was granted affiliation to the Royal Astronomical Society.

The Bulletin
Four issues a year of the Bulletin keep Members up to date with happenings within the Society and articles written by Members. Articles range from serious studies of dial design, research into types and the history of dials, and straightforward reports on dials worthy of comment. There is no doubt that the high standard of the Bulletin has been a factor in the growth of the Society.

Sundial Register
The cataloguing of fixed dials within the British Isles has continued steadily since the foundation of the Society. The details are bound in The Register of dials, and the fifth edition was produced in 2005. Information such as the location, type of dial, condition, technical information, mottoes, etc, is given. However, for some dials where security is important, their locations can be withheld at the request of the owner. The statistics are that over 8000 reports have been submitted for about 5600 dials, and steady production of new dials keeps members and the registrar busy.

Mass Dials
Sometimes called scratch dials, these are found on churches up and down the country. Quite a number date from Saxon times, and the study of these dials is carried out by a group within the Society. Again, a register of such dials is being prepared. This is a nationally important activity as many of these simple dials are disappearing through natural erosion of stone surfaces, especially the softer stones on which some dials are often carved.

Dial Restoration
Some dials can be restored quite easily if a new gnomon is needed, whereas a few may be better left alone. The Society can offer advice (almost always based on practical experience) and, within limited resources, may be able some offer some financial support.

Education
Informing the public has always been a strong point in the Society, and the Education Group

pioneered the Society's first publication, 'Make a Sundial'.

Publications

Occasional publications are produced by the Society. These include:

Make a Sundial. A book aimed specifically at schools which continues in demand.

BSS Sundial Glossary, explaining the terms used in dialling, (now in its second edition).

Biographical Index of British Sundial Makers, a booklet with details of known dial makers from the 7th century to 1920.

The Ancient Sundials of Ireland, a significant contribution in a specialist field that continues in modest demand.

Many of us want to make a sundial and the Society has a list of useful books that any practical person can follow. Failing that, the Society has quite a number of professional dial makers amongst its members, and a list, without favouring any special commercial activities, is available. For both lists apply to the Secretary.

Sundial Award Scheme

In order to promote the construction of good quality sundials, the Society set up an award scheme. It has been run at 5-yearly intervals, attracting some 20-30 entries each time, with classes for junior, amateur and professional makers. Substantial financial rewards are given, together with certificates and a plaque to be fixed near the dial.

Photographic Competition

Introduced for the first time in 2004 the competition has attracted some excellent entries. One of its aims is to improve the artistic quality of Member's photographs persuading them to upgrade their photographic skills, an important feature in dial recording.

The Internet

This has revolutionised the quest for information and the Society's website has much information together with links to other societies and related interest groups.

The address is www.sundialsoc.org.uk

Conferences & Meetings

Annual Conference. Social aspects are not neglected and the Society's primary event where members get together is the Annual Conference. This is a two-day residential event held in the spring at a different university conference centre. Each year it moves to a different part of the Country. Lectures are given on all aspects of dialling, there are demonstrations and exhibits of sundials and projects, and an afternoon tour. The lecturing highlight is a memorial lecture, (honouring the Founder Chairman), given by a leading expert of academic or international standing.

One Day Meetings. In the autumn a one-day meeting is held, with the emphasis on informality of displays and short talks.

Sundial 'Safaris'. Not so regular, but always successful, are week long tours. These have been to Scotland, Wales, Cornwall, Ireland, Austria, Germany, France, and Italy. The overseas tours have always been assisted by enthusiastic members of the local sundial societies.

The British Sundial Society is now a well established scientific society that is founded on the diverse interests of its professional and amateur members who come from all walks of life. In addition, members delight in exchanging ideas and information about dials and the Society has built up a reputation for being informative and sociable.

For more information about The British Sundial Society, contact the Secretary:

Douglas A Bateman, 4 New Wokingham Road, Crowthorne, Berkshire, RG45 7NR

INDEX

Refer to the Contents on pages v & vi for the regions of the British Isles and list of Authors.
Entries in CAPITAL letters refer to the main entries in the various Chapters.
Entries in *italics* signify the job or function of that person.

INDEX

'Whilst thou lookest; I fly'